NEW PARTICLES 1985

NEW PARTICLES 1985

edited by
Vernon Barger
David Cline
Francis Halzen

University of Wisconsin – Madison

World Scientific

Published by

World Scientific Publishing Co. Pte. Ltd.
P. O. Box 128, Farrer Road, Singapore 9128

Library of Congress Cataloging-in-Publication data is available.

NEW PARTICLES 1985

ISBN 9971-50-045-0

Printed in Singapore by Fu Loong Lithographer Pte Ltd.

CONTENTS

NEW PARTICLES 1985

TOP AND BOTTOM EXPECTATIONS AT THE CERN $p\bar{p}$ COLLIDER

R.J.N. Phillips
Rutherford Appleton Laboratory, Chilton, Didcot, Oxon, England

ABSTRACT

A brief summary is made - with illustrative calculations - of the expectations for b and t quark production and detection at the CERN $p\bar{p}$ collider.

INTRODUCTION

This talk is about the expected signals from t and b quarks (carrying quantum numbers top and bottom, or alternatively truth and beauty) at the CERN $p\bar{p}$ collider.

The expected production mechanisms are

$$\text{Electroweak} \begin{cases} q\bar{q} \to Z \to Q\bar{Q}, & Q = t,b \\ q\bar{q}' \to W \to t\bar{b}, & \end{cases}$$

$$\text{QCD} \begin{cases} O(\alpha_s^2) & q\bar{q},\ gg \to Q\bar{Q}, \\ O(\alpha_s^3) & q\bar{q},\ qg,\ gg \to Q\bar{Q}x, \quad x = q,\ \bar{q},\ g\ . \end{cases}$$

There may conceivably be diffractive channels too, where the incident p or $\bar{p}$ is excited to a high-mass N^* state with $N^* \to Q\bar{Q}X$, but these would presumably be confined to low p_T unlike the hard processes above.

Originally attention concentrated[1,2] on the $2 \to 2$ subprocesses, but more recently people have recognized[3-6] the importance of $2 \to 3$ channels[7] (especially those containing gluon fragmentation diagrams) which dominate over the former in some places. It happens that the $2 \to 2$ subprocesses do not contain the most favorable kinds of

Fig. 1
$2 \to 2$
subprocesses

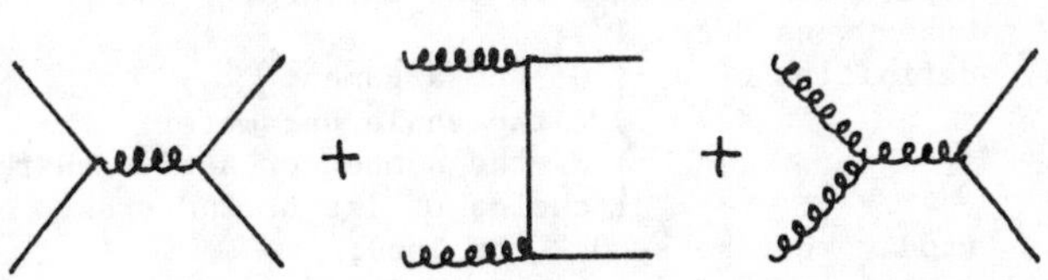

exchange: you could call them "accidentally suppressed" with $\sigma(s) \sim s^{-1} \ln s$. In contrast, the $2 \to 3$ processes include some with gluon exchange (favorable: $\sigma(s) \sim \ln s$) and with favorable color factors too. The $2 \to 3$ processes are important not because they are enhanced, but rather because the $2 \to 2$ channels happen to be

Fig. 2
Some 2 → 3
subprocesses

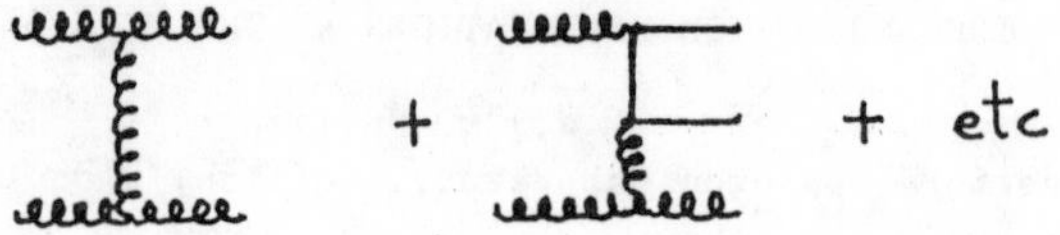

suppressed. It is not a breakdown in the perturbation expansion. It is somewhat like photon interactions, where e^+e^- pair production dominates over Compton scattering as soon as there is enough energy to get away from threshold, in spite of its higher order in α.

Note by the way that "flavor excitation"[8] is included among the 2 → 3 processes; note too that 2 → 3 amplitudes have soft and collinear divergences that need some cutoff.

Fig. 3
A typical flavor excitation $qQ \to qQ$ can be seen as arising from 2 → 3 (and higher order) processes.

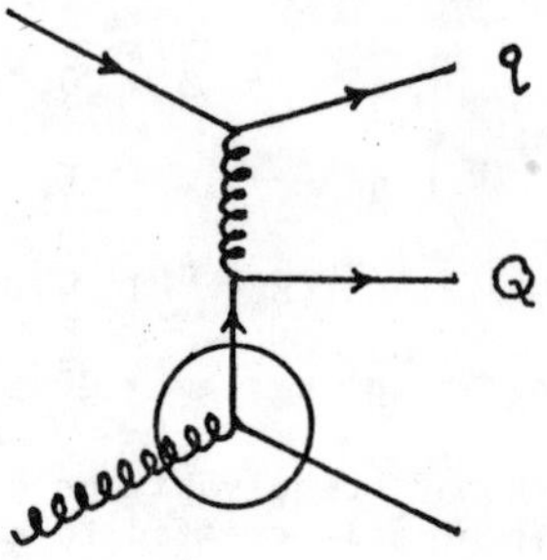

When the Feynman diagrams are specified, there still remain a number of inputs to choose:

(i) parton distributions in the incident p and $\bar{p}$ (today we take Duke-Owens Model 1[9]);

(ii) definition of α_s: { Q^2 the argument, Λ the scale parameter, N_f the number of active quark flavors, choice of 1st or 2nd order formula, (today we take $\hat{s}$, 0.2, 5, 1st);

(iii) cutoff for 2 → 3 subprocess (today we take $p_T(Q\bar{Q}) > 5$ GeV);

(iv) quark masses (today $m_b = 5.2$ GeV, $m_t = 40$ GeV, ref 10).

Uncertainties in these choices (and in K-factor enhancements from higher order loop diagrams: K=1 is taken here) make hadroproduction cross section calculations uncertain by perhaps a factor 2. The electroweak cross sections can be normalized to the W and Z data[11,12].

With the choices above and $\sqrt{s}$ = 630 GeV we get

$$
\begin{aligned}
\sigma(Z^o \rightarrow b\bar{b}) &= 0.2 \text{ nb} \\
\sigma(QCD : b\bar{b}) &= 5.0 \ \mu\text{b} \\
\sigma(QCD : b\bar{b}x) &= 0.8 \ \mu\text{b} \\
\sigma(W \rightarrow t\bar{b}) &= 1.2 \text{ nb} \\
\sigma(Z \rightarrow t\bar{t}) &= 0.1 \text{ nb} \\
\sigma(QCD : t\bar{t}) &= 0.4 \text{ nb} \\
\sigma(QCD ; t\bar{t}x) &= 0.1 \text{ nb}
\end{aligned}
$$

For an integrated luminosity around 360 nb^{-1}, this implies that about two million $b\bar{b}$ pairs and some eight hundred top quarks (singly or in pairs) have been produced. Should any trace of them have been seen?

HEAVY QUARK SIGNATURES

Leptons are the most popular signature since

$BF(b \rightarrow e) \simeq 0.12$ from experiment[13]
$BF(t \rightarrow e) \simeq 0.10$ from quark/lepton counting

and similarly for muonic decays. Fast leptons from b (or c) are usually in jets. There are strict kinematic constraints: e.g. for $B \rightarrow De\nu\ldots$ decays

$$
\begin{aligned}
\sin \Theta_{eD} &< (m_B^2 - m_D^2)/(2p_e m_D) \\
p_D &> \left[4m_D^2 p_e^2 - (m_B^2 - m_D^2)^2\right]/\left[4p_e(m_B^2 - m_D^2)\right]
\end{aligned}
$$

so that if p_e = 15 GeV we have $\Theta_{eD} < 25^o$ and p_D > 1.8 GeV as the extreme limits: most events lie well inside the limits. This expresses what we all know already, that fast leptons must have fast parents and hence the rest of the decay debris (plus the original $b \rightarrow B$ fragmentation debris) must be travelling in the same general direction. But when the lepton energy is low enough to be comparable with the energy released in the decay, these constraints become weak or vanish.

Thus fast leptons from b (or c) decays are accompanied by hadronic debris. This is made quantitative by defining a fiducial cone ΔR < something,

$$\Delta R = (|\Delta\phi|^2 + |\Delta\eta|^2)^{1/2}$$

around the lepton direction in azimuth ϕ and pseudorapidity η = ln cot(Θ/2) and by setting some threshold for the summed hadronic E_T within this cone. Leptons with ΣE_T > threshold are "non-isolated". Those with ΣE_T < threshold are "isolated". With an appropriate choice of cone and threshold, it can be arranged that most leptons expected from b or c are nonisolated, while most of those from t are isolated - which differentiates these signals.

To calculate lepton production we need quark → hadron fragmentation (another choice to be made: today it is the Peterson et al model[14] with parameter $\varepsilon = 0.5/m_Q^2$) and semileptonic decay distributions (V-A gives an adequate approximation).

ONE-LEPTON SIGNALS

At large p_T the production of b and c quarks is similar (mass effects do not matter when $p_T >> m$). However b → B fragmentation is harder than c → D and the B → ℓ spectrum is harder than D → ℓ for V-A quark matrix elements; also the primary B → ℓ branching fraction is a bit bigger than D → ℓ and b-decays have additional secondary b → c → ℓ contributions. As a result leptons from $b\bar{b}$ dominate over those from $c\bar{c}$ at large p_T.

Figure 4 shows UA1 data at $\sqrt{s}$ = 540 GeV from ref. 10; they may contain some background contamination and should perhaps be regarded as upper limits instead (private communications from UA1). These data are broadly consistent with the $b\bar{b}$ and $c\bar{c}$ calculations. At large p_T a substantial top-quark contribution may be present but the data here are sparse.

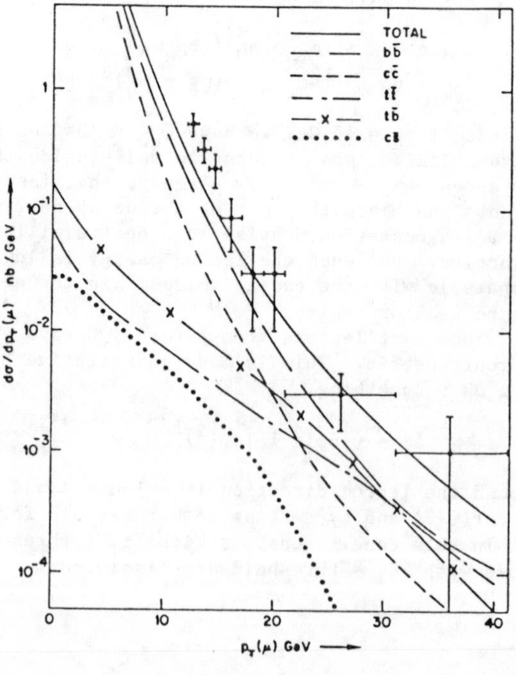

Figure 4.
Predicted single-muon distributions versus p_T, compared with UA1 data

DILEPTON SIGNALS

Consider non-isolated dileptons to pick out heavy quark contributions (Drell-Yan and ψ, Υ dileptons are mostly isolated). Expected $b\bar{b}$ and $c\bar{c}$ signals[4,15,16] dominate over top; there are both like-sign and unlike-sign dileptons.

$$c\bar{c} \to \ell^+\ell^- \quad (\text{via } c \to \ell^+, \quad \bar{c} \to \ell^-)$$

$$b\bar{b} \to \ell^+\ell^- \quad (\text{via } b \to c \to \ell^+, \quad \bar{b} \to \bar{c} \to \ell^-)$$

$$b\bar{b} \to \ell^\pm\ell^\pm \quad (\text{via } b \to \ell^-, \quad \bar{b} \to \bar{c} \to \ell^-$$
$$\text{or } b \to \ell^-, \quad (\bar{b}s) \to (b\bar{s}) \to \ell^-, \text{ "mixing"})$$

The mechanisms above give back-to-back dileptons from different primary parents. There is also another component, giving low-mass +− pairs from a common parent, usually appearing on the same side:

$$b \to c\ell^-\bar{\nu},\ c \to s\ell^+\nu \qquad \text{with } m(\ell^+\ell^-) < m_b .$$

Questions. Do data agree with QCD fusion calculations? Are there excess like-sign dileptons that would require B^o-$\bar{B}^o$ mixing?

Preliminary UA1 dimuon data[17] have the cuts

$$p_T(\mu_1) > 3 \text{ GeV} , \quad p_T(\mu_1) + p_T(\mu_2) > 10 \text{ GeV} ,$$

$$|\eta(\mu_1)| < 2 , \quad |\eta(\mu_2)| < 1.3 .$$

A typical calculation[4] at $\sqrt{s}$ = 630 GeV with these cuts gives the following cross sections in picobarns, assuming no B-$\bar{B}$ mixing.

	σ(+−,m<4)	σ(+−,m>4)	σ(±±)
$b\bar{b}$	21	290	100
$b\bar{b}x$	20	145	48
$c\bar{c}$		56	
$c\bar{c}x$		33	
$t\bar{t}$	1	8	4
$t\bar{t}x$		2	1
$W\to t\bar{b}$	2	10	9

With the CERN luminosity, plus chamber geometry and track quality acceptance factors[17] 0.45 x 0.58, these numbers imply about 65 heavy-flavor dimuons with like-sign/unlike-sign ratio 0.26. The UA1 results for non-isolated dimuons are 34 events with ±±/+− ratio 0.3±0.1. Within theoretical uncertainties, this total event rate is compatible with our expectations, and so is the like-sign/unlike-sign ratio.

B^o-$\bar{B}^o$ MIXING?

The most likely contributor is the $B^o(b\bar{s})$ state (for discussion and references, see e.g. ref. 4), where the mixing could even be maximal - i.e. $B^o(b\bar{s})$ forgets its origin and decays equally as a $(b\bar{s})$ or $(\bar{b}s)$ state. We neglect $B^o(b\bar{d})$ where the mixing should be much less. The effect experimentally depends on how many $(b\bar{s})$ states are produced per initial b-quark; at the very most this might be 33%, - which comes from assuming that $\bar{u}$, $\bar{d}$, $\bar{s}$ are picked up in the ideal ratio 1:1:1 and neglecting baryon formation.

This ideal maximum-mixing scenario would have a fraction $\varepsilon=1/6$ of the produced b-quarks decaying like $\bar{b}$. Then for the back-to-back dimuons, a fraction $2\varepsilon(1-\varepsilon)=0.28$ of the expected unlike-sign dimuons would actually come out as like-sign dimuons (and vice versa): see ref 4 for the formulae. The predicted like-sign/unlike-sign ratio would become 0.47.

There is no evidence for such an extreme scenario in the data above. However, the ideal assumption of $\bar{u}:\bar{d}:\bar{s}$ = 1:1:1 is probably exaggerated. It seems more reasonable to guess ratios 1:1:1/2 or even 1:1:1/3 which give $\varepsilon=1/10$ or 1/14 respectively and correspondingly smaller corrections.

The discussion here is a bit sloppy, but so are the data. A future careful analysis with future high-statistics data should take careful account of the precise isolation criterion and calculate exactly how much of the heavy-quark, Drell-Yan and other contributions fall into the isolated and non-isolated categories; hopefully there will also be accurate estimates of backgrounds from non-prompt and misidentified muons, etc.

ISOLATED SAME-SIGN DIMUONS

The UA1 results[17] include 7 isolated like-sign dimuons which may pose a problem. With their isolation criterion (both muons have ΣE_T < 4 GeV in cones ΔR < 0.7) relatively few heavy-flavor events are expected to be isolated.

Figure 5 shows the calculated fractions of $c\bar{c}$ and $b\bar{b}$ events in different categories that are isolated versus the ΣE_T threshold, for a fiducial cone ΔR < 0.7. The calculations are at the parton level

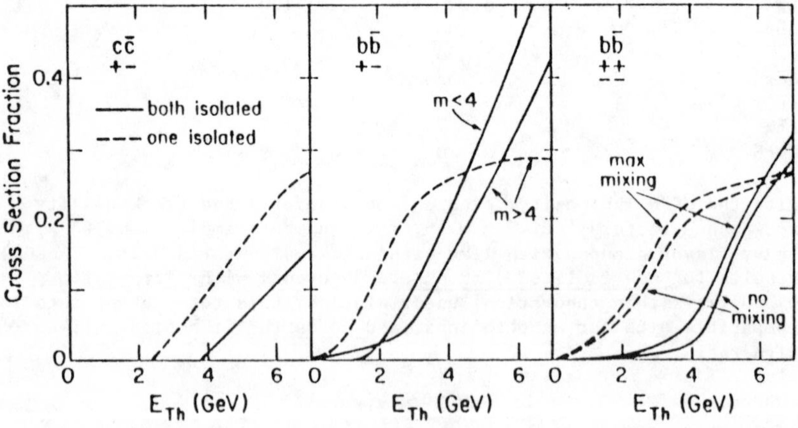

Figure 5

and therefore imperfect, but they show the expected qualitative features: $c\bar{c}$ events have the most collimated debris and are least isolated; $b\bar{b}$ unlike-sign events with m > 4 are mostly from two primary b-decays where the muon carries much of the parent momentum and the collimated debris is minimal, so they are more likely to be isolated than the like-sign events. ΣE_T contributions from the underlying event have been omitted, but they would make the isolated fractions even smaller.

Figure 5 suggests that only 10-20% of the like-sign events should be isolated - even with maximal mixing - say 1 or 2 events at most. It is not easy to understand these 7 observed events as $b\bar{b}$ (unless we are seeing a huge statistical fluctuation). It would require about half the like-sign events to be isolated. If a similar fraction of the unlike-sign pairs were isolated (actually we expect a larger fraction) they would leave no room for the Drell-Yan and T signals.

(At this meeting UA1 presents new data with looser cuts and more events: see talk by K. Eggert. They require separate discussion).

SIGNALS FOR TOP?

Dileptons are a possible place to look[4]. Although the $b\bar{b}$, $c\bar{c}$ signals are bigger, they are strongly correlated in azimuth near $\Delta\phi=0$ and 180°, whereas top is not: see Figure 6. At present the rate is too low but we may see top signals here eventually.

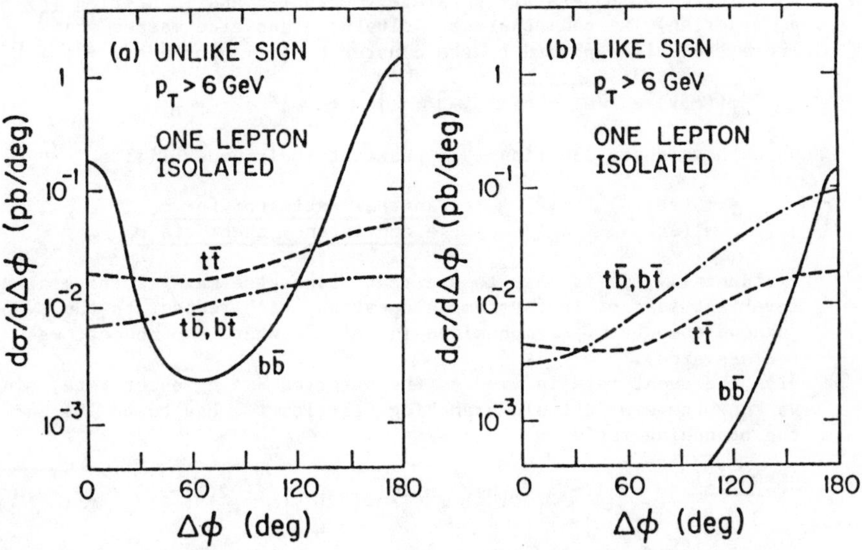

Figure 6. Azimuthal dependence of dilepton signals (ref 4).

Single-leptons offer a higher rate, but we need extra conditions to beat the $b\bar{b}$, $c\bar{c}$ backgrounds. First of all, require the lepton to have large p_T(cf. Fig. 4) and to be isolated. This can cut the background dramatically but it would be nice to have some further properties to identify top decay. For this, many people look to the $W \to t\bar{b}$ production channel with $t \to be\nu$ decay .

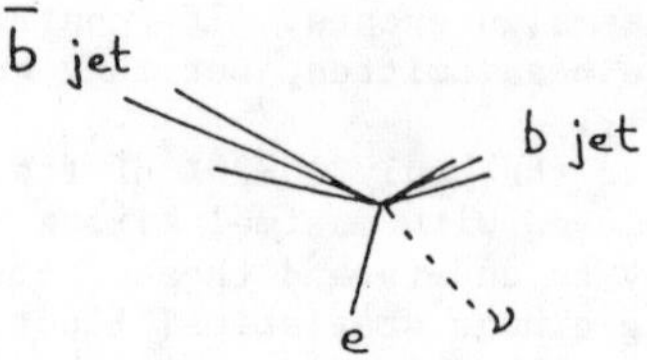

Fig. 7
Topology of $W \to t\bar{b}$
with $t \to be\nu$

This process has many nice specific features[18]

i) $e + 2$ jet event.

ii) e usually hard and isolated.

iii) $\bar{b}$ jet is usually the one with higher p_T. It has a Jacobian peak at $p_T(\bar{b}) \sim (m_W^2 - m_t^2)/(2m_W)$ <u>an estimator for m_t</u>.

iv) ν carries off missing p_T. If measured, we can construct the $e + \nu$ "transverse mass"[19] (the nearest thing to a true invariant mass when the neutrino p_L is unmeasurable).

$$m_T^2(e,\nu) = (e_T + \nu_T)^2 - \left|\vec{e}_T + \vec{\nu}_T\right|^2$$

with upper limit $m_T(e,\nu) < m_t - m_b$: <u>another estimator for m_t</u>. (For $b\bar{b}$ background events the upper limit would be $m_T < m_b - m_c \approx 3$ GeV; in principle a cut on m_T could further suppress this kind of thing).

v) If the b, $\bar{b}$ jets are separately measured and identified (eg by p_T ordering), we can construct "cluster transverse masses"[18,20] where the e is replaced by the cluster $c = (e + b)$ or $c = (e + b + \bar{b})$

$$m_T^2(c,\nu) = \left[\sqrt{m_c^2 + c_T^2} + \nu_T\right]^2 - \left|\vec{c}_T + \vec{\nu}_T\right|^2$$

which have sharp Jacobian-type peaks at their upper limits

$m_T(eb,\ \nu) \leqslant m_T$	<u>another estimator for m_t</u>
$m_t(eb\bar{b},\nu) \leqslant m_W$	<u>a consistency check via m_W</u>

Incidentally, it is easy to see that transverse mass is the minimum invariant mass of the system in question, with respect to the unknown momentum ν_L (emphasized in ref. 21 where the concept was rediscovered).

vi) The event rate is tied to the observed $W \to e\nu$ event rate, since we know the semileptonic branching fraction $t \to be\nu$ to be 10% and the branching ratio

$$\frac{W \to t\bar{b}}{W \to e\nu} = 3\left[1 - \frac{1}{2}(x_t + x_b) - \frac{1}{2}(x_t - x_b)^2\right]\sqrt{1 - 2(x_t + x_b) + (x_t - x_b)^2}$$

where $x_Q = m_Q^2/m_W^2$.

There are also some complications and problems.

(i) The full cascade decays $t \to b \to c \to s$ and $\bar{b} \to \bar{c} \to \bar{s}$ provide a sizeable probability of extra neutrinos from additional semileptonic decays (with unidentified charged leptons). The solution is to add them into the calculation; the extra neutrinos are mostly soft, so the corrections are not disastrous[22].

(ii) The missing p_T of neutrino(s) has big measurement errors. The solution is to fold the uncertainty into the calculation. In the

illustrations below, $p_T(\nu)$ has a 4 GeV r.m.s. error.
(iii) $b\bar{b}$, $c\bar{c}$ events are often e + 2 jet. But usually e lies in a jet, is not isolated.

Fig 8.
$b\bar{b}$, $c\bar{c}$ event topology

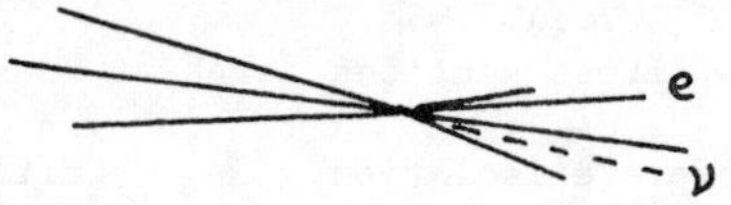

(iv) $b\bar{b}x$, $c\bar{c}x$ events with an additional hard parton x are a more serious background: they might give some e + 2 jet events in which the b-parent of e does not give an identifiable jet, and the e manages to pass the isolation cut. Solution: suppress these events by tight isolation cuts plus if necessary cuts on transverse masses or other dynamical features.

Fig 9.
$b\bar{b}x$, $c\bar{c}x$ possible event topology

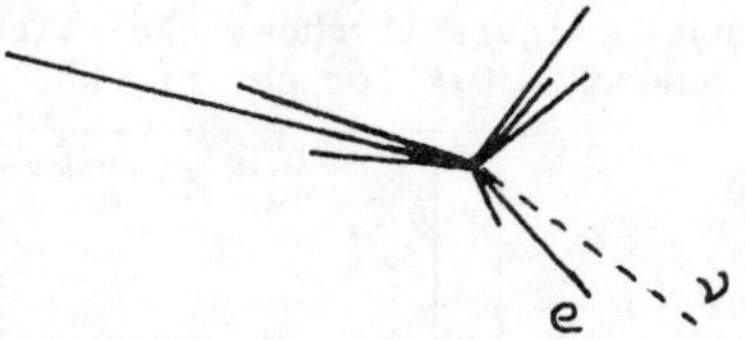

Altogether, $W \rightarrow t\bar{b}$ is a tightly constrained situation and seems a promising place to look for top. The strategy would be to select e + 2 jet events; to suppress backgrounds by cutting on $p_T(e)$, isolation and maybe m_T variables; to pull out the top signal by looking for Jacobian peaks in the cluster transverse masses and the recoil $\bar{b}$ jet p_T; and finally to cross-check that the absolute rate is correct.

The UA1 top search[10] makes the following cuts

Jets : $p_T(j_1) > 8$ GeV , $p_T(j_2) > 7$ GeV , $|\eta(j_i)| < 2.5$

Electron : $p_T(e) > 15$ GeV, $|\eta(e)| < 3$, $\Sigma E_T < 1$ GeV in $\Delta R < 0.4$,

which are used in the following calculations. (They also have

Fig 10

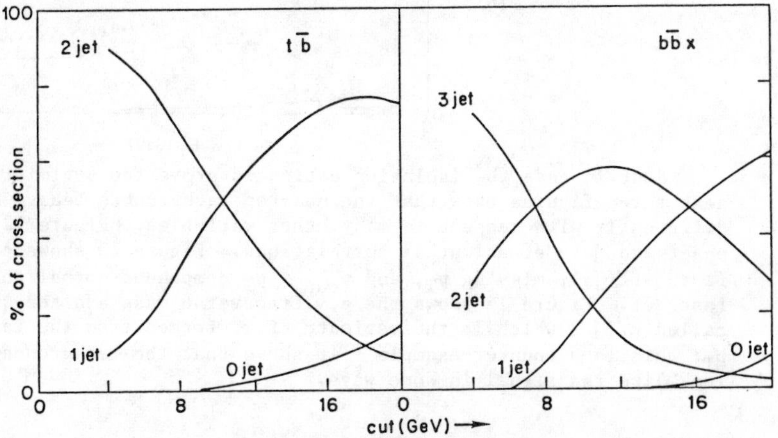

extra conditions plus somewhat different cuts on muons, that are not used here). Note first that the fraction of the cross section classified as 0-jet, 1-jet, etc depends absolutely on the jet cuts. Figure 10 shows how these fractions vary with the cut for the $t\bar{b}$ signal and $b\bar{b}x$ background, when we require $p_T(j_1)$ > cut, $p_T(j_n)$ > cut - 1 (for n = 2,3 ...).

The e + 2 jet cross sections in pb at $\sqrt{s}$ = 630 GeV are

	before e isolation	after isolation
$c\bar{c}$	20	-
$c\bar{c}x$	50	-
$b\bar{b}$	80	-
$b\bar{b}x$	54	-
$t\bar{t}$	10	9
$t\bar{t}x$	2	2
$t\bar{b}$	23	21

It seems that the isolation criteria are severe enough to remove the $b\bar{b}$, $c\bar{c}$ backgrounds. Figure 11 shows the calculated distributions of ΣE_T within the cone $\Delta R < 0.4$ for $t\bar{b}$, $t\bar{t}$, $b\bar{b}$, $b\bar{b}x$ contributions.

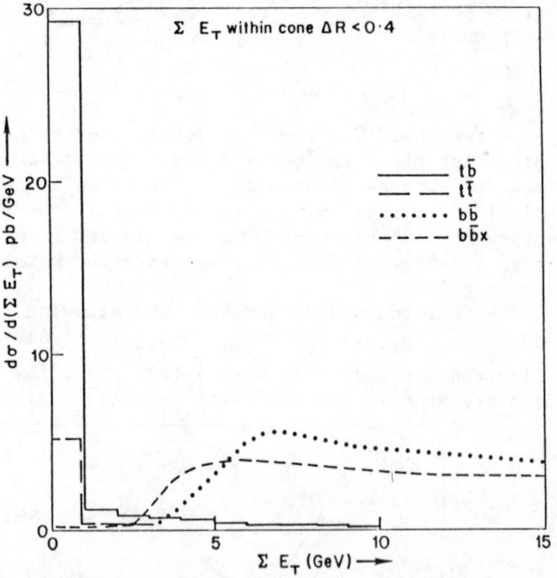

Figure 11. Dimuon isolation compared for various channels.

Just in case the isolation estimates prove too optimistic, the next three figures show that the non-top backgrounds behave quite differently with respect to many other variables. Figure 12 shows e-jet and jet-jet azimuthal correlations. Figure 13 shows p_T of the fast jet (j_1), missing p_T, and p_{out} = p_T component normal to the fast jet. Figure 14 shows the e,ν transverse mass and the quantity called $m_T(j_1)$ which is the estimate of m_t formed from the fast jet p_T; this is a counter-example - it shows that the backgrounds <u>can</u> look like the signal in some ways.

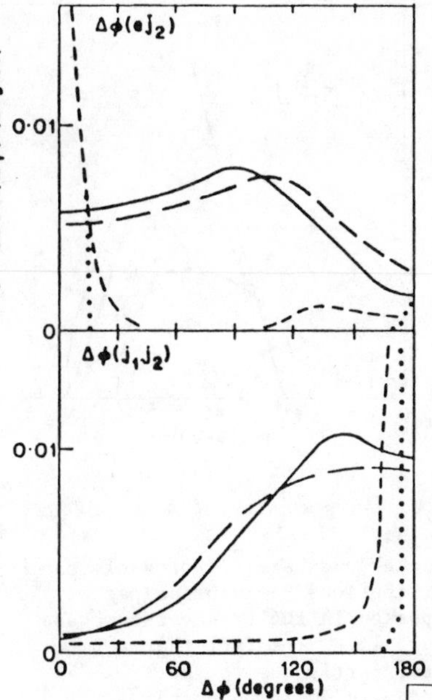

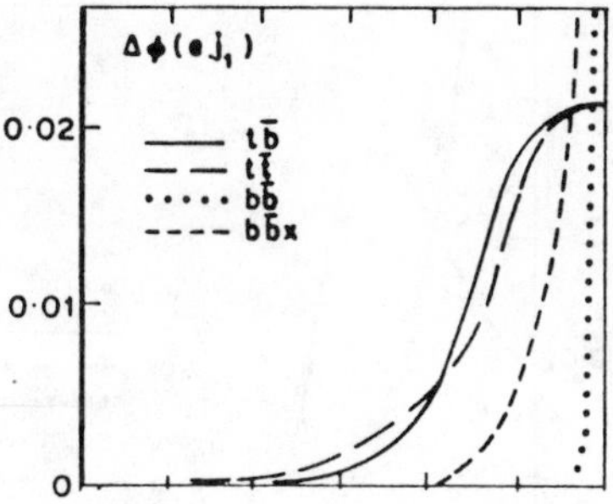

Figure 12.
Azimuthal correlations of
ej_2, ej_1, j_1j_2.

Figure 13.
p_T(jet 1), missing p_T,
p_{out}(e) and p_{out}(jet 2)
distributions.

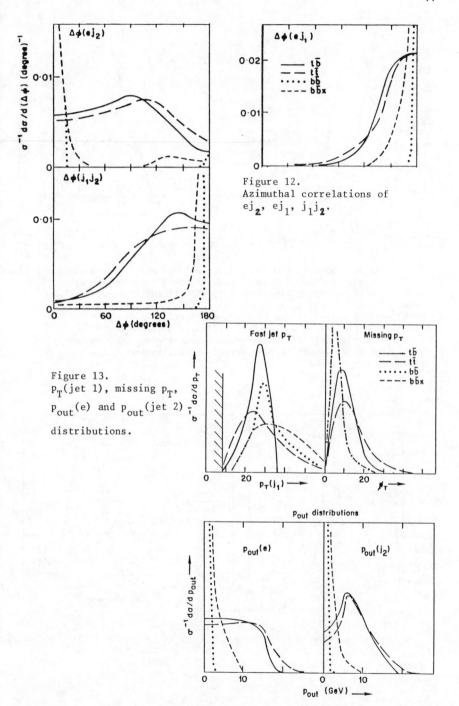

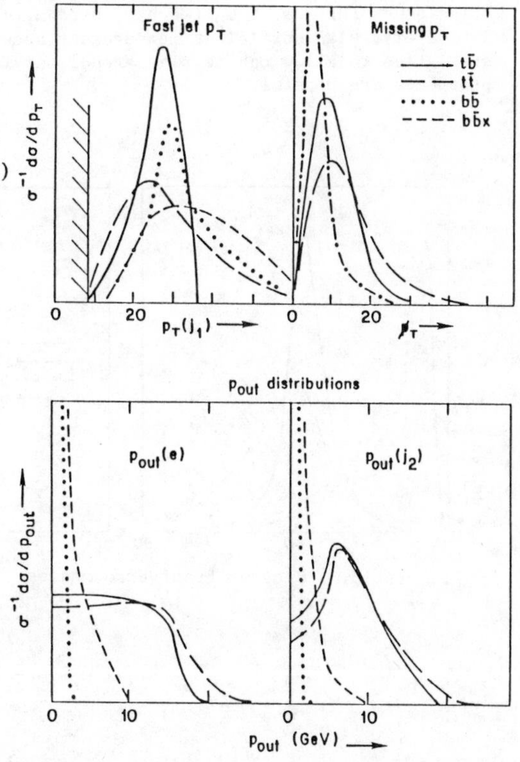

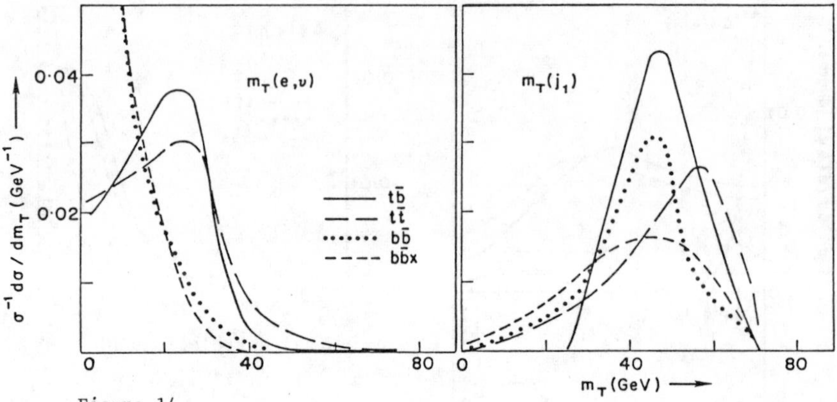

Figure 14.

Finally, supposing the non-top backgrounds are disposed of, can we hope to distinguish $t\bar{b}$ from $t\bar{t}$ contributions? In principle, yes. The $t\bar{t}$ signal is less sharply peaked in the transverse masses $m_T(ej_2, \nu)$ and $m_T(ej_1j_2, \nu)$: see Figure 15. The fact that $t\bar{t}$ peaks near $t\bar{b}$ is partly due to the cuts, partly due to the coincidence that $2m_t \approx m_W$ (we have used $m_t = 40$ throughout). Ultimately, with sufficient measurement accuracy and good enough statistics to allow one to play around varying the cuts, the prospects are hopeful.

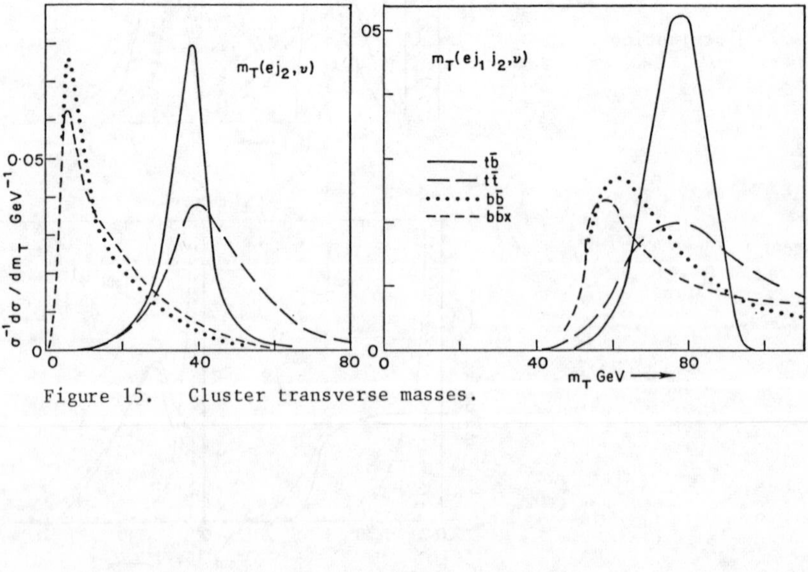

Figure 15. Cluster transverse masses.

Incidentally, some exotic backgrounds have recently been mooted. Hall and Raby[23] suggested that squark pair production (with one semileptonic and one hadronic decay giving $\tilde{q}\tilde{\bar{q}} \to q\bar{q}\ e\nu\tilde{\gamma}\tilde{\gamma}$) could fake the e + 2jet signal, but detailed calculation shows it could not[24]. Also Baer and Tata[25] wondered whether top decays via a light stop (e.g. $t \to \tilde{t}\tilde{\gamma} \to be\nu\tilde{\gamma}\tilde{\gamma}$) could confuse the issue but decided it could not. In particular, both ideas lead to too much missing p_T.

Experimentally, UA1 last year reported a handful of candidate $W \to t\bar{b}$ events[10]. Their distributions (including m_T) look about right but their rate is a bit high - above expectation by a factor 2.3 ± 0.9. See ref. 10 for all the details.

VERTEX DETECTION

In the future, vertex detectors will supply an important new ingredient. By tagging b,c decays they will greatly tighten the analysis, positively identify b or c jets, etc.

They also might allow direct topological identification[26] of top decays from cascades with a side-branch like Fig. 16.

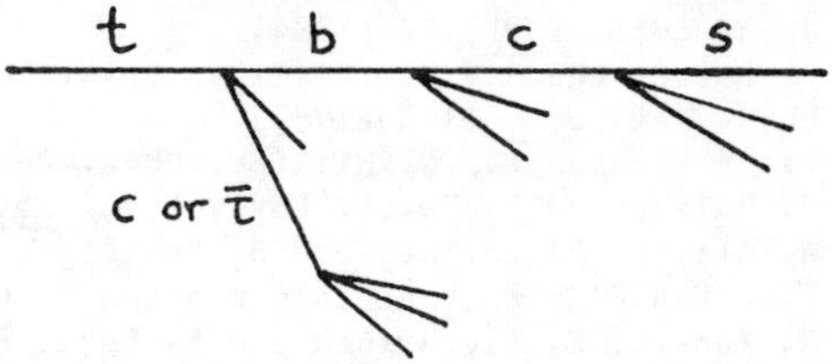

Figure 16.
t-decay chain
with a side-branch
at the first vertex

Surprisingly, as many as 44% of $t\bar{b}$ systems and 70% of $t\bar{t}$ systems should have one or more such side-branches at the first t or $\bar{t}$ decay. Although b-decay can have side-branches through $b \to c\bar{c}s$ or $b \to c\tau\bar{\nu}$ (with $\bar{c}$ or τ branching off), it cannot fake the full t chain. However, we need to identify all decay vertices and to establish their parentage (i.e. ordering), which requires a resolution of order 20μm or better[26].

Apparently side-branching backgrounds can come from charm plus beauty double-production $p\bar{p} \to b\bar{b}c\bar{c}$ ... : these may be distinguishable by dynamical features but will be interesting anyway in their own right.

CONCLUSIONS

The prospects are good for top and bottom physics at $p\bar{p}$ colliders. The hope is eventually to check the QCD/electroweak production mechanisms and weak decay processes of heavy quarks, to study $B^o/\bar{B}^o$ mixing, to confirm top and see if there is any new physics beyond.

The data so far are broadly similar to expectations but there is a long way to go: detailed analysis is only just becoming possible.

REFERENCES

1. M. Abud, R. Gatto and C.A. Savoy, Phys. Lett. 79B, 435 (1978); S. Pakvasa et al., Phys. Rev. D20, 2862 (1979); N. Cabibbo and L. Maiani, Phys. Lett. 87B, 366 (1979); R. Horgan and M. Jacob, Phys. Lett. 107B, 395 (1981); L.L. Chau, W.Y. Keung and S.C.C. Ting, Phys. Rev. D24, 2862 (1981); F.E. Paige, AIP Conf. Proc. No. 85, p. 168 (1981).
2. V. Barger, A.D. Martin and R.J.N. Phillips, Phys. Lett. 125B, 339, 343 (1983); R.M. Godbole, S. Pakvasa and D.P. Roy, Phys. Rev. Lett. 50, 1539 (1983); F. Halzen and D.M. Scott, Phys. Lett. 129B, 341 (1983); G. Ballocchi and R. Odorico, Phys. Lett. 136B, 126 (1984); K. Hagiwara and W.F. Long, Phys. Lett. 132B, 202 (1983); V. Barger et al., Phys. Rev. D28, 145 (1983), D29, 887, (1984); D.P. Roy, Z. Phys. C21, 333 (1984); L.M. Sehgal and P. Zerwas, Nucl. Phys. B234, 61 (1984); R. Odorico, Nucl. Phys. B242, 297 (1984); R. Horgan and M. Jacob, Nucl. Phys. B238, 221 (1984); J. Lindfors and D.P. Roy, Z. Phys. C24, 271 (1984).
3. V. Barger et al, Phys. Rev. D29, 1923 (1984); I. Schmitt et al, Phys. Lett. 139B, 99 (1984).
4. V. Barger and R.J.N. Phillips, Phys. Lett. 143B, 259 (1984), Phys. Rev. D32, 1128 (1985).
5. V. Barger and R.J.N.Phillips, Phys. Rev. D31, 215 (1985); F. Halzen and P. Hoyer, Phys. Lett. 154B, 324 (1985).
6. A. Ali, E. Pietarinen and B. Van Eijk, Eurojet Monte Carlo (see Van Eijk talk at this meeting or CERN-EP/85-121).
7. Z. Kunszt, E. Pietarinen and E. Reya, Phys. Rev. D21, 733 (1980); F.A. Berends et al, Phys. Lett. 103B, 124 (1981).
8. B. Combridge, Nucl. Phys. B151, 429 (1979); V. Barger et al, Phys. Rev. D24, 1428 (1981); R. Odorico, Phys. Lett. 118B, 425 (1982).
9. D.W. Duke and J.F. Owens, Phys. Rev. D30, 49 (1984).
10. UA1 collaboration: G. Arnison et al, Phys. Lett. 147B, 493 (1984).
11. UA1 collaboration: G. Arnison et al, Phys. Lett. 122B, 103 (1983), 126B, 398 (1983), 129B, 273 (1983), 134B, 469 (1984), 135B, 250 (1984), 147B, 241 (1985), CERN-EP/85-108.
12. UA2 collaboration: M. Banner et al, Phys. Lett. 122B, 476 (1983); P. Bagnaia et al, Phys. Lett. 129B, 130 (1983), Z. Phys. C24, 1 (1984).
13. K. Chadwick et al, Phys. Rev. D27, 475 (1983); H.J. Behrend et al, Z. Phys. C19, 291 (1983); E. Fernandez et al, Phys. Rev. Lett. 50, 2054 (1983); B. Adeva et al, Phys. Rev. Lett. 51, 443 (1983); G.G. Hanson, Proc. Brighton Conference (1983); G. Levman et al, Phys. Lett. 141B, 271 (1984); M. Althoff et al Z. Phys. C22, 219 (1984); W. Bartel et al, DESY 85-071 (1985).
14. C. Peterson et al, Phys. Rev. D27, 105 (1983).
15. E.W.N. Glover, F. Halzen and A.D. Martin, Phys. Lett. 141B, 429 (1984); R. Kinnunen, Z. Phys. C25, 167 (1984);
16. A. Ali and C. Jarlskog, Phys. Lett. 144B, 266 (1984).

17. UA1 collaboration: G. Arnison et al, Phys. Lett. 155B, 442 (1985); N. Ellis, report at Aosta meeting (1985).
18. V. Barger, A.D. Martin and R.J.N. Phillips, Phys. Lett. 125B, 343 (1983), 151B, 463 (1985).
19. W.L. Van Neerven, J.A.M. Vermaseren and K.J.F. Gaemers, NIKHEF report H/82-20 (1982); UA1 collaboration, Phys. Lett. 129B, 273 (1983).
20. V. Barger, A.D. Martin and R.J.N. Phillips, Phys. Lett. 125B, 339 (1983).
21. E. Berger et al, Phys. Lett. 140B, 259 (1984).
22. V. Barger, A.D. Martin and R.J.N. Phillips, Phys. Rev. D28, 145 (1983); D29, 887 (1984).
23. L. Hall and S. Raby, Phys. Lett. 153B, 433 (1985).
24. V. Barger, W.Y. Keung and R.J.N. Phillips, Phys. Rev. D32, 320 (1985).
25. H. Baer and X. Tata, CERN-TH. 4147/85.
26. V. Barger and R.J.N. Phillips, Nucl. Phys. B250, 741 (1985).

ENERGY LEVEL SPACING IN TOPONIUM AND THE STRONG INTERACTION COUPLING CONSTANT*

D.B. Lichtenberg, L. Clavelli, and J.G. Wills
Physics Department, Indiana University,
Bloomington, IN 47405

ABSTRACT

We use a potential model of a form motivated by quantum chromodynamics to calculate the energy level spacings in toponium. In the potential, we use a strong-interaction coupling strength estimated from various high-energy experiments. We find that, for a top quark mass of about 40 GeV, the energy difference of the first two levels in toponium is at least 700 MeV, or at least 100 MeV larger than the analogous energy differences in charmonium and bottomonium.

I. INTRODUCTION

The toponium ($t\bar{t}$) mass spectrum may be difficult to measure in the near future with hadron machines, and may have to wait for the high-energy electron-positron accelerators now under construction. It is nevertheless useful to point out the importance of measuring the mass spectrum of toponium for what it can tell us about the coupling strength α_s governing quantum chromodynamics (QCD). Other aspects of toponium are discussed in a talk by A.D. Martin at this conference.

Quark-antiquark potentials models have been fairly successful in leading to a qualitative understanding of the spectra of the charmonium ($c\bar{c}$) and bottomonium ($b\bar{b}$) families. However, as pointed out by a number of authors, and emphasized by Buchmüller and Tye,[1] the spectra of charmonium and bottonium do not determine the quark-antiquark potential very well outside a limited region of interquark separation r. In fact V is not really known from experiment outside the interval 0.1 fm $\leq r \leq$ 1 fm. Because of this circumstance, various authors have been able to obtain rather good fits to the heavy meson spectra with quite a wide variety of potentials, such as Coulomb plus linear,[2] logarithmic,[3] and power-law,[4] most of which are similar in the limited interval between 0.1 and 1 fm.

---

* Presented by the first-named author.

Of the potentials that have been considered in the literature, the ones most justified from QCD behave at small r like a Coulomb potential weakened logarithmically in accordance with assymptotic freedom.[5] Such potentials are able to fit the charmonium and bottomonium data as well as the others, but the data do not require the use of such potentials. For this reason, Buchmüller and Tye[1] and others have suggested that a measurement of the energy levels of toponium would give important information about the quark-antiquark potential at short distances. The reason for the sensitivity of the toponium system to the small-r behavior of the potential is the large mass m_t of the t quark, which the UA1 group finds is most probably in the range 30 GeV $\leq m_t \leq$ 50 GeV.

However according to Hagiwara et al.[7] there is a limitation on how well a measurement of the toponium mass spectrum determines the short-distance behavior of the potential, and, in particular, how much it constrains α_s. Basically, the argument of Hagiwara et al. is that the toponium mass spectrum is relatively insensitive to a simultaneous increase in α_s and m_t. Granting this insensititivity, a measurement of the toponium spectrum can still discriminate among the potentials suggested in the literature. For example, Moxhay and Rosner[8] find that various potentials in the literature lead to an energy difference ΔE_{21} between the 2S and 1S levels of as little as 520 GeV or as much as 1000 GeV.

In this talk we shall take a more modest approach of using an estimate of α_s obtained from other experiments to predict the energy differences between the first three levels of toponium. The estimate we use[9] for α_s is a rather small one compared to some others that have been made. We therefore regard our prediction that the energy difference ΔE_{21} is about 700 MeV (for m_t = 40 GeV) as a lower limit. We believe that a value of ΔE_{21} as high as 800 MeV or more is compatible with the α_s from other experiments. However, if the experimental value of ΔE_{21} should turn out to be significantly smaller than 700 MeV, we would not know how to explain the result in terms of asymptotically free QCD.

II. THE MODEL

We assume that, to a first approximation, the heavy mesons of the charmonium, bottomonium, and toponium families can be described by a flavor-independent central potential V(r) which depends only on the interquark separation r. We use the potential in a Schrödinger equation of the form

$$-\nabla^2\Psi_n + mV(r)\Psi_n = k_n^2\Psi_n, \qquad (1)$$

where m is the quark mass (m_c, m_b, or m_t), and Ψ_n and k_n^2 are the eigenfuctions and momentum-squared eigenvalues respectively. Although this equation is basically nonrelativistic, we use relativistic kinematics to obtain the meson masses M_n according to the formula

$$M_n = 2(m^2 + k_n^2)^{1/2} \qquad (2)$$

Especially in charmonium, we find that the masses given by Eq. (2) differ significantly from the masses obtained by using the nonrelativistic expression $M_n = 2m + k_n^2/m$.

In principle, our potential should contain a spin-dependent term which depends on flavor through the inverse square of the quark mass. Nevertheless, in this treatment we neglect spin-dependence, and assume our potential is the appropriate one to describe the triplet-S states of heavy quarkonia. This neglect of spin dependence is most serious for charmonium, and therefore we attach more importance to good agreement of our model with the observed bottomonium spectrum than with the observed charmonium spectrum.

We take for our potential a form

$$V(r) = -4\alpha_s(r)/(3r) + V_i + br \qquad (3)$$

where the first term, which is dominant at small r, arises from asymptotically-free QCD, the second term V_i is a phenomenological potential which may be important at intermediate r, and the third term is motivated by non-perturbative lattice QCD. In order to satisfy asymptotic freedom, $\alpha_s(r)$ should behave like (as $r \to o$)

$$\alpha_s(r) \longrightarrow -\frac{6\pi}{33-2n_f}\,\frac{1}{\ln \lambda r} \qquad (4)$$

where n_f is the number of quark flavors, and λ is related to the QCD scale parameter Λ by the approximate expression[10]

$$\lambda = \Lambda e^{\gamma} = 1.78\Lambda, \qquad (5)$$

where γ is Euler's constant. The form of $\alpha_s(r)$ given in Eq. (4) must be modified at larger r in order to be useful, because, in particular, the $\alpha_s(r)$ of Eq. (4) blows up when $\lambda r = 1$. We shall discuss later (in Section III) how we modify $\alpha_s(r)$ for finite r.

Let us now assume that we have suitably modified $\alpha_s(r)$ and ask whether we can get a good fit to the $c\bar{c}$ and $b\bar{b}$ spectra with $V_i = 0$. The answer is that we can, but the value of α_s (at some appropriate value of r, say r = 0.2 fm) needed to fit the data is appreciably larger than values obtained from other experiments.[2] We here adopt an alternative approach and attempt to fit the $c\bar{c}$ and $b\bar{b}$ data with a small α_s consistent with other experiments and with $V_i \neq 0$. To be specific, we take V_i to be of the form $V_i = a_1 \ln a_2 r$, where a_1 and a_2 are constants to be determined. We put constraints on these constants so as to satisfy a criterion suggested by Buchmüller and Tye,[1] namely, that in the interval of r where α_s is small, lowest-order QCD perturbation theory should give a good approximation to the potential. We take this to mean that for $r \leq 0.2$ fm, $|V_i + br|$ should be smaller than $|\alpha_s/r|$. We find that it is just barely possible to satisfy the constraints and get reasonably good fits to the $c\bar{c}$ and $b\bar{b}$ data.

III. THE FORM OF THE STRONG-INTERACTION COUPLING STRENGTH

One of us[9] has made a systematic study of $\alpha_s(Q^2)$, where Q is the four-momentum transfer, from various high-energy experiments. These include measurements of R (the ratio of hadrons to muons in electron-positron annihilation), measurements of jets, and measurements of decays rates (but not masses) of heavy quarkonia. In this same paper,[9] the second-order renormalization group equation for α_s was integrated to obtain the Q^2 dependence of that quantity, with α_s taken from experiment at one value of Q^2 ($\alpha_s = 0.11$ at 30 GeV). The function $\alpha_s(Q^2)$ determined in this way agrees rather well with the experimental values at a variety of Q^2. However, it should be pointed out that measurements of R at 30 GeV tend to give a somewhat larger value of α_s than the other methods considered.

The behavior of α_s as a function of Q^2 is often given in terms of the QCD scale parameter Λ. In particular, the value $\Lambda_{\overline{ms}}$ in the modified minimal-subtraction scheme is given. We wish to emphasize that because the effective number of quark flavors is a function of Q^2, straightforward integration of the renormalization group equations leads to a $\Lambda_{\overline{ms}}$ which is also a function of Q^2. We therefore focus directly on the coupling strength α_s rather than on the QCD scale parameter Λ.

Our procedure for finding $\alpha_s(r)$ starting from the numerical values of $\alpha_s(Q^2)$ obtained from the renormalization group equations is the following:

1. We approximate $\alpha_s(Q^2)$ by an analytic function of Q^2 which is finite in the entire interval $0 < Q^2 < \infty$. This approximate analytic form is not very well determined in the small Q^2 region, and the extrapolation to this region must be considered to be phenomenological.

2. We replace the four-dimensional Q^2 by the three-dimensional $\vec{Q}^2$ in $\alpha_s(Q^2)$ and numerically take the three-dimensional fourier transform of $\alpha_s(\vec{Q}^2)/\vec{Q}^2$. We identify this fourier transform (suitably normalized) with $\alpha_s(r)/r$.

3. We find an approximate analytic formula for $\alpha_s(r)$ which agrees rather well with the numerical value, is finite in the interval $0 < r < \infty$, and which reduces to the form of Eq. (4) as r goes to zero.

Assuming that there are only six quark flavors, our analytic expression is

$$\alpha_s(r) = \frac{2\pi}{7} \quad \frac{1}{1 + cr} \quad \frac{\lambda r - 1}{\ln \lambda r}, \qquad (6)$$

where c is a constant and λ is the following function of r:

$$\lambda = c_1 + c_2 r^{1/2}, \qquad (7)$$

with c_1 and c_2 additional constants. The values of c_1, c_2, and c obtained from this procedure are

$$c_1 = 11 \text{ MeV},\ c_2 = 137 \text{ MeV fm}^{-1/2},\ c = 0.05 \text{ fm}^{-1}. \qquad (8)$$

It should be emphasized that c_1, c_2, and c are determined from a procedure which has nothing to do with the energies of the quarkonia spectra. They are not parameters to be adjusted in a fit to the energy levels of the $c\bar{c}$ and $b\bar{b}$ systems.

Because $\alpha_s(Q^2)$ is not well determined at small Q^2, our expression (6) for $\alpha_s(r)$ should be regarded as purely phenomenological in the large-r region. Since the value of c affects our expression for α_s only at large r, it is not well determined. For definiteness we use $c = 0.05$ fm^{-1}, but our results are not very sensitive to the value of c.

RESULTS AND DISCUSSION

In fitting the charmonium and bottomonium triplet-S wave spectra, we varied the quark masses m_c and m_b, and the potential parameters a_1, a_2, and b, subject to the constraint that in the small-r region the term $-4\alpha_s/(3r)$ should be dominant. It is not easy to satisfy this contraint with $\alpha_s(r)$ given by Eqs. (6)-(8). It appears that our model "prefers" a somewhat larger $\alpha_s(r)$. Nevertheless, we obtained a fit which is satisfactory in the sense that its quality is about the same as that obtained with most other potential models. The values of the parameters turn out to be

$$a_1 = 333 \text{ MeV},\ a_2 = 5.08 \text{ fm}^{-1},\ b = 605 \text{ MeV fm}^{-1},$$
$$m_c = 1.55 \text{ GeV},\ m_b = 4.73 \text{ GeV}.$$

With all parameters of the potential fixed, we are able to calculate the mass spectrum of toponium for various values of the mass of the t quark. Of primary interest are the energy differences ΔE_{21} and ΔE_{32}, and we exhibit these in Table I for t-quark masses of 30, 40, and 50 GeV. For comparison we also show the experimental values of the analogous energy differences in charmonium and bottomonium.

Table I. Predicted energy-level splittings in toponium compared with the experimental splittings in charmonium and bottomonium. The notation ΔE_{nm} means the energy difference between the nS and mS levels.

System	m_t (GeV)	ΔE_{21} (MeV)	ΔE_{32} (MeV)
$t\bar{t}$	30	673	327
$t\bar{t}$	40	702	332
$t\bar{t}$	50	725	336
$c\bar{c}$ (experiment)		590	344
$b\bar{b}$ (experiment)		563	332

We see that the energy difference between the first excited state and the ground state of toponium is about 100 MeV larger than the analogous differences in charmonium and bottomonium. The reason for this is that the ground-state energy of toponium is significantly lowered because the toponium wave function is pulled into the Coulomb part of the potential. The Coulomb term in the potential is not so important for the excited states, and we see that the energy difference Δ_{32} is about the same in our model for all heavy quarkonia.

We should emphasize again that we have obtained this result using an unusually small value of $\alpha_s(r)$. A somewhat larger value would still be consistent with other high-energy experiments and would increase the calculated value of ΔE_{21} in toponium. However, we do not know of any mechanism consistent with asymptotically free QCD which would lead to a value of ΔE_{21} for toponium which is significantly smaller than the value shown in Table I.

ACKNOWLEDGEMENTS

One of us (DBL) is grateful to M.G. Olsson for an enlightening discussion. This work was supported in part by the Department of Energy.

REFERENCES

1. W. Buchmüller and S.-H.H. Tye, Phys. Rev. D24, 132 (1981).
2. E. Eichten et al., Phys. Rev. Lett. 34, 369 (1975).
3. M. Machacek and Y. Tomozawa, Ann. Phys. (NY) 110, 407 (1978);
   C. Quigg and J.L. Rosner, Phys. Lett. 71B, 153 (1977).
4. A. Martin, Phys. Lett. 93B, 338 (1980).
5. A representative list of such potentials is given in Ref. 1.
6. G. Arnison et al., Phys. Lett. 147B, 493 (1984).
7. K. Hagiwara, S. Jacobs, M.G. Olsson, and K.J. Miller, Phys. Lett. 131B, 455 (1983).
8. P. Moxhay and J.L. Rosner, Phys. Rev. D31, 1762 (1985).
9. L. Clavelli, Intl. Symposium on High Energy e^+e^- Interactions, Nashville, Tenn. (Apr., 1984).
10. W. Celmaster, H. Georgi, and M. Machacek, Phys. Rev. D17, 879 (1978).

DETECTING TOPONIUM

A.D. Martin
Department of Physics, University of Durham, England

ABSTRACT

After surveying the properties of toponium, we discuss the possibility of detecting $(t\bar{t})$ states at the $p\bar{p}$ colliders and at the forthcoming high energy e^+e^- colliders.

THE TOPONIUM SPECTRUM

The flavour-independent potential model approach has been very successful in accounting for the spectrum of $c\bar{c}$ and of $b\bar{b}$ bound states, and there is the exciting possibility that the much more massive toponium $(t\bar{t})$ system, when it is observed, will probe the potential in the short-distance regime where perturbative QCD is applicable. Although the $Q\bar{Q}$ potential has not, as yet, been derived from first principles and although there is no complete understanding of why flavour independence works so well, there are general expectations for the form of the potential. As implied above, at short distances it is expected to have the "QCD-like" form

$$V(r) \sim -\frac{4}{3}\frac{\alpha_s}{r} \sim -\frac{C}{r\ln(1/\Lambda^2 r^2)} \quad \text{as} \quad r \to 0 \tag{1}$$

where the constant C is known, whereas at large distances it is expected to exhibit linear confinement

$$V(r) \sim Kr \quad \text{as} \quad r \to \infty .$$

It is natural to identify K with the string tension, which is given in terms of the Regge slope by $K = 1/2\pi\alpha' \simeq 0.2\ \text{GeV}^2$.

It turns out that the values of r relevant to the states of the $c\bar{c}$ and $b\bar{b}$ spectra are in the intermediate region, $0.1 \lesssim r \lesssim 1$ fm, where neither of the above limits are applicable. This is well-illustrated by Fig. 1 which is taken from Buchmüller and Tye[1]. It compares a QCD-like potential with two purely empirical forms

$$V(r) = A + Br^{0.1} \quad \text{(Martin potential}^2\text{)} \tag{2}$$

$$V(r) = -\frac{a}{r} + br \quad \text{(Cornell potential}^3\text{)} . \tag{3}$$

All these potentials describe the $c\bar{c}$ and $b\bar{b}$ spectra equally well and, indeed, are virtually indistinguishable in the intermediate region where we see they have essentially a logarithmic dependence on r.

Where does toponium fit into this discussion? The toponium spectrum will be much richer even than that for $b\bar{b}$, since it can be readily shown[4] that the number of narrow S states of a $Q\bar{Q}$ system (which lie below the threshold for OZI-allowed decays) grows with

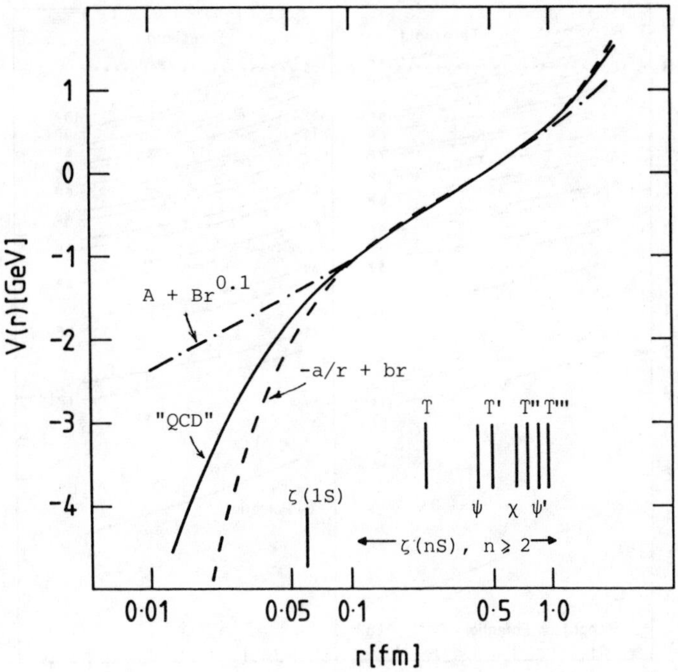

Fig. 1 Potentials which successfully describe the $c\bar{c}$ and $b\bar{b}$ spectra. Quarkonia states are displayed at their mean-square radii. The figure is adapted from a figure in ref. 1.

quark mass as

$$n \simeq \frac{1}{4} + \text{const.}\sqrt{m_Q} \ , \qquad (4)$$

independent of the form of the potential. For example, if the constant is determined from the known $b\bar{b}$ spectrum, then we find $n = 10$ taking[5] $m_t = 40$ GeV. That is there will be ten narrow S-wave $t\bar{t}$ states below the $T\bar{T}$ threshold, where T is the lightest meson containing a t quark.

Fig. 2 shows the S and P wave toponium levels for two different 'QCD-like' potentials and Fig. 3 establishes the notation we shall use for the states. Alternative notations, which exist in the literature, for the ζ state are θ, ψ_t, V or $(t\bar{t})_{1S}$. The fine and hyperfine splittings are anticipated to be at most 10 MeV and are going to be almost impossible to resolve experimentally.

Only the spacings between the lowest toponium levels depend on the form of the potential; compare, for example the sets of levels shown in Fig. 2. The reason is evident from the values of the mean

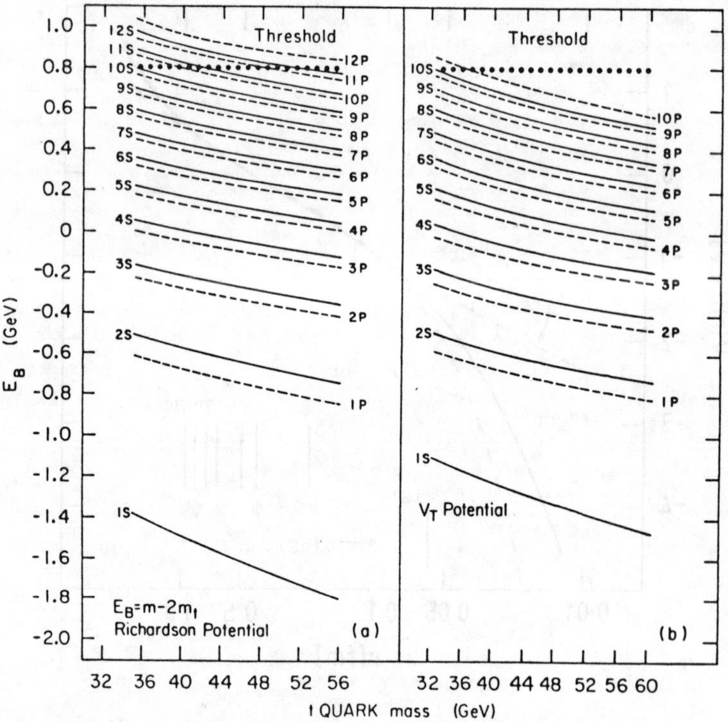

Fig. 2 S and P wave toponium levels as a function of the top quark mass for two different 'QCD-like' potentials[6,7]. The figure is taken from ref. 8.

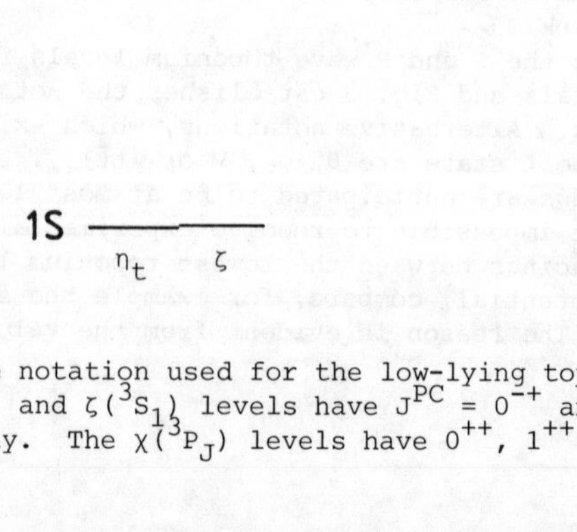

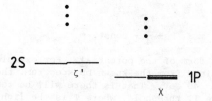

Fig. 3 The notation used for the low-lying toponium levels. The $\eta_t(^1S_0)$ and $\zeta(^3S_1)$ levels have $J^{PC} = 0^{-+}$ and 1^{--} respectively. The $\chi(^3P_J)$ levels have 0^{++}, 1^{++} and 2^{++}.

square radii of the $\zeta(nS)$ states indicated on Fig. 1. Only $\zeta(1S)$ is an effective probe of the very short distance regime. Thus a measurement of the difference[1] $E(2S) - E(1S)$ offers the best opportunity to determine the QCD scale parameter Λ, although in ref. 9 it is argued that the constraint on Λ may not be quite so tight as it appears at first sight. The ζ leptonic width is another quantity which is sensitive to $V(r)$ at short distances, see Fig. 4. To lowest-order the ζ leptonic widths are given by

$$\Gamma_{ee}(nS) = \frac{4\alpha^2 e_Q^2}{M_\zeta^2} |R_{nS}(0)|^2 \tag{5}$$

where $e_Q = 2/3$ and the radial wave function $R_{nS} = \sqrt{4\pi}\,\psi_{nS}$. Here and in Fig. 4 we have omitted the intermediate Z contribution which of course dominates Γ_{ee} if $M_\zeta \simeq M_Z$. For a 'QCD-like' potential the leptonic width $\Gamma(\zeta \to \gamma \to ee) \simeq 5$ keV, whereas had the more singular Coulomb behaviour been appropriate then $\Gamma \propto M_\zeta$ and hence Γ would be nearly an order of magnitude larger for $M_\zeta = 80$ GeV. The toponium production cross-sections are proportional to Γ_{ee} and so the value is crucial.

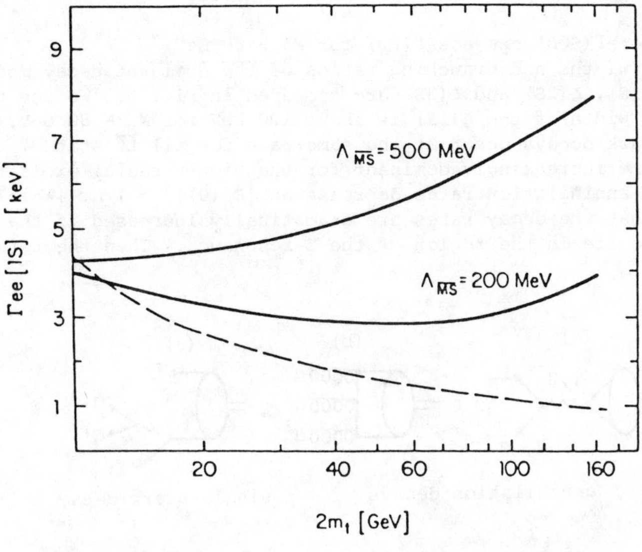

Fig. 4 The leptonic width $\Gamma(\zeta \to \gamma \to e^+e^-)$ as a function of m_t for a 'QCD-like' potential with two different values of Λ. The dashed line is for the Martin potential of eq. (2). The figure is taken from ref. 1.

TOPONIUM DECAY MODES

Since the production and decay of toponium are closely correlated, it is relevant to survey the decay characteristics of the various $t\bar{t}$ states.

The main decay modes of the $\zeta(nS)$ states are shown in Fig. 5. Modes (a) and (b) proceed through $t\bar{t}$ annihilation and depend on the wave function at the origin. Examples are $\zeta \to e^+e^-$ of eq. (5) and the decay into three gluon jets, which to lowest order is given by

$$\Gamma(\zeta \to 3g) = \frac{40(\pi^2-9)}{81\pi\, M_\zeta^2} \alpha_s^3 \, |R_{nS}(0)|^2 \tag{6}$$

and which decreases with increasing M_ζ primarily due to the factor α_s^3. Recall that the decays of the ψ and Υ states are dominated by such annihilation diagrams. However for the massive $\zeta(t\bar{t})$ states there is a significant probability that one quark decays, with the other remaining as a spectator, as shown in diagram (c). The rate[10,6] for these single quark decays (SQD) increases as m_t^5,

$$\Gamma(\mathrm{SQD}) \simeq 18\Gamma(t \to be\nu) \simeq 18 \frac{G_F^2 \, m_t^5}{192\pi^3} \tag{7}$$

and in fact $\Gamma(\mathrm{SQD})$ exceeds $\Gamma(3g)$ for $M_\zeta > 60$ GeV.

The widths and branching ratios of the dominant decay modes of the $\zeta(1S)$, $\zeta(2S)$ and $\zeta(3S)$ are compared in Fig. 6. We see that the total width of the $\zeta(1S)$ is about 100 keV for $M_\zeta \simeq 80$ GeV. Single quark decays occur at the same rate for all $t\bar{t}$ states. They thus become increasingly dominant for the higher radial excitations since the annihilation rates decrease as $|R_n(0)|^2 \sim 1/n$. We also observe that the decay rates are dramatically increased if the ζ happens to lie in the region of the Z resonance. Then the total ζ

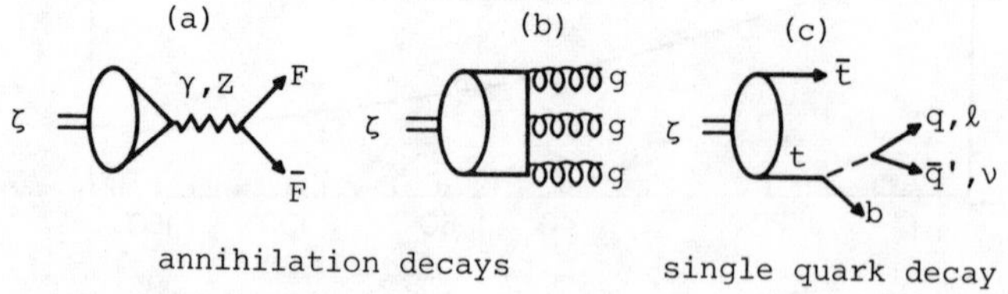

Fig. 5 The main decay modes of the ζ. Diagram (a) represents the fermionic decays $\zeta \to q\bar{q}$, $\ell^+\ell^-$, $\nu\bar{\nu}$; for $\zeta \to b\bar{b}$ there is also a W-exchange diagram. Other possible decay modes, with smaller branching fractions, are $\zeta \to gg\gamma$, $\zeta \to H\gamma$ (if the Higgs mass $M_H < M_\zeta$) and for $n \geqslant 2$ the E_1 radiative transitions $\zeta(nS) \to \chi\gamma$.

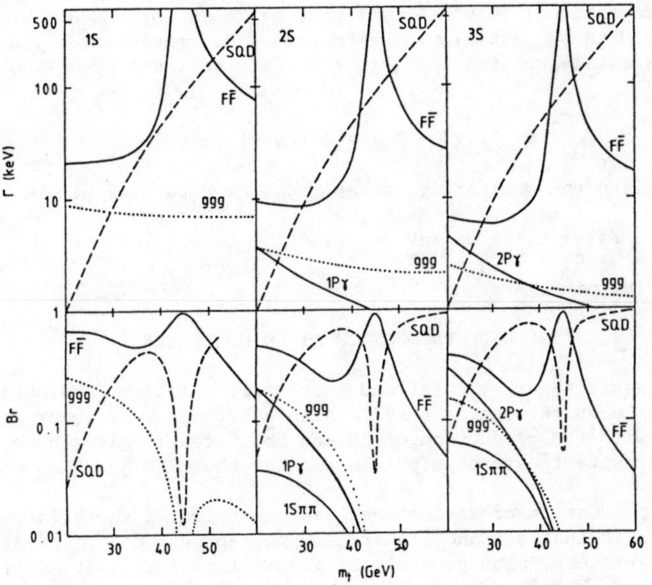

Fig. 6 The partial widths and branching fractions of the dominant decay modes of the $\zeta(1S)$, $\zeta(2S)$ and $\zeta(3S)$ states depending on the mass of the top quark (taken from ref. 6).

width could be as large as 20 MeV, and the decays dominantly proceed through diagram 5(a) with an intermediate Z. In this case the branching ratios simply reflect the vector couplings of the Z to the various $F\bar{F}$ states.

The decays of the η_t $(^1S_0)$ states are straightforward in comparison to the various competing decay modes of the ζ $(^3S_1)$ states. η_t dominantly decays into two gluon jets. To lowest order

$$\Gamma(\eta_t \to 2g) = \frac{8\alpha_s^2}{3M^2} |R_S(0)|^2 \simeq 2 \text{ MeV} \tag{8}$$

throughout the mass region of interest. Thus the single quark decays of the η_t are negligible unless $M > 120$ GeV.

For the χ_J states the annihilation amplitudes are dependent on the derivative of the P wave function at the origin. In lowest order

$$\Gamma(\chi_2 \to 2g) = \frac{4}{15}\,\Gamma(\chi_0 \to 2g) = \frac{128\alpha_s^2}{5M^4} |R_P'(0)|^2 \tag{9}$$

and are proportional to v^2/c^2, whilst $\chi_1 \rightarrow 2g$ is forbidden. With increasing M these partial widths therefore decrease rapidly, unlike the approximately constant behaviour of $\Gamma(\eta \rightarrow gg)$, and by $M \simeq 80$ GeV we have

$$\Gamma(\chi_2 \rightarrow 2g) \simeq 4 \text{ keV} . \qquad (10)$$

For comparison the dominant χ_2 decay modes have partial widths

$$\Gamma(\chi_2 \rightarrow \zeta\gamma) \simeq 25 \text{ keV} , \quad \Gamma(\text{SQD}) \simeq 50 \text{ keV} , \qquad (11)$$

at this value of M.

TOPONIUM PRODUCTION AT $p\bar{p}$ COLLIDERS

A classic paper by Baier and Rückl[11] is entitled "Hadronic collisions: a quarkonium factory". Although their work is primarily concerned with the production of $c\bar{c}$ and $b\bar{b}$ states it gives hope that the more massive $t\bar{t}$ states might be seen at the CERN and FNAL $p\bar{p}$ colliders.

The production cross-sections can be estimated rather directly. The picture is that a t and $\bar{t}$ quark are pair-produced in a hard scattering process, most probably by gluon-gluon fusion at collider energies, and that the probability that the $t\bar{t}$ pair form a bound state is determined by the appropriate quarkonium wave function. To leading order there is therefore a direct relationship between the production and decay of $t\bar{t}$ states, particularly for the η, χ_0 and χ_2 states which couple directly to two gluons, see Fig. 7(a).

The subprocess cross-section for producing one of these onia states, $\mathcal{O} = \eta, \chi_0$ or χ_2, can be written in the form[11,12]

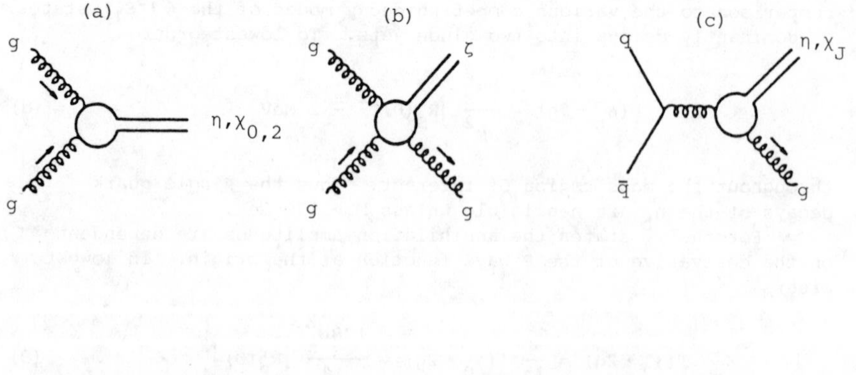

Fig. 7 Lowest-order QCD diagrams for the hadroproduction of the $\eta(^1S_0)$, $\zeta(^3S_1)$ and $\chi_J(^3P_J)$ toponium states.

$$\hat{\sigma}(gg \to \mathcal{O}) = \frac{(2J+1)\,\pi^2}{8M^3}\,\Gamma(\mathcal{O} \to gg)\,\delta\left(\frac{\hat{s}}{M^2} - 1\right)$$

where M and J are the mass and spin of the state, and where the partial width is given by eq. (8) or (9). The hadroproduction cross-section is obtained by folding $\hat{\sigma}$ with the QCD-evolved gluon distributions evaluated at $Q^2 = M^2$. For production at $p\bar{p}$ (or pp) colliders we have

$$\sigma(p\bar{p} \to gg \to \mathcal{O}) = (2J+1)\;\Gamma(\mathcal{O} \to gg)\left[\frac{\pi^2\tau}{8M^3}\int_\tau^1 \frac{dx}{x}\,D_g(x,Q^2)\,D_g\!\left(\frac{\tau}{x},Q^2\right)\right] \tag{12}$$

where $\tau = M^2/s$ and D_g is the gluon distribution in a proton. Higher-order QCD effects will give some enhancement, usually parametrized in terms of a semi-empirical multiplicative factor K, which for toponium should be at most a factor of 1.5. The factor in square brackets, which depends only on the onia mass M and the collider energy $\sqrt{s}$, is shown in Fig. 8. Given the onia width the $p\bar{p}$ (or pp) production

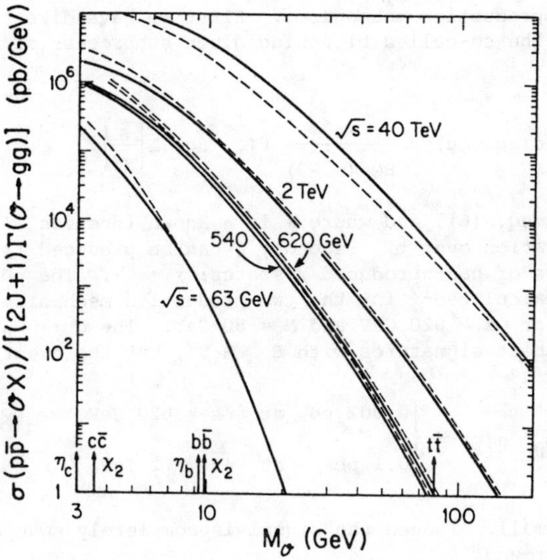

Fig. 8 The universal cross-section to width ratio for gluon-gluon fusion production of onia states versus mass for several different $\bar{p}p$ (or pp) collider energies, taken from ref. 12. Solid (dashed) curves are obtained using Duke-Owens[14] (Eichten et al.[15]) structure functions.

cross-section can be read directly off this figure. For example, with $\Gamma(\eta_t \to gg) = 2$ MeV (see eq. (8)) and M = 80 GeV, we have[12,13]

$$\sigma(\eta_t) \simeq \begin{cases} 2 \text{ pb} & \text{at } \sqrt{s} = 620 \text{ GeV} \\ 50 \text{ pb} & \text{at } \sqrt{s} = 2 \text{ TeV}. \end{cases} \tag{13}$$

The predicted event rate is very low, even though η_t is the toponium state with the largest production cross-section. Presumably the decay $\eta_t \to \gamma\gamma$ is the best signature, giving a pair of photons, essentially back-to-back in the transverse plane, with invariant mass equal to M_η. Taking $B(\eta_t \to \gamma\gamma) \sim 0.2\%$ gives

$$B_{\gamma\gamma}\ \sigma(\eta_t) \simeq \begin{cases} 0.004 \text{ pb} & \text{at } \sqrt{s} = 620 \text{ GeV} \\ 0.1 \text{ pb} & \text{at } \sqrt{s} = 2 \text{ TeV}, \end{cases} \tag{14}$$

and so η_t detection looks impossible. χ_t production is even smaller. For χ_2, for example, the expected width, eq. (10), implies $\sigma(\chi_2) \simeq \sigma(\eta_t)/100$.

Turning now to the $^3S_1$ state, ζ, we have to distinguish two different hadroproduction mechanisms. First we have direct ζ production via the so-called bleaching gluon subprocess of Fig. 7(b) with

$$\hat{\sigma}(gg \to \zeta g) = \frac{9\pi^2}{8M^3(\pi^2-9)}\ \Gamma(\zeta \to 3g)\ I\left(\frac{\hat{s}}{M^2}\right)$$

with Γ given by eq. (6), and where I is a known function[12,16] resulting from the integration over $\hat{t}$. Secondly ζ can be produced from the radiative decays of hadroproduced χ states; $\chi \to \zeta\gamma$. The $p\bar{p} \to \zeta X$ cross-section calculated[12] for the two production mechanisms are each about 0.03 pb for $\sqrt{s} = 620$ GeV and M = 80 GeV. The $\zeta \to \mu^+\mu^-$ or e^+e^- decays are the best signatures with $B_{\mu\mu} \simeq 5\%$, but the event rates

$$B_{\mu\mu}\ \sigma(\zeta) \simeq \begin{cases} 0.002 \text{ pb} & \text{at } \sqrt{s} = 620 \text{ GeV} \\ 0.1 \text{ pb} & \text{at } \sqrt{s} = 2 \text{ TeV} \end{cases} \tag{15}$$

are extremely small. Indeed the signal is completely swamped by Drell-Yan background

$$q\bar{q} \to \gamma, Z \to \mu^+\mu^- \text{ or } e^+e^-$$

as shown in Fig. 9.

The η_t and ζ cross-sections are shown as a function of collider energy in Fig. 10, along with those for the production of the analogous $c\bar{c}$ and $b\bar{b}$ states. The sad conclusion is that it will be virtually impossible to detect the toponium states at the CERN and

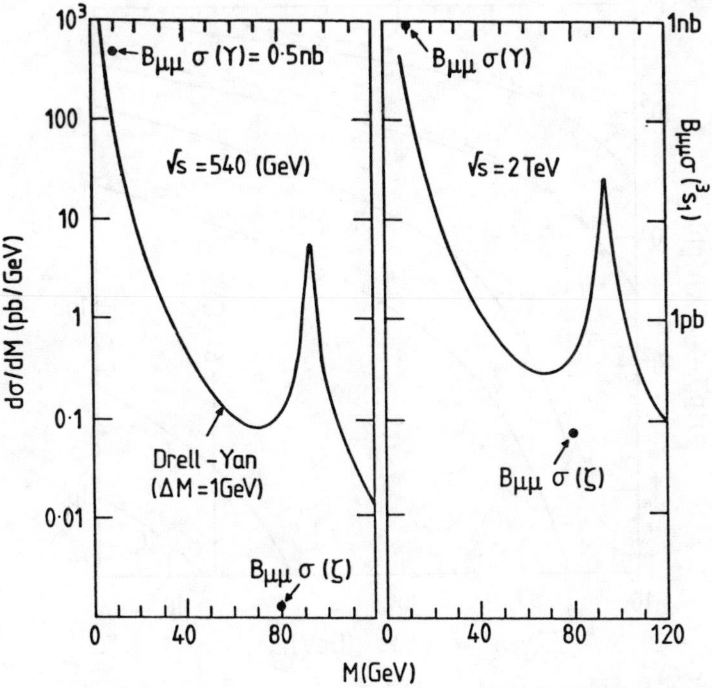

Fig. 9 The cross-section for $p\bar{p} \to (\zeta \text{ or } \Upsilon) \to \mu^+\mu^-$ at $\sqrt{s}$ = 540 GeV and $\sqrt{s}$ = 2 TeV compared with the continuum from the Drell-Yan process $p\bar{p} \to \gamma, Z \to \mu^+\mu^-$ assuming a mass resolution of ΔM = 1 GeV.

FNAL $p\bar{p}$ colliders (see eqs. (14) and (15)). The situation is entirely different for η_b and Υ, however. From Fig. 10 we predict

$$\sigma(\eta_b) \simeq 400 \text{ nb}$$
$$B_{\mu\mu}\,\sigma(\Upsilon) \simeq 0.5 \text{ nb} \tag{16}$$

at $\sqrt{s}$ = 640 GeV. How best can we exploit the enormous η_b event rate? The mode $\eta_b \to \gamma\gamma$ would require the clean identification of photons of about 5 GeV transverse momentum. Other possible characteristic modes, such as $\eta_b \to \Lambda\bar{\Lambda}$, have extremely small branching ratios. On the other hand the $\Upsilon \to \mu^+\mu^-$ signal is particularly clean, and moreover from Fig. 9 we see that it lies above the Drell-Yan background. Indeed it has already been observed with a cross-section compatible with eq. (16).

The estimates of the quarkonium hadroproduction cross-sections have been based on gluon-gluon fusion and it is relevant to ask if there are significant contributions from other parton subprocesses. Experimental information is available for charmonium production. Fig. 11 compares the gluon-gluon fusion prediction for

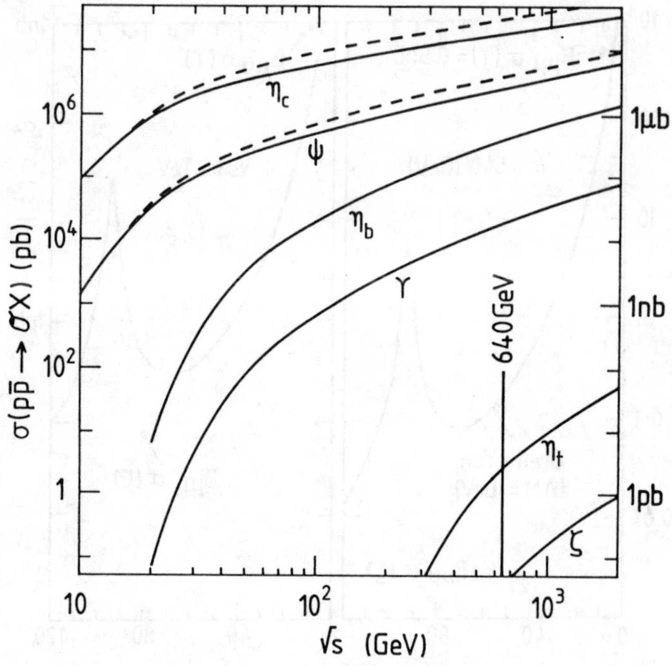

Fig. 10 The predicted cross-sections for the production of $^1S_0$ and $^3S_1$ heavy quark bound states versus the p̄p (or pp) collider energy (adapted from ref. 12). The solid (dashed) curves are obtained using Duke-Owens[14] (Eichten et al.[15]) structure functions.

J/ψ production with the data. From gg fusion we expect equal J/ψ production in pp and pp̄ collisions. However at low $\sqrt{s}$ it is found[18] that σ(pp̄) > σ(pp), though by $\sqrt{s} \simeq 20$ GeV the ratio has dropped to

$$\sigma(\bar{p}p)/\sigma(pp) = 1.46 \pm 0.25$$

consistent with the increasing importance of gg fusion as the relevant region of integration in eq. (12) moves to smaller x values, namely $x \simeq M/\sqrt{s}$. Data are also available for χ production[19,20]. Recent FNAL measurements[20] of χ_J production in both π^- Be and p Be interactions at $\sqrt{s} = 20$ GeV are shown in Fig. 12. For the proton-initiated reactions the ratio $\sigma(\chi_1)/\sigma(\chi_2)$ is small, $\sigma(\chi_2) \simeq 300$ nb, and about half of the J/ψ's are produced via radiative χ decay; these features of the data are all compatible with the predictions based on gluon-gluon subprocesses alone. On the other hand for an incident π^- we have $\sigma(\chi_1) \simeq \sigma(\chi_2)$, whereas $\sigma(\chi_1)$ should vanish at leading order, $O(\alpha_s^2)$. A plausible explanation is that the $O(\alpha_s^3)$ subprocess $q\bar{q} \to \chi g$ of Fig. 7(c) also contributes and is more important for an incident π^- than an incident proton, since we have both valence $\bar{q}$ and q available. In summary, the expectation is that gg-initiated subprocesses completely dominate at low $M/\sqrt{s}$, but that

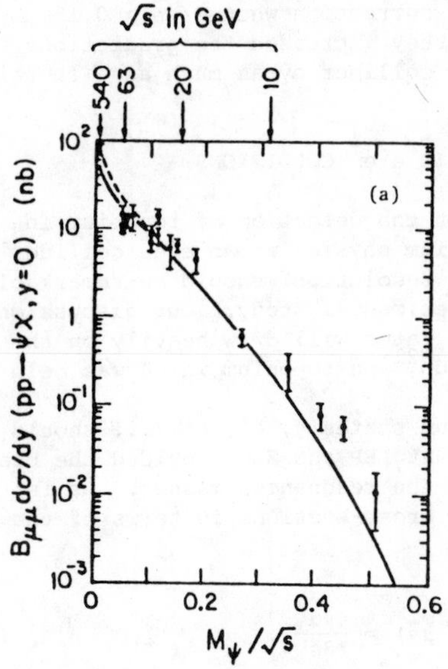

Fig. 11 A comparison of the $(gg \to \chi \to \psi\gamma) + (gg \to \psi g)$ model predictions with the data on J/ψ production in pp collisions (taken from ref. 12). An enhancement factor of K = 2 is included in the predictions.

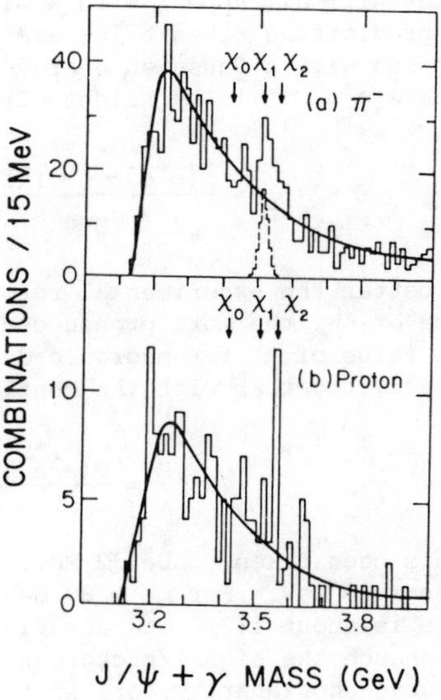

Fig. 12 χ_J production in π^- and proton collisions on Be at 200 GeV/c, taken from ref. 20. The absence of χ_0 production is consistent with the low $\chi_0 \to \psi\gamma$ branching fraction.

$q\bar{q}$ processes may make a significant correction when $M/\sqrt{s} \gtrsim 0.15$, although it would be surprising if they increased the predictions for toponium production at the CERN collider by as much as a factor of two.

TOPONIUM PRODUCTION IN e^+e^- COLLISIONS

We have been pessimistic about the detection of toponium in hadronic collisions. However toponium physics at an e^+e^- collider of sufficient energy (and good beam resolution) should be remarkably rich and should reward detailed experimental study. Our discussion of the expected structure and event rates will draw heavily on the results obtained by a LEP study group[21] on toponium which was held recently at CERN.

If $m_t \simeq 40$ GeV the $^3S_1$ toponium states ζ, ζ', ζ'', ... should be directly seen in e^+e^- collisions at LEP and SLC provided the beam energies can be accurately tuned to the resonances masses. As is usual in e^+e^- physics we shall give cross-sections in terms of the standard QED cross-section

$$\sigma_0 \equiv \sigma(e^+e^- \to \gamma \to \mu^+\mu^-) = \frac{4\pi\alpha^2}{3s}, \tag{17}$$

which at $\sqrt{s} = 80$ GeV has the value $\sigma_0 = 13.6$ pb. The narrow toponium states will all appear with a width characteristic of the beam spread. The predictions given below are made by folding the production cross-sections with a Gaussian energy spread of rms δW. For example, in the process $e^+e^- \to f\bar{f}$ the height of a resonance peak is given by[22,21]

$$R^{\zeta}_{peak} = \frac{\sigma(e^+e^- \to \zeta \to f\bar{f})}{\sigma_0} = \frac{9\pi}{2\alpha^2} \frac{\Gamma_{ee}}{\sqrt{2\pi}\ \delta W} B(\zeta \to f\bar{f}) . \tag{18}$$

The better the experimental resolution, that is the smaller is the value of δW, the more pronounced will be the resonance peak. The peak value of R^{ζ} for hadronic decays is shown in Fig. 13 as a function of mass, together with the continuum background contribution

$$R^{\gamma,Z} = \frac{\sigma(e^+e^- \to \gamma, Z \to \text{hadrons})}{\sigma_0} . \tag{19}$$

δW has been taken to be 32 MeV, which is the expected LEP resolution at $\sqrt{s} = 80$ GeV. For $m_\zeta \sim 80$ GeV we see the signal-to-background ratio is about 1:2. Characteristics of $\zeta(t\bar{t})$ decay may be used[23] to enhance the signal/background, such as requiring multijet events of large acoplanarity, or, in the case of semileptonic t, $\bar{t}$ decays, isolated energetic leptons.

Presumably the LEP and SLC e^+e^- colliders will first concentrate on the Z resonance and so a measurement of $\Gamma(Z \to t\bar{t})/\Gamma(Z \to \mu^+\mu^-)$ should yield a reasonably accurate value of m_t, see Fig. 14. However even if the mass of the top quark is known to an accuracy of one or two GeV, a lot of machine time will be required

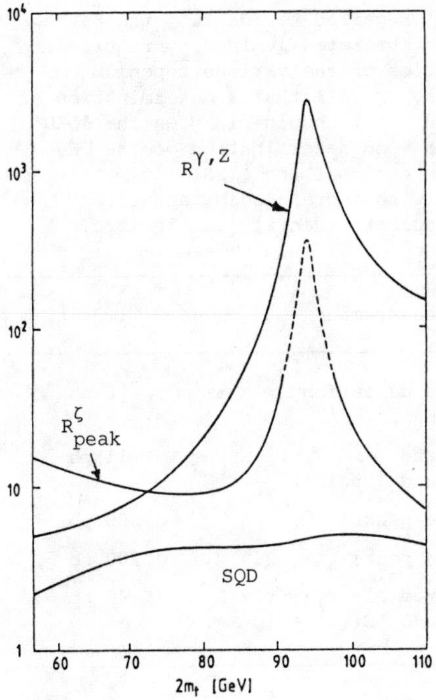

Fig. 13 The peak value of the cross-section ratio $\sigma(e^+e^- \to \zeta \to \text{hadrons})/\sigma_0$ of eq. (18), but including radiative corrections, as a function of M_ζ. The single-quark-decay component of this ratio is the curve labelled by SQD. The continuum contribution, $R^{\gamma,Z}$ of eq. (19) is also shown. The dotted part of the curve shows the region where the interference between ζ and the continuum would be important. The figure is due to Buchmüller and Kühn[21].

Fig. 14 The dependence of the $Z \to t\bar{t}$ decay rate on m_t. The dashed line gives the ratio in the absence of QCD radiative corrections. The figure is taken from ref. 24.

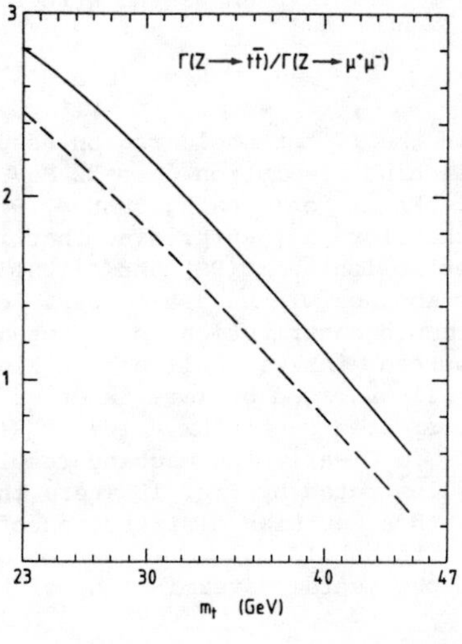

for the fine scan (in steps of δW) required to identify the narrow ζ states. Table 1 shows the e^+e^- integrated luminosity required to identify and determine the properties of the various toponium states in ideal running conditions. Recall that a determination of the properties of $\zeta(1S)$, $\zeta(2S)$ and $\chi(1P)$ should pin down the 'QCD' potential at small r and provide a good determination of the QCD scale parameter Λ. From Fig. 6 we see that the $\zeta(2S) \to \chi\gamma$ branching ratio decreases rapidly with increasing M_ζ and if $M_\zeta \gtrsim 90$ GeV it will be very difficult to identify the 1P level.

$t\bar{t}$ state	accuracy of measurement	$\int dt\, L$
1S	To find a 3 standard deviation ζ peak in a 4 GeV scan	10 pb^{-1}
1S,2S	$\Delta(M_{\zeta'}-M_\zeta) \simeq 40$ MeV and $\Delta\Gamma_{ee}/\Gamma_{ee} \simeq 10\%$ (ζ) and 20% (ζ')	11 pb^{-1}
3S-8S	To find 3σ resonance peaks	200 pb^{-1}
9S,10S	To find 3σ resonance peaks	200 pb^{-1}
1P	γ_1,γ_2 coincidence from $\zeta' \to \gamma_1\chi \to \gamma_1\gamma_2\zeta$ with $\Delta E(\gamma_1) \simeq 5$ MeV and $\Delta E(\gamma_2) \simeq 10$ MeV	40 pb^{-1}

Table 1. The LEP study group[21] estimates of the e^+e^- integrated luminosity required for toponium studies if $M_\zeta \simeq 80$ GeV and $\delta W = 32$ MeV. [An integrated luminosity of 1 pb^{-1} is equivalent to about 30 hours of running time at the planned luminosity of 10^{31} cm^{-2} s^{-1}].

Fig. 15 shows the structure expected for

$$R = (e^+e^- \xrightarrow{\gamma,Z} t\bar{t} \to \text{hadrons})/\sigma_0 \qquad (20)$$

in the $T\bar{T}$ threshold region assuming that $m_t = 40$ GeV and that the machine resolution $\delta W = 32$ MeV. Recall that the $\zeta(nS)$ resonance peaks sit on a background of $R^{\gamma,Z} \simeq 20$ (see Fig. 13). It is therefore not surprising that it will take a lot of machine time to establish the $\zeta(9S)$ and $\zeta(10S)$ states! Above threshold the resonance widths are comparable to the level spacing and so a smooth contribution is expected, even before allowing for the beam spread. As we shall see in Fig. 16, the resonance contribution is well-averaged by (or, in other words, is locally dual to) the QCD-corrected continuum value of $R(t\bar{t})$.

Clearly the machine resolution is crucial. This is well-illustrated by Fig. 16 where the narrow resonances have been smeared with a Gaussian distribution of width 60 MeV. The position of the $T\bar{T}$ threshold can no longer be observed. Indeed, in this case the experimental determination of the position of the threshold would be

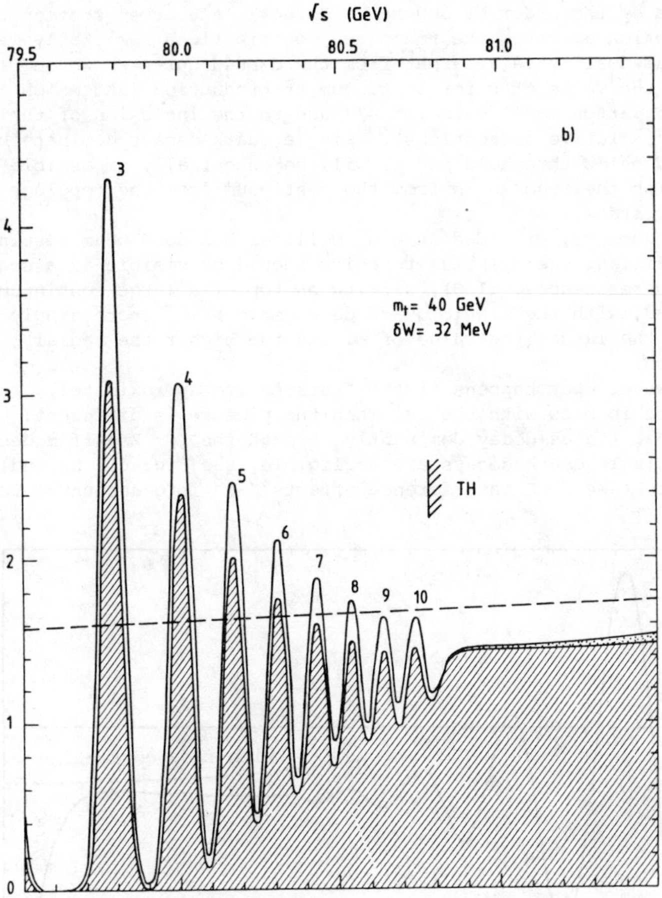

Fig. 15 The expected behaviour of R of eq. (20) in the region of the $T\bar{T}$ threshold for m_t = 40 GeV, which shows the ζ(3S), ... ζ(10S) resonance peaks each with a width characteristic of the beam spread (taken to be δW = 32 MeV). The shaded area indicates the fraction of single quark decays. The small axial-vector current contribution just above threshold is indicated by the dotted area. The dashed curve corresponds to the (massless) parton value of $R(t\bar{t})$; the curve including QCD corrections and m_t effects, but without the intermediate Z contribution, is shown in Fig. 16. The figure is taken from ref. 24.

uncertain by the order of 500 MeV. Moreover the cross-section for $t\bar{t}$ production averages the resonance contributions remarkably well. This local duality is apparent from the dashed curve shown on Fig. 16 which is the value of R for continuum $t\bar{t}$ production (and which lies below the parton model value of 4/3 due to the inclusion of the order α_s radiative corrections). Single quark decays dominate both above and below threshold and it will be essentially impossible to distinguish the resonances from the continuum from the topology of the final state.

In summary, provided an e^+e^- collider has good beam resolution (and sufficient energy) then toponium should be visible as a sequence of narrow resonances, $\zeta(nS)$, sitting on top of a large continuum background, with the single quark decay mode being increasingly dominant the larger the value of M_ζ and the higher the radial excitation.

However what happens if the ζ states are approximately degenerate in mass with the Z? Then the picture is different. First the ζ states decay dominantly through the $\zeta \to Z \to f\bar{f}$ modes, and the single quark decays are negligible, see Fig. 6. Secondly we have to take $Z-\zeta$ interference effects[25-28] into account. To

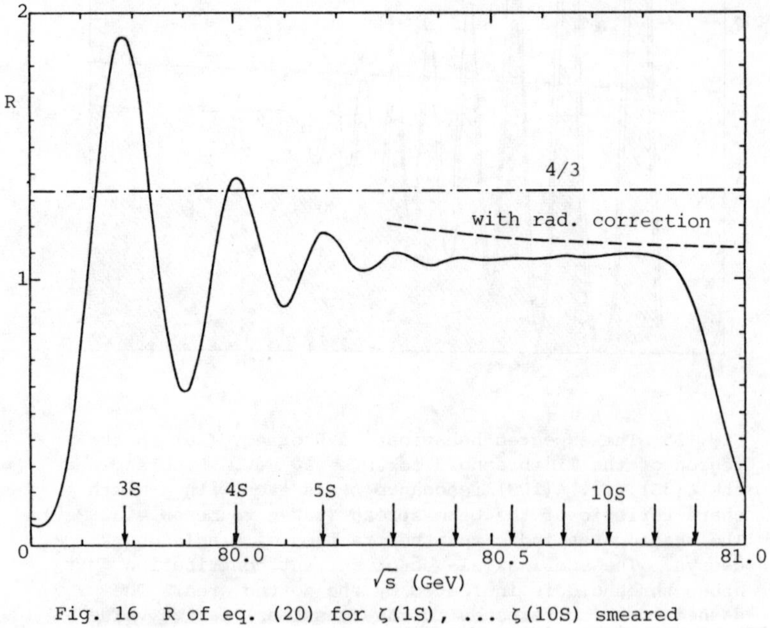

Fig. 16 R of eq. (20) for $\zeta(1S)$, ... $\zeta(10S)$ smeared with a Gaussian distribution of width δW = 60 MeV, assuming m_t = 40 GeV. For simplicity the Z contribution has been omitted. The dashed curve is the QCD-corrected value of $R = \sigma(e^+e^- \to \gamma \to t\bar{t} \to \text{hadrons})/\sigma_0$. The figure is taken from ref. 24.

consider Z - ζ mixing, we note $\Gamma_Z >> \Gamma_\zeta$ and that $M_Z\Gamma_Z$ is much larger than the effective ζ - Z coupling. It is straightforward to show[8,27] that the mixing angle is then small and lowest-order perturbation theory is inadequate.

As an example of Z - ζ interference, consider the process $e^+e^- \to f\bar{f}$ in the case when $M_\zeta \simeq M_Z$. The process may proceed either directly via the Z or via the ζ

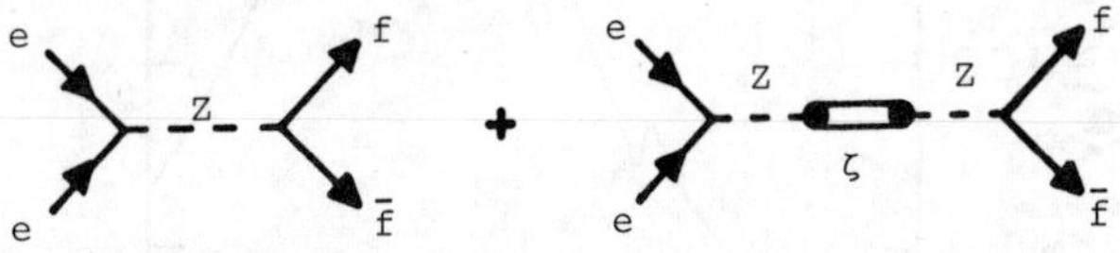

and so the amplitude has the form

$$A(s) \simeq \frac{1}{D_Z(s)} + \frac{1}{D_Z(s)}\frac{g_{\zeta Z}^2}{D_\zeta(s)}\frac{1}{D_Z(s)} \tag{21}$$

where $D(s) \equiv s - M^2 + iM\Gamma$. In the special case when $M_\zeta = M_Z = M$ we have, on resonance

$$A(s=M^2) \simeq \frac{1}{iM\Gamma_Z} + \frac{1}{iM\Gamma_Z}\frac{g_{\zeta Z}^2}{iM\Gamma_\zeta}\frac{1}{iM\Gamma_Z} \tag{22}$$

and hence

$$|A|^2 = |A_{continuum} - A_\zeta|^2 , \tag{23}$$

where the two contributions are in phase. Thus the ζ appears as a narrow dip on top of the Z resonance peak[26-29]. It is easy to see that a ζ state just below (above) the Z peak will appear as a peak-dip (dip-peak) structure. A typical result, with allowance for beam spread, is shown in Fig. 17. The ζ(1S) appears as a peak-dip structure whereas ζ(4S) and ζ(5S) are just visible as dips on top of the Z.

If M_ζ is sufficiently different from M_Z the interference effects disappear. $D_Z(s)$ is now essentially real and from eq. (21) we obtain

$$|A(s=M_\zeta^2)|^2 \simeq |A_{continuum} - iA_\zeta|^2$$
$$\simeq |A_{continuum}|^2 + |A_\zeta|^2 .$$

The inclusion of the $e^+e^- \to \gamma \to f\bar{f}$ continuum contribution does not change the argument. We therefore have an incoherent sum leading to the narrow ζ peaks sitting on top of the γ,Z continuum as shown in Fig. 15.

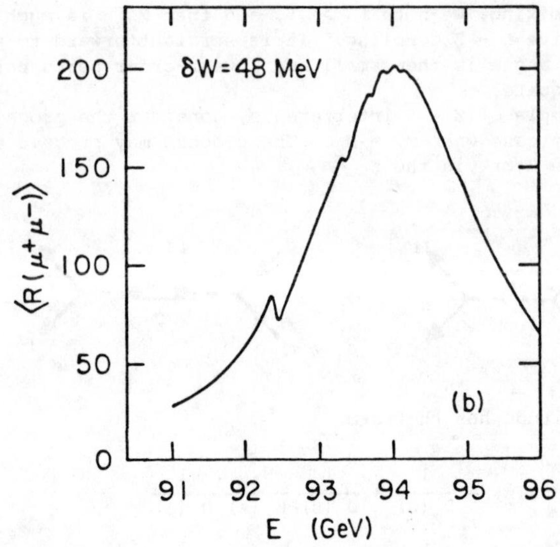

Fig. 17 ζ - Z interference effects in the ratio $\sigma(e^+e^- \to \mu^+\mu^-)/\sigma_0$, assuming $2m_t = M_Z = 94$ GeV and a beam spread $\delta W = 48$ MeV. The figure is from ref. 8.

Several years ago[30] it was pointed out that there was an interesting possibility of directly producing the $^3P_1$ toponium states (with $J^P = 1^{++}$) through the axial part of the neutral current. The rate, however, is proportional to the square of the derivative for the P wave function at the origin which is strongly suppressed for large M and, with the anticipated beam spread, the optimal signal-to-background ratio is only of the order of 0.5%[31,21]. Imposing topology cuts on the decays and improving the experimental resolution will help, but detection of the $\chi(^3P_1)$ states in this way will be extremely difficult.

Once toponium is found it opens the door to searches for new particles. The classical example is the search for the neutral Higgs scalar using the Wilczek mechanism[32] of Fig. 18. The $\zeta \to H\gamma$ branching ratio is sizeable due to the large Higgs-$t\bar{t}$ coupling. In lowest order the ratio

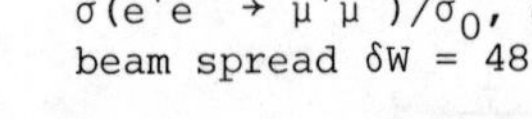

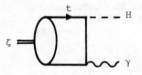

Fig. 18 The decay $\zeta \to H\gamma$.

$$\frac{\Gamma(\zeta \to H\gamma)}{\Gamma(\zeta \to \gamma \to \mu^+\mu^-)} = \frac{1}{8\sin^2\theta_W} \frac{M_\zeta^2}{M_W^2}\left(1 - \frac{M_H^2}{M_\zeta^2}\right)$$

which is approximately 0.5 if $M_\zeta \simeq M_W$ and $M_H < M_\zeta$. Radiative corrections[33,34] will somewhat reduce this ratio, particularly if

$M_H \lesssim M_\zeta$ when the decay photon is soft. However the expected event rate for $M_\zeta \simeq 80$ GeV is such that this mechanism offers the best opportunity to find the Higgs if it lies in the mass range up to M_ζ. Indeed nature would be kind if $M_\zeta \simeq 80$ GeV as this appears the optimum value for a Higgs search. For the larger values of M_ζ, $90 \lesssim M_\zeta \lesssim 100$ GeV, the ζ would decay via an intermediate Z and the $\zeta \to H\gamma$ mode would be swamped by $\zeta \to f\bar{f}$.

Another interesting possibility is to use toponium decays to search for supersymmetric particles. The different possible scenarios that may occur have been recently reviewed in ref. 31.

CONCLUSIONS

We have discussed the possibility of detecting toponium in both $p\bar{p}$ and e^+e^- collisions. At the hadron colliders the $^1S_0$ toponium state, η_t, is expected to have the largest production cross-section. The best signature is $\eta_t \to \gamma\gamma$, but the event rate appears to be much too small for detection at the CERN and FNAL $p\bar{p}$ colliders. The rate for the production and leptonic decay of the $\zeta(^3S_1)$ state (that is for $p\bar{p} \to \zeta \to \mu\mu$) is equally small and, moreover, is swamped by the Drell-Yan background.

However the rich toponium spectrum should be able to be observed in e^+e^- collisions at LEP and SLC. A sequence of ten or so narrow $\zeta(nS)$ peaks is expected, sitting on top of a sizeable continuum. The beam resolution is crucial; the high n radial excitations, just below the threshold for open top production, are only 80 MeV apart and single quark decay is dominant both for the resonances and for open top. If $M_\zeta \simeq M_Z$ the picture is different. The main toponium decays are then $\zeta \to f\bar{f}$ and simply reflect the vector couplings of the Z. In this case interesting Z - ζ interference effects would occur on the Z resonance peak; in particular a ζ state occurring exactly at the top of the Z resonance would appear as a narrow dip.

In summary, a mass of 40 GeV for the top quark would be ideal. It is large enough for toponium to probe the short-distance behaviour of the 'QCD' potential and to give the possibility of Higgs detection (via $\zeta \to H\gamma$) over the sizeable mass range up to $M_H \lesssim 2m_t$. Yet it is small enough for toponium to be sufficiently far from the Z to prevent the Z resonance dominating its properties.

ACKNOWLEDGEMENTS

It is a pleasure to thank W. Buchmüller and J.H. Kühn for providing information obtained by the LEP study group, V. Barger for an enjoyable collaboration while studying the hadroproduction of toponium and M. Olsson for useful discussions.

REFERENCES

1. W. Buchmüller and S.-H.H. Tye, Phys. Rev. D24, 132 (1981).
2. A. Martin, Phys. Lett. 93B, 338 (1980).
3. E. Eichten et al., Phys. Rev. D17, 3090 (1978); D21, 203 (1980).

4. C. Quigg and J.L. Rosner, Phys. Rep. 56, 167 (1979).
5. UA1 collaboration, Phys. Lett. 147B, 493 (1984).
6. J.H. Kühn and S. Ono, Z. Phys. C21, 395 (1984); (E) C24, 404 (1984).
7. J.L. Richardson, Phys. Lett. 82B, 272 (1979).
8. S. Güsken, J.H. Kühn and P.M. Zerwas, SLAC-PUB-3580 (1985).
9. K. Hagiwara, S. Jacobs, M.G. Olsson and K.J. Miller, Phys. Lett. 131B, 455 (1983).
10. J.H. Kühn, Acta Phys. Pol. B12, 347 (1981).
11. R. Baier and R. Rückl, Z. Phys. C19, 251 (1983).
12. V. Barger and A.D. Martin, Phys. Rev. D31, 1051 (1985).
13. S. Geer and R. Raja, Phys. Lett. 150B, 223 (1985).
14. D. Duke and J. Owens, Phys. Rev. D30, 49 (1984).
15. E. Eichten, I. Hinchliffe, K. Lane and C. Quigg, Rev. Mod. Phys. 56, 579 (1984).
16. W.-Y. Keung, Proc. of the Z^0 Physics Workshop, Cornell report 485 (1981).
17. UA1 collaboration, K. Eggert, these proceedings.
18. J. Baider et al., Z. Phys. C20, 100 (1983).
19. Y. Lemoigne et al., Phys. Lett. 113B, 509 (1982).
20. D.A. Bauer et al., Phys. Rev. Letts. 54, 753 (1985).
21. W. Buchmüller, J.H. Kühn, A. Martin, G. Coignet, F. Richard, L. Rolandi and H. Takeda, LEP study group on toponium.
22. J.D. Jackson, S. Olsen and S.-H.H. Tye, Proc. of Snowmass meeting, p175 (1982).
23. S. Goggi and G. Penso, Nucl. Phys. B165, 429 (1980).
24. S. Güsken, J.H. Kühn and P.M. Zerwas, Phys. Lett. 155B, 185 (1985).
25. F.M. Renard, Z. Phys. C1, 225 (1979).
26. J.H. Kühn and P.M. Zerwas, Phys. Lett. 154B, 448 (1985).
27. P.J. Franzini and F.J. Gilman, SLAC-PUB-3541 (1985).
28. A. Martin, CERN preprint TH.4138 (1985).
29. L.J. Hall, S.F. King and S.R. Sharpe, Harvard preprint A012 (1985).
30. J. Kaplan and J.H. Kühn, Phys. Lett. 78B, 252 (1978).
31. J.H. Kühn, CERN preprint TH.4083 (1984).
32. F.A. Wilczek, Phys. Rev. Lett. 39, 1304 (1977).
33. M.I. Vysotsky, Phys. Lett. 97B, 159 (1980).
34. J. Ellis, K. Enqvist, D.V. Nanopoulos and S. Ritz, CERN preprint TH.4143 (1985).

NEUTRINO COUNTING AT CERN COLLIDER

N.G. Deshpande
Institute of Theoretical Science, University of Oregon
Eugene, OR 97403

ABSTRACT

The measured ratio of production cross section of W to Z, $R = \sigma^{W^+ + W^-} B(W \to e\nu)/\sigma^Z B(Z \to e^+e^-)$, puts stringent limits on the number of neutrinos N_ν. The effect of supersymmetric extension of the standard model is to strengthen these limits further. Improvement in the measured value of R could not only limit N_ν, but also remove from contention some extensions of the standard model.

Improving bounds on the number of light neutrinos (N_ν), which bears directly on the number of families, is clearly of great importance. Hitherto, the best limits have come from either cosmology, whose limit $N_\nu \leqslant 4$ from the cosmological He-to-H abundance ratio can perhaps be evaded,[1] or terrestrially from the lack of observation of $K^+ \to \pi^+\nu\bar{\nu}$, which yields[2] only a limit of $N_\nu < 1400$. The discovery of the W and Z weak bosons[3] provides new ways of putting rather stringent limits on N_ν.

One such limit can be obtained from the experimental upper bound on the total width of the Z boson. If we allow the possibility of further generations and new particles that Z can decay into (including new non-standard channels), the bound on N_ν is

$$N_\nu - 3 \leqslant \frac{\Gamma_Z(\text{expt}) - \Gamma_Z(\text{stand})}{(0.182\ \text{GeV})} \qquad (1)$$

where Γ_Z(stand) is the QCD-corrected theoretical width for three families in the standard model and 0.182 GeV is the partial width for $Z \to \nu_i \bar{\nu}_i$.

$$\Gamma\ (\text{stand}) = \frac{G_F M_Z^3}{12\sqrt{2}\,\pi} \Big\{ 3(1 + a_W)$$

$$+ 3\left(1 + \frac{\alpha_s}{\pi}\right)\left[\, 3d_W + u_W(2 + \rho_t) \,\right] \Big\} \qquad (2)$$

where $a_W = 1 - 4x_W + 8x_W^2$, $u_W = 1 - 8x_W/3 + 32x_W^2/9$, $d_W = 1 - 4x_W/3 + 8x_W^2/9$ and ρ_t is the phase space correction for t quark. We assume $M_Z = 94$ GeV, $x_W = 0.22$, $\alpha_s(M_Z) = 0.12$ and $M_t = 40$ GeV, and find Γ_Z(stand) = 2.83 GeV. Using the experimental bounds[4] Γ_Z(expt) < 8.1 (7.1) GeV at 90% confidence limit (C.L.) we obtain $N_\nu < 31$ (26). It is possible to improve greatly on this limit by use of measurements of $W \to e\nu$ and $Z \to e^+e^-$ production.

The method[5] is to obtain the best bound on the ratio

$$R = \frac{\sigma^{W^+ + W^-} B(W \to e\nu)}{\sigma^Z B(Z \to e^+e^-)} \qquad (3)$$

which is directly observable. This ratio can be expressed in terms of theoretically calculable quantities as

$$R = \frac{\Gamma(W \to e\nu)}{\Gamma(Z \to e^+e^-)} \frac{\Gamma_Z}{\Gamma_W} \frac{\sigma^{W^+ + W^-}}{\sigma^Z} \qquad (4)$$

The first ratio in Eq. (4) can be calculated accurately. The width ratio depends on the accessible decay channels. In the standard model with three generations Γ_Z/Γ_W is about 1. With more generations Γ_Z

increases because of additional $Z \to \nu\nu$ decays while Γ_W is unaffected because of lack of phase-space for new decays; we shall assume that the fourth- or higher-generation charged leptons have $m_L > M_W$, though this assumption can be relaxed. When the model is extended to include light supersymmetric particles, both Γ_W and Γ_Z increase, with the Γ_Z increase still dependent on N_ν. The results depend on the particular model of supersymmetry and the masses of the particles. In popular supersymmetry models Γ_Z increases by a larger fraction than Γ_W and tends to strengthen the bounds on N_ν. This not only has the effect of limiting N_ν but it also provides tests of the validity of the supersymmetry model. We shall discuss two different models of supersymmetry including the grand unified model discussed by Ellis and Sher[6] which could provide one explanation[7] of anomalous missing p_T events.

This brings us to the discussion of the third factor in Eq. (4), i.e., the production ratio of W to Z at collider energies. The production of W and Z is governed by structure functions of the quark and antiquark distributions in the proton. To obtain these at Sp$\bar{p}$S collider energies of $\sqrt{s} \simeq 540$ GeV and 630 GeV and at $Q^2 = M_W^2$ or M_Z^2 we use the most recent calculation of Altarelli et al.[8] which includes QCD corrections and employs the renormalization-group-improved parton distributions of Duke and Owens[9] with $\Lambda_{QCD} = 0.2$ GeV. Although the expected theoretical uncertainties on the individual W and Z production cross sections are of the order of 30%, they cancel in the ratio and one finds

$$\frac{\sigma^{W^+ + W^-}}{\sigma^Z} = 3.3 \pm 0.2. \tag{5}$$

This uncertainty reflects an estimate of uncertainties due to as yet uncalculated higher-order effects and changes in Λ_{QCD} within the experimentally allowed range. It is furthermore important to note that the above ratio is rather insensitive to the energy (in the Sp$\bar{p}$S range), and to the choice of the specific structure functions. We have tried the set of Gluck, Hoffman, and Reya[10] and the set of structure functions of Ref. 9 with $\Lambda_{QCD} = 0.4$, and the results were always within the range in Eq. (4). If we choose the set of structure functions of Eichten et al.[11] (EHLQ) with $\Lambda_{QCD} = 0.2$ GeV the ratio is slightly lower than 3.0 and outside the error in Eq. (4). However, the EHLQ parametrization does not describe the experimental ratio $d_v(x)/u_v(x)$ derived from $F_2^{ep}(x)/F_2^{ep}(x)$ data, thus underestimating W production at the Sp$\bar{p}$S, which renders it inappropriate to our study. Let us also note that at the values of x most relevant to our analysis ($x \simeq$ 0.1-0.2) the discrepancy between the EHLQ d/u ratio and the data of Abramowicz et al.[12] is the most pronounced, while all the other parametrizations are consistent with the data. Whether d/u ratios from higher-Q^2 data are consistent with the low-Q^2 data of Bodek et al.[13] is still inconclusive. An eventual disagreement would indicate the presence of some higher-twist effect in SLAC data, and could result in an increase of the error estimate in Eq. (4). We have also checked for the approximate stability of the ratio under changes (within an admissible region) of M_W, M_Z, x_W, m_t, Kobayashi-Maskawa angles, the poorly known heavy quark distributions, and the assumption of SU(3) symmetry for sea quarks. We may therefore use it with reasonable reliability.

If the only extension of the standard model is more families, with the only light particles being neutrinos, we obtain

$$R = (2.715)(3.3 \pm 0.2)\,\frac{\Gamma_Z(\text{stand}) + (N_\nu - 3)\Gamma_{Z \to \nu\bar{\nu}}}{\Gamma_W(\text{stand})} \qquad (6)$$

where we have evaluated $\Gamma(W \to e\nu)/\Gamma(Z \to e^+e^-)$ using M_W = 83 GeV, M_Z = 94 GeV, and x_W = 0.22. In Fig. 1 we plot R as a function of N_ν. In evaluating Γ_Z and Γ_W we have used m_b = 4.5 GeV and m_t = 40 GeV. The QCD corrections are included with α_s = 0.12 at $Q^2 = M_W^2$. We find Γ_W = 2.82 GeV and Γ_Z = 2.83 GeV. The upper limits on R at 90% C.L. from existing data[4] are drawn to show that interesting limits on N_ν can be expected from forthcoming higher statistics data.

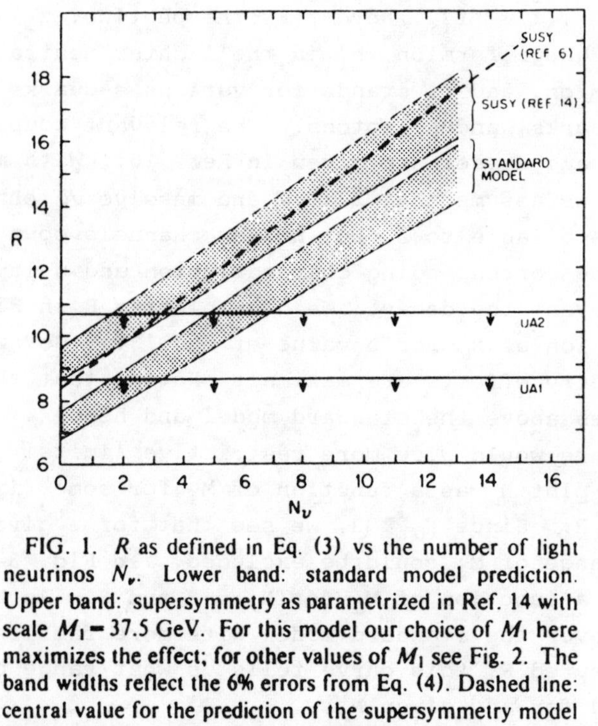

FIG. 1. R as defined in Eq. (3) vs the number of light neutrinos N_ν. Lower band: standard model prediction. Upper band: supersymmetry as parametrized in Ref. 14 with scale $M_1 = 37.5$ GeV. For this model our choice of M_1 here maximizes the effect; for other values of M_1 see Fig. 2. The band widths reflect the 6% errors from Eq. (4). Dashed line: central value for the prediction of the supersymmetry model of Ref. 6 with W gauge fermion mass $M_1 = 37.5$ GeV. Horizontal full lines: upper limits (90% C.L.) from Ref. 4. These limits include systematic as well as statistical errors.

If the extension of the standard model is supersymmetric, how will this bound behave? In order to answer this question we study two models of supersymmetry.

Model A.—We take the model of supergravity of Arnowitt, Chamseddine, and Nath[14], with SU(2) x U(1) breaking at tree level.[15] We shall use the simpler parametrization of Barger et al.,[16] which has only one scale parameter, which we take to be the mass M_1 of ω_1, the lighter mixture of the Higgs fermion and the W gauge fermion. The extra channels for W and Z decays are $W^+ \to \omega_1^+ + \tilde{\gamma}$, $\omega_1^+ + z_1$, $\omega_1^+ h_1$, and $\tilde{f}_i \tilde{f}_j$ and $Z \to \omega_1^+ + \omega_1^-$, $z_1 + h_1$, $\tilde{f}_i + \tilde{f}_i$. Here $\tilde{\gamma}$ is the photino, z_1 is the lighter Z gauge fermion, h_1 is the lighter neutral Higgs fermion, and $\tilde{f}_i$ stands for various s-quarks (scalar quarks) and s-leptons. The relevant coupling constants and masses are given in Ref. 16. With more families, we assume that $Z \to \nu\bar{\nu}$ and massive $\nu\bar{\nu}$ channels are also available to Z, but no new channels open for W because the corresponding charged lepton and s-lepton masses prevent the decay. We have plotted R in Fig. 1 as a function of N_ν for a value of $M_1 \simeq 37.5$ GeV where supersymmetry effects are maximal. Notice that the result lies above the standard model and hence measurements would give more restrictive limits. In fig. 2 we plot N_ν as a function of M_1 for some fixed values of R. Since $N_\nu \geqslant 3$, we see that for a given R, a whole range of M_1 could be excluded. In Fig. 3 we plot R as a function of M_1 for $N_\nu = 3$ and 8. As $M_1 \to \infty$, we recover the standard model with more families. For a measured R, this curve tells us what range of M_1 is allowed for any given N_ν.

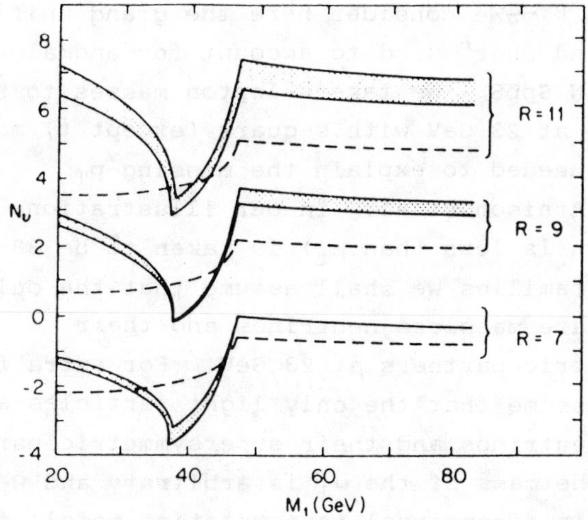

FIG. 2. The number of light neutrinos N_ν vs M_1 for the supersymmetry models of Ref. 14 (shaded regions between full lines) and Ref. 6 (dashed lines) for $R = 7, 9, 11$. 6% errors are shown only for the first model.

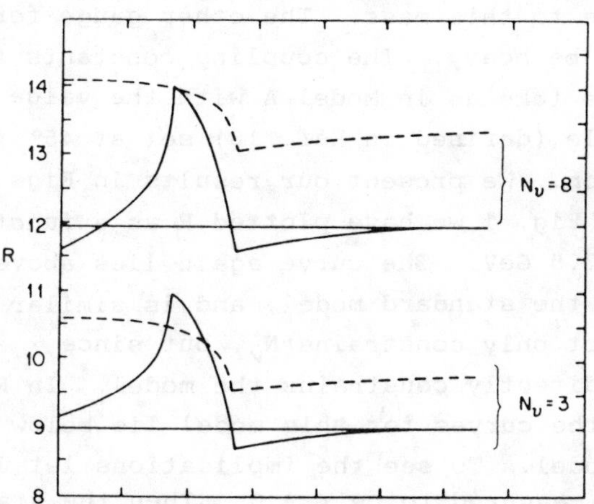

FIG. 3. R as defined in Eq. (3) vs M_1 for the standard model (horizontal lines), the supersymmetry models of Ref. 14 (full lines; M_1 is the supersymmetry scale), and Ref. 6 (dashed lines; M_1 is the W gauge fermion mass), for $N_\nu = 3$, 8. 6% errors are not shown.

Model B.—We consider here the grand unified model of Ellis and Sher[6] used to account for anomalous events at the CERN $Sp\bar{p}S$. We take s-lepton masses to be degenerate at 23 GeV with s-quark (except t) masses at 40 GeV as needed to explain the missing-p_T events of Arnison et al. In our illustrations the $\tilde{t}$ mass (which is less than m_t) is taken to be 33 GeV. For extra families we shall assume that the only light particles are massless neutrinos and their supersymmetric partners at 23 GeV. For extra families we shall assume that the only light particles are massless neutrinos and their supersymmetric partners at 23 GeV. The mass of the ω_1 is arbitrary and we take it to be M_1; in dimensional transmutation models the lightest ω_1 is expected to have a mass of order 70 GeV. The decays allowed for W and Z other than s-fermion channels are $W+ \rightarrow \omega_1^{\pm} + \tilde{\gamma}$ and $Z \rightarrow \omega_1^{+}\omega_1^{-}$. We assume that the $\tilde{\gamma}$ mass is 7 GeV though the results are insensitive to this mass. The other gauge fermions are assumed to be heavy. The coupling constants for these channels we take as in Model A with the value of the mixing angle (defined in Ref. 16) set at 45° for illustration. We present our results in Figs. 1, 2, and 3. In Fig. 1 we have plotted R as a function of N_ν for M_1 = 37.5 GeV. The curve again lies above the results of the standard model, and is similar to Model A. This not only constrains N_ν, but since $N_\nu \geqslant 3$ it directly constrains the model. In Fig. 2 we note that the curves for this model lie below the standard model. To see the implications let us assume that R was measured to be < 9.0. Then the standard model and Model A with M_1 > 47 GeV are compatible with $N_\nu = 3$; however, Model B would be ruled out in the complete range of M_1. In Fig. 3 we plot R vs M_1 for fixed N_ν for this model. It can be seen that for N_ν =

3, R > 9 for the full range of M_1 for this model. Model A and the standard model are seen to be compatible from this figure though the result is not as restrictive as Fig. 2.

We see that the measurement of R not only provides for restrictions of the number of neutrinos, but can help distinguish extensions of the standard model. This is a powerful way of testing models because one does not have to study exclusive modes. Once the test suggested here is performed and supersymmetry is found to be consistent with the data, one can go to exclusive channels and search for specific rates and distributions.[17,18] The interesting explanation of anomalous events at $Sp\bar{p}S$ thus can be subjected directly to test by measurement of R. Improvements in the accurate measurements of this value will pay rich dividends.

ACKNOWLEDGEMENTS

This work was done in collaboration with V. Barger, F. Halzen and G. Eilam, and supported by Department of Energy grant DE-FG-06-85ER 40224.

REFERENCES

1. M. Turner, in Proceedings of the Telemark Neutrino Mass Miniconference, edited by V. Barger and D. Cline (unpublished), p. 73.

2. Y. Asano et al., Phys. Lett. 107B, 159 (1981).

3. G. Arnison et al. (UA1 Collaboration), Phys. Lett. 126B, 298 (1983); P. Bagnaia et al. (UA2 Collaboration), Phys. Lett. 129B, 130 (1983).

4. J. Hansen, in Proceedings of the Aspen Winter Physics Conference, 13-19 January 1985, edited by M.M. Block (to be published); O. Vuillemin, ibid.

5. This method was proposed by F. Halzen and K. Mursula, Phys. Rev. Lett. 51, 857 (1983); D. Cline and J. Rohlf, unpublished; K. Hikasa, Phys. Rev. D 29, 1939 (1984).

6. J. Ellis and M. Sher, Phys. Lett. 148B, 309 (1984); M. Sher, private communication; C. Kounnas, A. Lahanas, D. Nanopoulos, and M. Quiros, Nucl. Phys. B336, 438 (1984), and Phys. Lett. 132B, 135 (1983).

7. V. Barger, K. Hagiwara, and W.-Y. Keung, Phys. Lett. 145B, 147 (1984); J. Ellis and H. Kowalski, Nucl. Phys. B246, 189 (1984); A. R. Allen, E.W. N. Glover, and A.D. Martin, Phys. Lett. 146B, 247 (1984).

8. G. Altarelli, R.K. Ellis, M. Greco, and G. Martinelli, Nucl. Phys. B246, 12 (1984); R.K. Ellis, Fermilab Report No. Conf-84/96-T, 1984 (to be published).

9. D.Duke and J. Owens, Phys. Rev. D 30, 49 (1981).

10. M. Gluck, E. Hoffman, and E. Reya, Z. Phys. D 313, 119 (1982).

11. E. Eichten, I. Hinchliffe, K. Lane, and C. Quigg, Rev. Mod. Phys. 56, 579 (1984).

12. H. Abramowicz et al., Z. Phys. C17, 283 (1983).

13. A. Bodek et al., Phys. Rev. D20, 1471 (1979).

14. R. A. Arnowitt, A.H. Chamseddine, and P. Nath, Phys. Rev. Lett. 50, 232 (1983), and Northeastern University REports No. 2588, No. 2597, and No. 2600,

1983 (unpublished).

15. S. Weinberg, Phys. Rev. Lett. 50, 387 (1983); J. Ellis, J.S. Hagelin, D.V. Nanopooulos, and M. Srednicki, Phys. Lett. 127B, 233 (1983).

16. V. Barger, R.W. Robinett, W.-Y. Keung, and R.J.N. Phillips, Phys. Rev. D 28, 2192 (1983).,

17. H. Baer, J. Ellis, D.V. Nanopoulos, and X. Tata, CERN Report No. TH.4059/84, November 1984 (unpublished).

18. V. Barger, W.-Y. Keung, and R.J.N. Phillips, University of Wisconsin Reports No. PH/219, No. 223, and No. 228, 1984 (unpublished).

BARYON PRODUCTION AT THE CLEO DETECTOR

Talk Presented by

Mac D. Mestayer
Vanderbilt University, Nashville, Tenn.

First, let me say that CLEO has no new particles to report contrary to the title of this conference, but we do have some new results on 'old' particles and some persistent mysteries. In this talk I will discuss three topics: 1) the baryon/meson ratio from Υ decays compared with this ratio from continuum e^+e^- annihilation data, 2) a study of Λ's and protons produced in B meson decays, in particular, the ratio of production rates and their momentum spectra, and 3) CLEO's observation of the Λ_c through its decay to $\Lambda\pi^+\pi^+\pi^-$ and measurement of its fragmentation function which I will compare to that of charmed mesons.

All three of these topics fall under the general heading of baryon production in e^+e^- annihilation. This is no accident. Studies of hadronization processes which produce baryons complement our knowledge of meson hadronization. One reason is that baryon momentum spectra are softened less by the cascade of decays from parent to daughter particles than are meson momenta. This is a kinematic effect due to the higher mass of baryons and is demonstrated well in a recent publication[1]. A more important reason for comparing baryon to meson hadronization is that it provides a comparison of quark-antiquark with quark-two quark forces. I said 'two quark' rather than 'diquark' because one of the topics our data addresses is whether diquarks (that is, tightly bound two quark states) exist as dynamical entities and constitute a significant portion of the baryonic wave function.

The first subject which I will discuss is the enhancement of the baryon/meson ratio in Υ decays compared to continuum e^+e^- annihilation events. This is an old result, having been presented at the Paris conference[2] and having been updated in conjunction with other hyperon studies[3]. Despite its antiquity it is perhaps still unexplained, as we will see.

The basic result is illustrated in Figs. 1a and 1b; which are plots of the Ξ, Λ, and K^0_s cross sections for the Υ and continuum data sets, respectively. As the figure shows, the relative production rates of Ξ and Λ baryons compared to the K^0_s rate, is higher for the Υ data sample than from the continuum. Table I summarizes the averages of these ratios.

Table I : Baryon Rates from Υ and Continuum

	Υ	Continuum
Λ / K^0_s :	0.180 ± 0.030	0.074 ± 0.014
Ξ / K^0_s :	0.015 ± 0.004	0.005 ± 0.001.

For a summary of CLEO's hadronization studies, I refer you to a recently published article which dealt with many aspects of hadronization.[4]

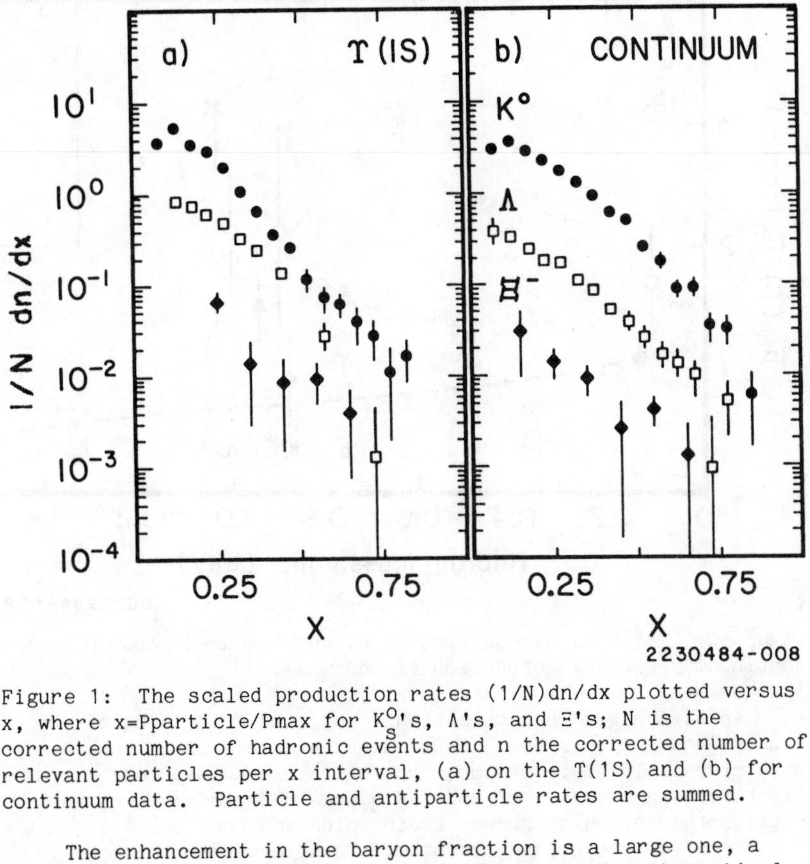

Figure 1: The scaled production rates (1/N)dn/dx plotted versus x, where x=Pparticle/Pmax for K^0_s's, Λ's, and Ξ's; N is the corrected number of hadronic events and n the corrected number of relevant particles per x interval, (a) on the ϒ(1S) and (b) for continuum data. Particle and antiparticle rates are summed.

The enhancement in the baryon fraction is a large one, a factor of 2.5 to 3 in size, and I will describe two theoretical approaches used in calculating this effect. The first is that of the LUND model[5] in which a color string mediates the force between quarks and/or gluons and the breaking of the string accounts for the hadronization process. If the string breaks through the formation of a $q\bar{q}$ pair a meson results; if a diquark pair breaks the string then a baryon will be produced. In this picture there are two color strings connected to each gluon and one to each quark, so ϒ decays are envisioned as a loop of string with three 'corners' (the gluons), whereas $q\bar{q}$ events will have a single string connection. Because the string ends are quarks in the $q\bar{q}$ events there are two less positions available for the string to break into diquark-antidiquark pairs than in the three gluon decay. This effect leads to an increase in the baryon to meson production ratio from the T decays, but the predicted[6] rise in the Λ/K^0_s ratio of about 30 - 40% is too low to account for the observed enhancement.

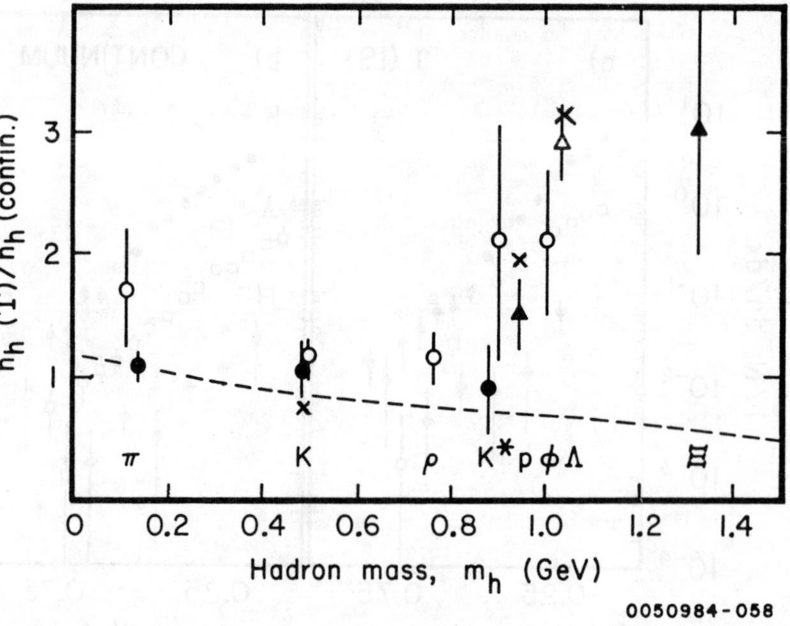

Figure 2: For each hadron species h, the measured ratio of T and continuum yields as a function of the mass.

An alternative hadronization model, the color singlet cluster model[7,8] has been worked on by a number of physicists. In their scheme, the initial state hadronizes by first forming a shower of colored quarks and gluons which then realign themselves to screen their color and form color singlet "clusters" of energy and momentum. These clusters then decay into pairs of hadrons which can then further decay. Their low order QCD calculations of the quark gluon fragmentation process predict that the three gluon decay of the T resonance will produce a greater proportion of higher mass color neutral clusters than will the fragmentation of $q\bar{q}$ jets. Since higher mass clusters tend to hadronize into higher mass hadrons, a greater proportion of higher mass hadrons, for instance, baryons, are produced. It should be pointed out that this effect is not thought to be due to the properties of gluon jets but rather to the three gluon nature[8] of T decays; in other words, two gluon decays of resonances are expected to produce baryons at roughly the same rate as $q\bar{q}$ jet events. The second point to note about this model is that the enhancement of particle production from the Υ compared to continuum production depends only on the mass of the produced particle, not on whether it is a baryon or meson.

In order to study this effect we compared the production rates for several types of hadrons from Υ decays relative to their rate from the continuum. In Fig. 2 we plot the ratio of production rates as a function of the mass of the particle and see a clear rise with increasing mass.

In an independent fragmentation model in which gluons fragment like quarks, the ratios would follow the dashed line. Also indicated as x's on the graph are the results of a particular theoretical calculation[9], which agree well with the data. It is not yet clear though how well the color singlet cluster model reproduces the flavor and spin dependence of baryon production. These are important hurdles which these models must clear if they are to be considered viable alternatives to the 'string' hadronization models. Alternatively, there may be some way for the 'string' model builders to incorporate low order QCD shower processes into their calculations in order to better model the observed high mass enhancement from three gluon decays.

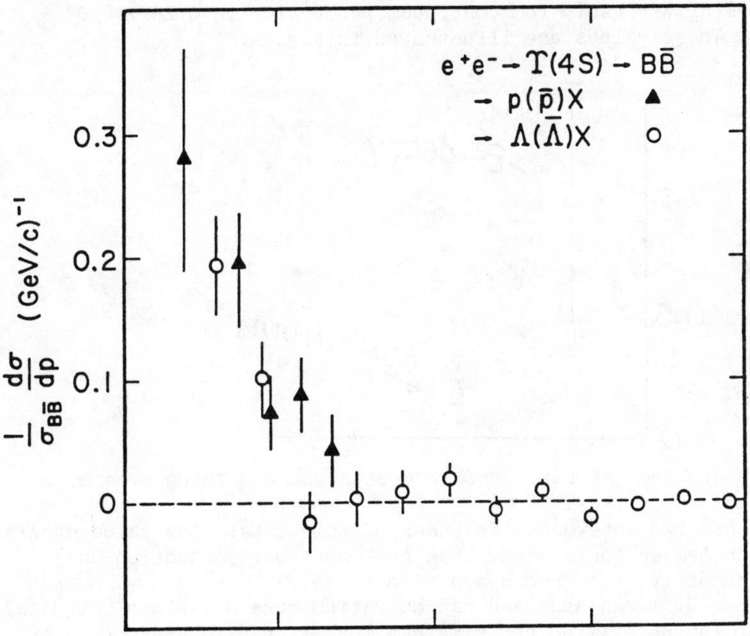

Figure 3: Corrected proton (and antiproton) and Λ (and $\bar{\Lambda}$) momentum spectra from $B\bar{B}$ decay.

The second topic in this talk is the production of baryons in the decay of B mesons, interesting because it will have bearing on determining the relative importance of 'spectator' and 'non-spectator' decays. These studies have yielded many interesting results in the charmed sector and there is reason to believe that they will be as informative in the b sector.

Once again, this data is not new[10], but the data have perhaps not been fully appreciated by phenomenologists studying the systematics of weak decay processes. The experimental results are summarized in fig. 3, where we have plotted the production cross sections for Λ's and for protons which come from the decay of B mesons.

The two cross sections look very similar, both in size and in shape. After subtracting from the proton yield those which come from Λ decays, we obtain the branching fraction for B mesons to decay into Λ's of $3.0 \pm 1.0 \pm 0.6\%$, and for protons, $3.0 \pm 0.6 \pm 0.9\%$.

If we divide these numbers to get the ratio of production rates, we obtain the following:

$$(B \rightarrow \Lambda X) / (B \rightarrow P X) = 1.0 \pm 0.4 \pm 0.3$$

where the errors shown are statistical and systematic. The equivalent Λ/P ratio for continuum production is $0.20 \pm 0.04 \pm 0.04$. These values are illustrated in Fig. 4.

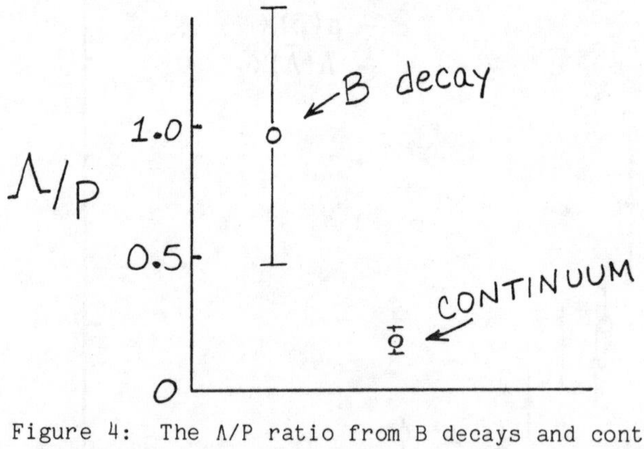

Figure 4: The Λ/P ratio from B decays and continuum events.

There are two noteworthy features of this data: the value of Λ/P is much larger for B decay than for continuum production and, unfortunately, the errors are also large.

How well can standard hadronization models explain the data? First, let us examine the simplest model, that is, non color-mixed spectator decays in which the b quark decays into a c quark (which combines with the spectator quark to form a D meson) and into a virtual W boson which then hadronizes into the observed baryons (i.e. $B \rightarrow D$ + baryons). The first thing to note is that in this particular application the W cannot fragment into a $c\bar{s}$ pair because the mass of the lightest pair of resulting baryons ($\Lambda_c + \Lambda$) is 3.44 GeV/c^2 which is greater than the available energy of 3.41 GeV. Therefore, we would expect the W to fragment into a $u\bar{d}$ quark pair, and so should produce baryons at a rate comparable with that from e^+e^- annihilations at a center of mass energy of 3.4 GeV and with a similar strangeness content.[11] The overall proton production rate of 0.03 per event is compatible with data from low energy continuum production but, as we have seen, the Λ/P ratio is much higher than the same ratio on the continuum.[12]

The high Λ/P ratio is easily explained if there are substantial contributions from color-mixed non-spectator processes such as $B \rightarrow \Lambda_c \bar{N}$.[13] In light of the large errors on the data we cannot make definite conclusions but we are currently analysing a data sample of B decays which is a factor of two larger and if the large Λ/P ratio persists then this would be evidence for a sizeable non-spectator contribution to B decay. This would be very interesting and similar to the current situation in D decay.

I would like to note in passing an interesting alternative explanation for this effect[14], that is, that the W boson might have a direct coupling to a diquark anti-diquark pair in direct analogy to its coupling to $q\bar{q}$ pairs. If, furthermore, spin 0 diquarks dominate then we would expect the two subprocesses,

$$W \rightarrow ud\,\bar{d}\bar{s} \quad \text{or} \quad W \rightarrow us\,\bar{d}\bar{s},$$

to be the predominant ones. The high strangeness content of these processes accounts for the relatively large number of Λ's present in B decays. At any rate, if future data corroborates the large Λ/P ratio it will be evidence for sizeable contributions from processes other than the simplest spectator diagrams and will be a very interesting result.

I now turn to the third and final topic of my talk, CLEO's measurement of the fragmentation function of a charmed baryon, the Λ_c. The result comes from an analysis of three years' accumulation of data, a sample which includes some 400,000 hadronic events. We observed a Λ_c signal through its decay into $\Lambda\pi^+\pi^+\pi^-$. By making certain assumptions about the overall Λ_c production rate, we were able to calculate the branching ratio into this particular mode.

The Λ_c search began by first identifying Λ candidates. They were detected by reconstruction of their decay vertex into a proton and a π^-. We required that the momenta of the daughter tracks exceed 100 MeV/c, that the secondary vertex be further than 8 mm from the e^+e^- beam line, that the Λ mass be within 5 MeV of its known value, and that the two track combination not be consistent with being a K^o_s decay.

Once we had a good Λ candidate, we combined it with three charged tracks (assumed to be pions) and calculated the invariant mass of the combination. We required that the cosine of the angle between the Λ direction and that of each of the pion's be greater than -0.4, and that the momentum of the Λ_c candidate exceed 2.5 GeV/c. This momentum cut drastically reduces the background but, because of the stiff fragmentation function of the Λ_c, does not reduce our reconstruction efficiency by much.

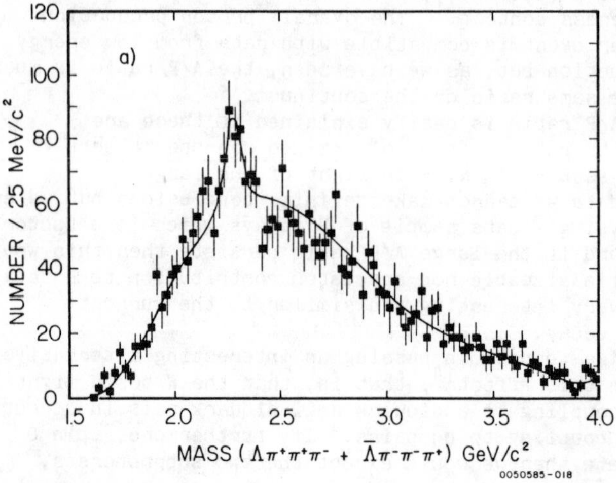

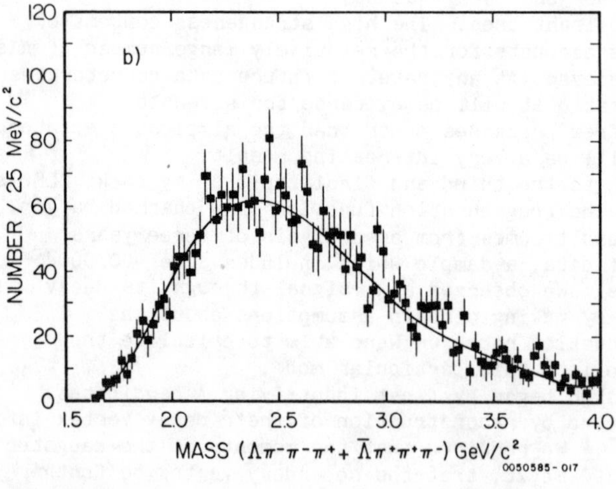

Figure 5: a) The $\Lambda\pi^+\pi^+\pi^-$ (+ $\bar{\Lambda}\pi^-\pi^-\pi^+$),"right-charge", invariant mass with the background and Gaussian fit overplotted. b) The $\Lambda\pi^-\pi^-\pi^+$ (+ $\bar{\Lambda}\pi^+\pi^+\pi^-$),"wrong-charge", invariant mass with the background fit overplotted.

In Fig. 5a, I have plotted the mass of $\Lambda\pi^+\pi^+\pi^-$ (and $\bar{\Lambda}\pi^-\pi^-\pi^+$) candidates. An enhancement is observed at the known Λ_c mass of 2.283 GeV. Figure 5b is a mass plot for "wrong charge" combinations (i.e.

$\Lambda\pi^-\pi^-\pi^+$ and $\bar{\Lambda}\pi^+\pi^+\pi^-$) and does not show a peak. The curves overplotted in the two figures are the results of fitting the two distributions to a common background polynomial plus, in the case of the "right charge" case, a Gaussian peak. The results of this fit are the following:

Mass = 2.287 ± 0.011 ± 0.005 GeV/c^2

FWHM = 94 ± 25 MeV/c^2

Number = 108 ± 28 Λ_c's,

where the mass has been corrected for the energy loss due to charged particles traversing the material of our detector (a 10 MeV effect).

For Λ_c candidates whose mass fell within 50 MeV of the peak position we histogrammed the distribution in scaled momentum, X = P/Pmax. We subtracted a fit to the wrong charge X distribution from the right charge distribution in order to correct for random (charge-symmetric) background. From Monte Carlo studies we determined that the Λ_c reconstruction efficiency was 0.06 and constant above X = 0.3. We divided the background subtracted X spectrum by this efficiency to obtain the branching ratio times differential cross section for Λ_c production which I show in fig. 6.

Note that the X spectrum peaks at a large value of X which is in sharp contrast to the steeply falling spectra of non-charmed hadrons but similar to the spectra observed for charmed mesons (D's and D*'s). This hard spectrum is expected since the Λ_c carries the original charmed quark produced from the e^+e^- annihilation. I would now like to compare this data with charmed meson fragmentation functions in order to compare the dynamics of "popping" an anti-quark out of the vacuum with the similar probability for producing two pairs of quarks (or possibly, a pair of diquarks). If a baryon is produced by popping a diquark pair out of the vacuum, then the baryon fragmentation function should look just like that of a meson, modulo differences due to a form factor of the diquark.

The solid curve in Fig. 6 represents the Peterson formula[15] for the fragmentation function of mesons containing a heavy quark. In this formulation, the parameter ε is the mass squared plus perpendicular momentum squared of the spectator quark divided by the square of the charmed quark mass. We have set ε = 0.14 which gives the best fit to our D^* X spectrum[16] in order to study differences between this data and the meson case. As is evident, this form provides a good fit to the Λ_c data, yielding a χ^2 of 4.0 for 4 degrees of freedom. If we allow ε to be a free parameter we obtain ε = 0.21 ± 0.08.

We have also overplotted as a dashed curve in Fig. 6 a formula suggested by DeGrand[17] as a hadronization function for charmed baryons. It is a generalization of the one for mesons, modified to include the independent production of two quark-antiquark pairs from the vacuum. We again fix the value for ε to be the same as in the meson case, that is, 0.14, which yields a χ^2 of 9.7 for 4 degrees of freedom. If we allow ε to vary, we obtain ε = 0.02 ± 0.01, which is too small to agree with reasonable values of the spectator quarks' masses.

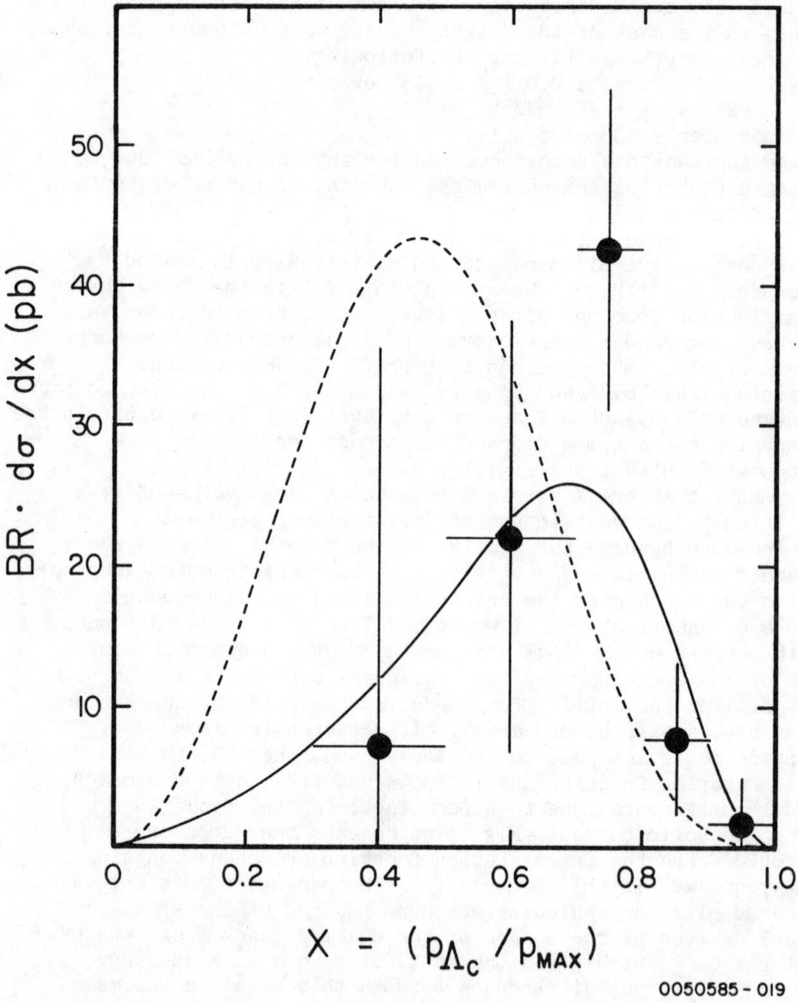

Figure 6. The branching fraction $BR(\Lambda_c \to \Lambda\pi^+\pi^+\pi^-)$ times the differential cross section $d\sigma/dx$ for Λ_c production. The solid curve is a theoretical prediction of Peterson, et al. (Ref. 15), while the dashed curve is the prediction of DeGrand (Ref. 17).

The x dependence of Λ_c hadronization is sensitive to the underlying dynamics, and we have used these two formulations to compare the spectrum expected from a one step diquark approach (Peterson formula) with a two step, independent quark production model (DeGrand formula). The Peterson fomula fits the data better, lending support to models in which baryons are formed by combining a quark with an effective two quark bound state (a diquark).

To measure the total number of Λ_c's produced, we use the Peterson formula with $\varepsilon = 0.14$ to correct for the unseen part of the spectrum (a correction of 39%). After dividing by the total continuum cross section[18] we find the yield of Λ_c's per hadronic event which decay into $\Lambda\pi^+\pi^+\pi^-$ to be 0.0044 ± 0.0011 ± 0.0015, where the errors are statistical and systematic, respectively.

If we assume that 40% of the continuum hadronic events are due to $c\bar{c}$ production then there are 0.8 c (+ $\bar{c}$) quarks per hadronic event. Therefore, the number of Λ_c's per charmed quark which decay into $\Lambda\pi^+\pi^+\pi^-$ is 0.0055 ± 0.0014 ± 0.0019. If we use the estimate from MARK II data[19] that there are 0.2 Λ_c's produced per c quark, we find the branching fraction for $\Lambda_c \to \Lambda\pi^+\pi^+\pi^-$ to be 2.8% ± 0.7% ± 1.1%. We have included a 20% uncertainty on the number of Λ_c's produced per c quark in the systematic error.

CONCLUSIONS

1) Three gluon decays produce 2.5 to 3 times as large a percentage of baryons as do continuum $q\bar{q}$ events. This effect may be due solely to the higher mass of baryons compared to mesons.

2) The Λ/P ratio from B meson decays is considerably higher than from continuum events, albeit with large errors, and is evidence for a sizeable contribution from non-spectator decays.

3) We have measured for the first time the Λ_c fragmentation function and found it to be very similar to that of charmed mesons. Its shape is more consistent with predictions from diquark fragmentation models than from independent quark production models when ε is fixed by charmed meson data. We have also made the first measurement of the $\Lambda_c \to \Lambda\pi^+\pi^+\pi^-$ branching fraction.

REFERENCES

1 H. Aihara et al., Phys. Rev. Lett. 53, 130 (1984).
2 F. M. Pipkin, Proceedings of the XXI Conference on High Energy Physics (1982).
3 M. S. Alam, Phys. Rev. Lett. 53, 24 (1984).
4 R. Giles, et al., Phys. Rev. D 29, 1285 (1984).
5 For a good review, see B. Andersson et al., LU TP 83-10.
6 B. Andersson et al., LU TP 84-9.
7 A. Casher et al., Phys. Rev. D 20, 179 (1979). T. Gottschalk, Nucl. Phys. B214, 201 (1983). B. Andersson, G. Gustafson, T. Sjostrand, LU-TP 84-9 (1984). G. Marchesini, B. R. Webber, Nucl. Phys. B 238, 492 (1984).
8 R. D. Field, Phys. Lett. 135, 203 (1984).

[9] Private communication, R. D. Field; for further details see F. Morrow, Ph. D. Thesis, Cornell University, 1984 (unpublished).

[10] M. S. Alam et al., Phys. Rev. Lett. 51, 1143 (1983).

[11] At center of mass energy of 3 GeV, $u\bar{u}$ and $d\bar{d}$, jets account for 5/6 of all continuum events, with $s\bar{s}$ events comprising the remainder.

[12] G. S. Abrams et al., Phys. Rev. Lett. 44, 10 (1980).

[13] I. I. Bigi, Phys. Lett. 106B, 510 (1981).

[14] Private communication, M. Hempstead.

[15] C. Peterson et al., Phys. Rev. D 27, 105 (1983).

[16] P. Avery et al., Phys. Rev. Lett. 51, 1139 (1983).

[17] Thomas A. DeGrand, Phys. Rev. D 26, 3298 (1982).

[18] Because the data were collected at energies both above and below B meson threshold, we calculate the average "continuum" cross section by taking the cross section below the $\Upsilon(4S)$ and extrapolating it to higher energies accounting for the 1/s dependence; therefore, the "continuum" cross section does not include $e^+e^- \to b\bar{b}$ events.

[19] In Ref. 17, DeGrand uses the observed step in baryon production in e^+e^- annihilation at Λ_c threshold, as reported by G. S. Abrams et al., op. cit., to obtain this estimate.

DISTINCTIVE SIGNATURES FOR NEW PARTICLES IN e^+e^- ANNIHILATION AND Z-DECAY

Yoshihisa Kitazawa[†,1] and Wu-Ki Tung[*,††,1,2,3]
[1]Enrico Fermi Institute, University of Chicago, Chicago, IL
[2]Argonne National Laboratory, Argonne, IL
[3]Illinois Institute of Technology, Chicago, IL

ABSTRACT

New particles and new interactions reveal themselves most clearly where standard model contributions are negligibly small. A prominent example with this advantage is the one-lepton inclusive longitudinal structure function (W_L) in e^+e^- annihilation and Z-decay. We discuss general features of this approach and present structure functions for four types of new particles (heavy charged fermion, e.g. new sequential lepton or top quark; heavy neutral lepton; and supersymmetric scalar lepton, i.e. slepton), along with the (small) standard model 'background'. The x-dependence of W_L provides a distinct signature of the identity of the new particle. Extensions of this approach are discussed.

INTRODUCTION

With the discovery of the W- and Z-bosons, the emphasis in particle physics has shifted toward the search for new particles and interactions. These searches are quite difficult in general because signals due to new phenomena are hard to distinguish from 'backgrounds' due to various standard-model processes. One commonly used technique to enhance the signal for new physics is to impose p_T-cuts to observed final-state particles. These p_T-cuts, while extremely useful, are clearly not the answer to all problems.

We are thus motivated to consider physical quantities which vanish (or nearly vanish) in the standard model, but which get sizable contributions from new particles and new interactions if they exist. One prominent example, examined in detail in this report, is the 1-lepton longitudinal structure function in e^+e^- annihilation, particularly at the Z-resonance.

The process we have in mind is $e^+e^- \to \ell^\pm + X$ where X denotes unspecified final state particles. At current e^+e^- storage rings as well as the anticipated SLC and LEP, this process is expected to

---

*Permanent Address: Illinois Institute of Technology.

†Robert R. McCormick Fellow, work supported in part by the NSF, PHY-83-01221, and the DOE, DE AC02-82ER-40073.
††Work supported by the NSF, PHY 83-05283.

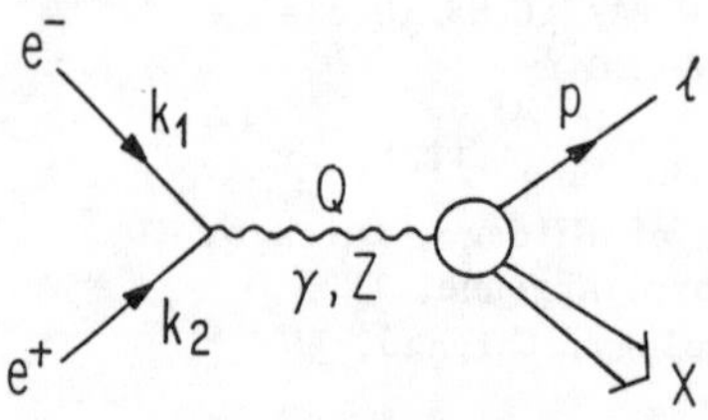

Fig.1 One-lepton inclusive production in e^+e^- scattering

proceed via γ^* and Z intermediate states as shown in Fig. 1. The inclusive differential cross-section can be expressed in terms of 'structure functions', as in deep inelastic scattering. With the exception of a possible contribution from the top quark (which is of intrinsic interest by itself), standard model processes yield only transverse structure functions corresponding to helicity ±1 for the gauge boson, due to a general theorem analogous to the Callan-Gross[1] relation for deep inelastic scattering. The longitudinal structure function W_L is negligibly small in standard model processes, and is sensitive to the presence of any elementary scalar particles as well as heavy leptons and quarks (including the top quark). In addition to signal the existence of new particles and new interactions, the functional dependence of W_L on the scaled momentum variable x contains much information on the mass and spin of the underlying new elementary particle. This report explores the general features of this approach to new particle search and present detailed case studies of several sources of possible new physics.[2]

1. INCLUSIVE CROSS-SECTION AND STRUCTURE FUNCTIONS

Our study applies to all one-particle inclusive final states in e^+e^- scattering via γ^* or Z^0 intermediate states. For the sake of clarity, we shall restrict ourselves to the special case of one-lepton inclusive process via Z^0 intermediate state in much of the following discussion. Generalizations to include the γ^* intermediate state and one-hadron or one-jet inclusive final states will be discussed in appropriate places.

The differential cross-section (normalized to that of $e^+e^- \to \mu^+\mu^-$) for this process due to the s-channel Z resonance term is given by

$$\frac{1}{\sigma_0}\frac{d\sigma}{dx\cos\theta} = \frac{3}{8}\frac{x}{\sin^4 2\theta_w}\frac{Q^4}{(Q^2-M_z^2)^2+M_z^2\Gamma_z^2}L^{\mu\nu}W_{\mu\nu} \ . \qquad (1)$$

Here θ_w is the Weinberg angle, $\sigma_0 = 4\pi\alpha^2/3Q^2$, $x = 2p\cdot Q/Q^2$ = fractional energy of ℓ in the CM frame, θ is the CM scattering angle of ℓ, and Q is the total CM energy. $L_{\mu\nu}$ is the initial state lepton tensor given by

$$L^{\mu\nu} = 2(1 + \delta^2)(k_1^\mu k_2^\nu + k_2^\mu k_1^\nu - g^{\mu\nu} k_1 \cdot k_2)/Q^2$$
$$+ 4\delta(i\varepsilon^{\mu\nu\lambda\sigma} k_{1\lambda} k_{2\sigma}/Q^2) \; . \tag{2}$$

where $\delta = 1 - 4\sin^2\theta_w$. $W_{\mu\nu}$ is the final state tensor given by

$$W_{\mu\nu} = \frac{1}{4\pi} \sum_x (2\pi)^4 \delta^4(q - p - p_x) \langle 0|J_\mu(0)|p,x\rangle\langle p,x|J_\nu^\dagger(0)|0\rangle \tag{3}$$

where J_μ is the current which couples to the Z-boson. Defining invariant structure functions $W_i(x,Q^2)$, $i = 1,2,3,\ldots$ by

$$W_{\mu\nu} = -W_1 g_{\mu\nu} + W_2 p_\mu p_\nu/Q^2 + W_3 i\varepsilon_{\mu\nu\lambda\sigma} p^\lambda Q^\sigma/Q^2$$
$$+ \text{(terms with } Q^\mu \text{ or } Q^\nu) \; , \tag{4}$$

we deduce

$$L^{\mu\nu} W_{\mu\nu} = 2W_1 + \frac{x^2}{4} \sin^2\theta \, W_2 - 2\delta \cos\theta \, xW_3$$
$$= (1 + \cos^2\theta) \, W_T + \sin^2\theta \, W_L - 2\delta \cos\theta \, xW_3 \tag{5}$$

where

$$W_T(x,Q^2) \equiv W_1 \qquad \text{(the transverse structure function)}$$
$$W_L(x,Q^2) = W_1 + \frac{x^2}{4} W_2 \qquad \text{(the longitudinal structure function)} \tag{6}$$

$\delta = 1 - 4\sin^2\theta_w$, and terms of order δ^2 and m_ℓ^2/Q^2 have been neglected.

The three structure functions which appear in Eq. (5) can be inferred, in principle, from the angular distribution at fixed x and Q. On the Z-resonance peak, W_3 will be hard to pin down because of the small parameter δ which multiplies the angular factor. (W_3 is measurable when the γ^*- Z interference term is non-negligible.) On the other hand, W_L and W_T have angular factors of comparable magnitudes but very distinctive shapes. Therefore, both can be quite easily measured. We are particularly interested in the longitudinal structure function because it is extremely sensitive to new particles and new interactions. We particularly emphasize the Z-resonance region because the expected large cross-section will make the separation of W_L and W_T quite practical.

The angular distribution formula (Eq. 5) does not apply to higher order QED processes (such as bremsstrahlung off the initial state leptons or two-photon processes) or to t-channel exchange processes. Both are strongly peaked in the forward-back direction, and represent a (calculable) small background over a large range of the CM scattering angle. Except for elastic scattering, which is easy to discriminate, all such processes also only occur in higher orders of the electro-weak coupling in the standard model, hence are expected to be insignificant in the absence of new physics. Any measured departure from this distribution beyond the standard-model background can only arise from some 'new physics', hence will be of a great deal of interest. We shall not, however, be concerned with departure from Eq. (5) in this study. In the Z-resonance region, one can be particularly confident that this formula holds.

2. VANISHING OF W_L IN STANDARD MODEL PROCESSES

For the sake of definiteness, we shall refer to the observed final state lepton as the 'electron', although the same results hold also for e^+ and for $\mu^\pm$. In the standard model, this final state electron can come from the following sources: (i) It can arise from direct pair-production by the Z; (ii) it can be the decay product of the τ-lepton; (iii) it can be the decay product of a c- or a b-quark; and (iv) it can be the decay product of the top-quark, if the latter indeed exists. Case (i) is distinguished from all the rest by the unambiguous kinematic feature $x = 1$. We shall not consider it further. The basic mechanism for the other channels is depicted in Figure 2, where f-$\bar{f}$ stands for an elementary, almost on-shell fermion-antifermion pair. For the moment, we do not need to specify the precise way in which f fragments (decays) into the observed lepton ℓ. In the energy range below the Z-resonance (relevant for physics at PETRA, PEP, and TRISTAN), we must take into account contributions from the s-channel γ^* intermediate state and the Z^0-γ^* interference term. The form of the angular distribution stays the same (i.e. Eq. 5), but the overall factors (cf. Eq. 1) and the structure functions are different. We shall present results in this energy range later, but shall dispense with the detailed formulas.[3]

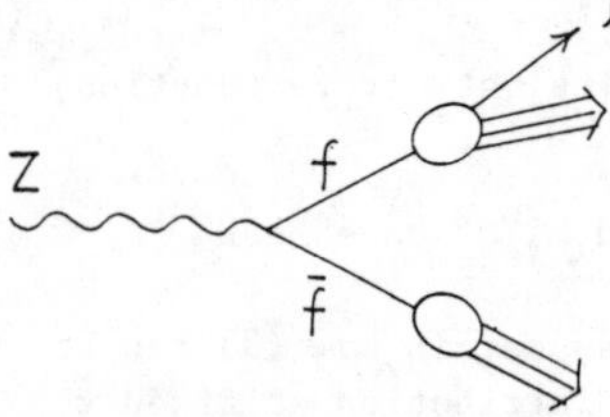

Fig.2 Parton picture for 1-lepton production

Subject to order α_s QCD radiative corrections and to small finite-mass effects of order m_f^2/Q^2, cases (ii) and (iii) give vanishing contribution to the longitudinal structure function[1].

This general result follows from the fundamental vector-axial-vector coupling of the gauge bosons to the elementary fermions, from the dominance of the cross-section by almost on-shell partons for hard collisions, and from the smallness of m_f^2/Q^2 for all elementary fermions (except the top quark). It is completely independent of the fragmentation (or decay) mechanism of the elementary fermion into the observed lepton.

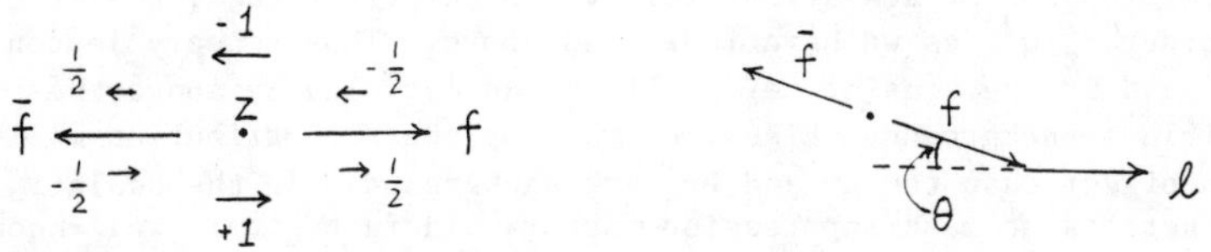

(a) Helicity of f-$\bar{f}$ pair and the Z-boson

(b) Parton axis and outgoing lepton direction

Fig. 3

The physical content of this general theorem is illustrated in Fig. 3. In part (a), we show the fermion-antifermion produced back-to-back in the CM frame of the Z-boson. Because the V-A nature of the coupling, the (almost) zero-mass fermion-antifermion pair carry opposite helicities. Hence their spins are aligned along the momentum of either the fermion or the anti-fermion. Thus, the initial state Z-boson must carry angular momentum ±1 along the direction of the parton-pair. In part (b), we show the fragmentation of one of the parton into the observed lepton ℓ. Since the helicity label "L" and "T" on the structure functions refer to an axis along the momentum of the observed particle, we must take into account the angle θ between the parton and the lepton. However, if the parton is almost on-shell and massless, the two must be collinear (i.e. $\theta \approx 0$). Thus the parton-level conclusion from part (a) remains valid at the level of physical particles. When the parton mass is non-negligible, the opening angle θ is of the order m_f^2/Q^2 ($\simeq m_f^2/M_z^2$). We then expect that $W_L \sim O(m_f^2/M_z^2)$.

In the Z-resonance region, the only substantial contribution to W_L from standard model processes at the tree level comes from the top-quark, if it exists. Thus, the quantity W_L/W_T provides a practically background free method of observing the top quark and other new particles/new interactions beyond the standard model. Additional advantages of using the inclusive structure functions for searching for new particles are: (i) the signal is sensitive to whole classes of new phenomena, rather than specific channels, (ii) the x-dependence of the structure functions provides the means to distinguish between various possibilities, and (iii) the one-lepton inclusive cross-section is among the easiest ones to measure from the experimental point of view; it represents one of the first levels of the chain of event selection.

3. MECHANISMS FOR NON-VANISHING W_L

Sizable contributions to W_L/W_T can arise from the production of heavy fermions or of scalar particles. In the first case, the effect is of order m_f^2/Q^2, as we have indicated above. Thus a heavy lepton with a mass in the tens of GeV's will stand out clearly above the negligible τ-background. Likewise, the top-quark contribution will be much bigger than the c- and b-quark background. In the scalar case, there is no mass-suppression factor. In fact, it is well-known that in the $M \to 0$ limit, scalar particle production alone contributes only to the longitudinal structure function. The validity of this result can be easily seen by referring back to Fig. 3. For scalar partons, the net angular momentum along the direction of motion must be zero; hence the Z-boson must be longitudinally polarized. The significance of a scalar parton signal therefore depends on the relative strength of scalar vs. fermion production, which, in turn, is determined by their couplings to the gauge boson.

A third source of non-vanishing W_L is QCD correction effects on light-quark pair production from the Z. For instance, if a gluon is emitted off one of the partons (Fig. 4), then that fermion leg can be off-shell, hence vitiating the previous argument on the vanishing of W_L even if the quark mass is negligibly small. The price to pay for this mechanism is an extra strong-interaction running coupling constant $\alpha_s(Q^2)/\pi$. Preliminary study indicates that this 'QCD background', although not totally negligible, does not pose a serious problem for new particle search. The reason is that it has rather different x-dependence compared to that of new particle production, hence can be separated from the latter. We remark that the study of this QCD effect is of value on its own right. It provides a useful way, along with photon structure functions and the longitudinal structure function in deep inelastic scattering, to test

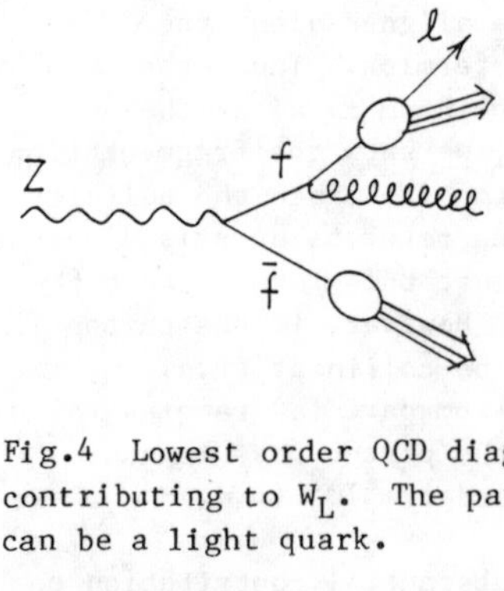

Fig.4 Lowest order QCD diagram contributing to W_L. The parton can be a light quark.

the validity of the theory in a quantitative way. We are currently studying this problem in greater detail.[3]

To illustrate the general features of new particle production, we present three concrete calculations: the production and decay of heavy charged fermions, of heavy neutral fermions, and of a charged scalar particle. For definiteness, we use fermion couplings of the standard model for the first two cases, and scalar lepton couplings of supersymmetric models for the third case. Most results can be taken as representative of general classes of possible new particle production and new interactions.

4. FEATURES OF HEAVY CHARGED FERMION PRODUCTION

We have calculated in detail the one-lepton inclusive functions due to heavy charged fermions such as a new sequential lepton or the top quark as well as the τ-, b-, and c-background. For this purpose, we specify the decay mechanism of f in Fig. 2, and use the standard model couplings appropriate for sequential heavy fermions as illustrated in Fig. 5. The method of calculation used is that of Ref. [4]. Although the heavy fermion line in Fig. 3 can be virtual, thus allowing us to probe, in principle, any mass value m_f; we shall assume $m_f < Q/2$ so that on-shell production at higher rate is possible. The calculation is straightforward, but quite tedious. Results can be obtained in closed form. The formulas are rather lengthy and involved, hence will not be reproduced here.[3] In practice, one can expand the relevant quantities as power series in the parameter $\rho = (m_f/M_W)^2$. The leading term in W_L is one order higher in ρ than that of W_T, as expected from the general theorem stated in Sec. 2. These series converge quickly. To give an indication of the behavior of these structure functions, we present the leading order formulas for W_T and W_L:

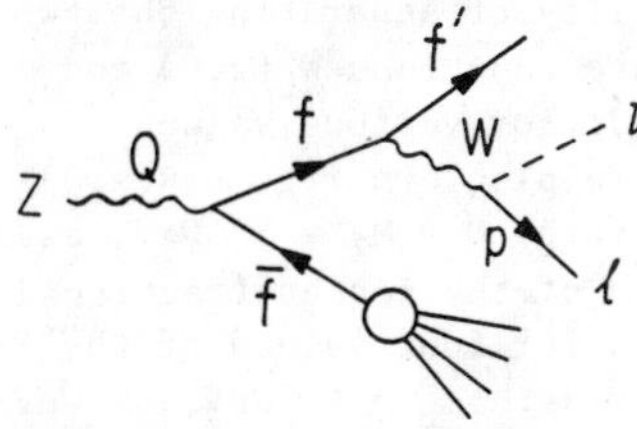

Fig.5 Heavy fermion pair production contribution to 1-lepton inclusive structure function.

$$W_T = 2g_L^2(1 - x^3) + 3g_R^2(1 - x)^2(1 + 2x)$$

$$W_L = \frac{m_f^2}{Q^2}\frac{1 - x}{x}\left[g_L^2(4 + x + x^2) + g_R^2(8 + x + x^2 - 4x^3) + g_L g_R 2x(5 + 5x - 4x^2)\right] \qquad (7)$$

where $g_L = T_3^f - q^f \sin^2\theta_W$, $g_R = -q^f \sin^2\theta_W$ if the standard model couplings to the Z-boson is assumed. A common factor independent of the variable x is omitted in both equations. These formulas apply in the range $m_f^2/Q^2 < x < 1$. In the small x region ($x < m_f^2/Q^2$), both structure functions decrease to zero as $x \to 0$ due to phase space restrictions.

We now present some numerical results obtained from the exact formulas. Let us define the ratio

$$R(x,Q) = W_L(x,Q)/W_T(x,Q) \tag{8}$$

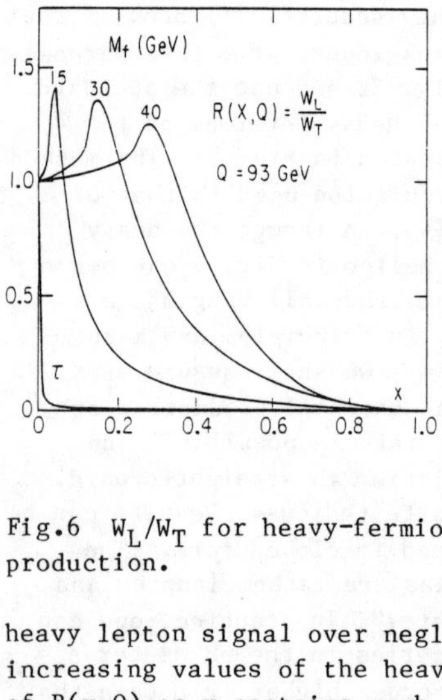

Fig.6 W_L/W_T for heavy-fermion production.

which is important in determining the angular distribution of the observed lepton. In particular, it quantifies the deviation of the angular distribution from standard $1 + \cos^2\theta$ behavior (hence the feasibility of separating the two structure functions $W_L(x,Q)$ and $W_T(x,Q)$), for various values of x. We plot, in Fig. 6 R(x,Q) at CM energy $Q = M_Z = 93$ GeV, as a function of the lepton fractional energy x for four values of the lepton mass: $m_L = 40$ GeV, 30 GeV, 15 GeV, and $m_L = m_\tau = 1.8$ GeV. The last curve represents the main background from the known lepton production. We see from these results the dramatic emergence of heavy lepton signal over negligible background (τ-curve) for increasing values of the heavy lepton. In addition, the dependence of R(x,Q) on x contains much information on the characteristics of the new particle. The shape of these curves, with peaking in the lower half of the x-range, is characteristic of fermions coupled to gauge bosons in the conventional way. (Cf. scalar particle production to be discussed later.) The position of the peak is sensitive to the mass of the lepton and increases with the value of the mass.

Although the particular curves presented in Fig. 6 correspond to couplings of a heavy lepton of the sequential type, entirely similar results are obtained for a heavy quark (top) of the corresponding masses. The shape of the curves are not very sensitive to the change of coupling constants (Cf. Fig. 7). The main background for top production comes from b- and c-quark production. The contributions of these processes to W_L are again strongly suppressed by the mass factor m_f^2/Q^2, rendering them insignificant as illustrated for the τ-lepton in Fig. 6. As pointed out earlier, there are first-order QCD contribution to W_L for quarks, but this effect is not a serious background for new particle production.

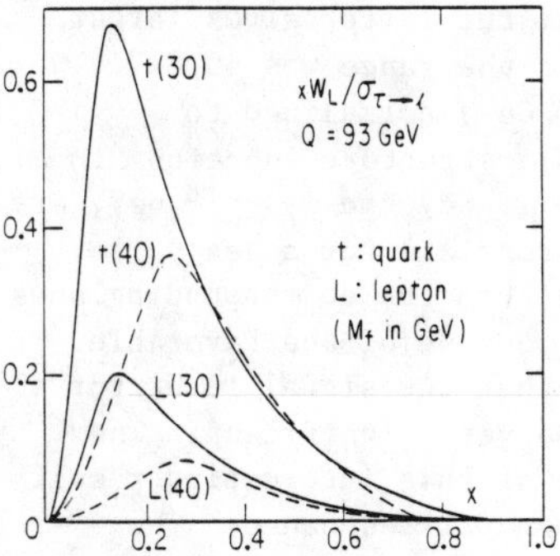

Fig.7 $xW_L(x)/\sigma_{\tau\to\ell}$ for heavy fermion production

The ability to measure the 'signal' in the W_L function depends on the absolute magnitude of this quantity, in addition to its value relative to W_T. In Fig. 7 we examine the function $xW_L(x,Q = M_Z)$ normalized to $\sigma_{\tau\to\ell}$, the integrated cross-section for a single light lepton due to τ-production. (For inclusive one-lepton production with fractional energy $x < 1$, $e^+e^- \to \tau^+\tau^- \to \ell + X$ is the simplest calculable electroweak process; hence it will be used as the standard normalization process in subsequent discussions.) We present results for a heavy lepton and a heavy quark, each at two mass values $m_f = 30$ GeV and $m_f = 40$ GeV. We see that the absolute rate for heavy quark production is much higher, as expected, mainly due to the color factor. We see here an even clearer display of two important features mentioned earlier: (i) the shape of the curves at a given mass is quite similar for a lepton and a quark; and (ii) the position and width of the distribution is very sensitive to the fermion mass parameter m_f. We also remark that: (i) the peaking of these curves at low x is partially due to 3-body decay phase-space constraints (cf. discussion on scalar electron production later); (ii) two potential handles on discriminating heavy lepton and heavy quark production are the absolute value of the W_L peak and the absence or presence of accompanying hadrons for the observed lepton.

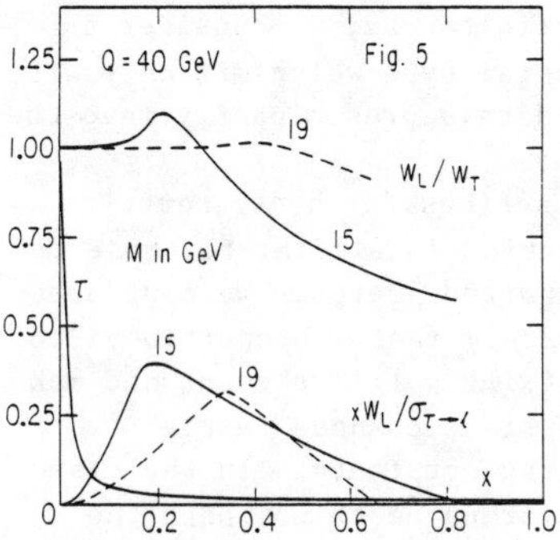

Fig.8 W_L/W_T and $xW_L/\sigma_{\tau\to\ell}$ at Q = 40 GeV for heavy fermion of masses 15 and 19 GeV.

As mentioned earlier, our analysis applies at all e^+e^- colliding energies, including currently available ones. When we are off the Z-resonance peak, the virtual photon intermediate state also has to be taken into account. It turns out that the latter gives larger contributions to W_L than the virtual Z in the range $Q < 60$ GeV. We present typical results on $xW_L(x, Q = 40 \text{ GeV})$ normalized to $xW_T(x, Q = 40 \text{ GeV})$ and to the integrated τ-structure functions in Fig. 8. Two values of the fermion, $m_f = 15$ GeV and $m_f = 19$ GeV, are shown. The couplings used are those appropriate for a heavy lepton. These distributions are broader than the corresponding ones at $Q = M_Z$ for the same mass ratio m_f^2/Q^2 because of the favorable coupling to the virtual photon. We see that the signal in W_L for a new particle, if it exists, will again be very significant. The practical limitation in taking advantage of this interesting result lies with the low total cross-section when off-resonance. The limited statistics represents, of course, the ultimate hindrance to any method of new particle search, inclusive or exclusive. Our approach has the advantage of utilizing the maximum possible lepton-final-state sample in an environment which is in principle almost background free. We would expect, therefore, that it be quite competitive compared with other methods of new particle search.

5. FEATURES OF HEAVY NEUTRAL LEPTON PRODUCTION

Heavy neutral leptons (neutrino) of either the Dirac or the Majorana type can be pair-produced in e^+e^- annihilation via the Z-pole. It is commonly assumed that heavy neutrinos decay through mixing with lighter generations. For definiteness, we consider a left-handed heavy neutrino of the sequential type which mix only with the observed lepton generation; thus it decays predominantly into the latter (Cf. Fig. 9).

The calculation of the structure functions for heavy neutrino production is similar to those for the charged lepton. We must add an overall factor proportional to the mixing matrix element and make appropriate changes in the coupling constants. In the case of Majorana neutrino, both the coupling scheme and the overall decay constant must take into account the non-conservation of lepton number, hence the possibility of its decaying into both the lepton and the anti-lepton. The results of this calculation are shown in Fig. 10a for the Dirac type neutrino and in Fig. 10b for the Majorana type

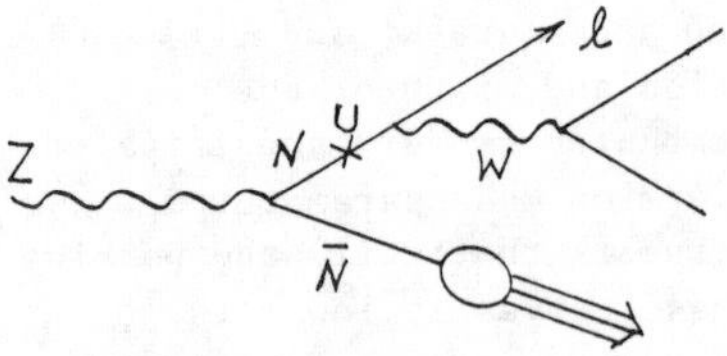

Fig.9 Heavy neutral lepton production and decay via mixing with light lepton

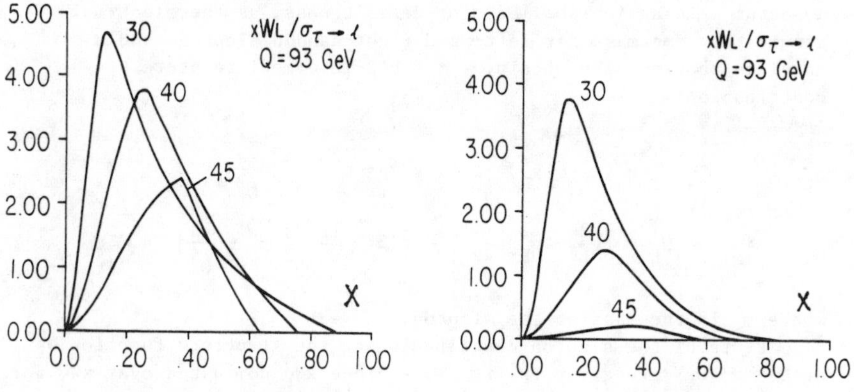

(a) Dirac-type (b) Majorana type

Fig. 10 xW_L from heavy neutral lepton production

respectively. The longitudinal structure function at Q = 93 GeV is plotted against x for three values of the assumed neutrino masses: 30, 40, and 45 GeV. Again, these curves are normalized with respect to $\sigma_{\tau\to\ell}$. We see that the shape of these curves are similar to those for the heavy charged fermion. The relatively large overall size of the effect arises from the assumption that the heavy neutrino decays exclusively into the observed lepton. We notice that the longitudinal structure function due to Dirac neutrinos are larger than those due to Majorana neutrinos. This is partially due to the fact that in our model calculation, the former couples to the Z like $\gamma_\mu(1 - \gamma_5)$ while the latter couples like $\gamma_\mu\gamma_5$. The cross-sections approach each other in the zero-mass limit as they should.

6. FEATURES OF NEW SCALAR PARTICLE PRODUCTION

As pointed out in Sec. 3, elementary scalar particle production contribute predominantly to the longitudinal structure function. We consider here a specific example of the production of a pair of scalar leptons (sleptons) expected in supersymmetric theories[5]. Many features of our results are, however, typical of scalar particle production. The production and decay mechanism is depicted in Fig. 11 where the double lines denote supersymmetric partners of ordinary particles. The detected lepton with

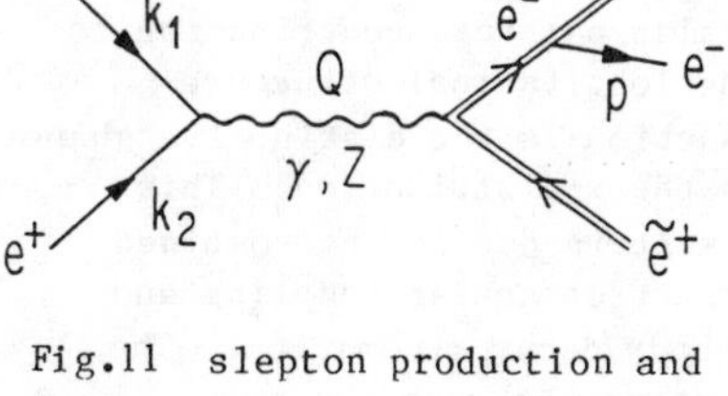

Fig.11 slepton production and decay

momentum p is again labelled, for definiteness, as the electron. We assume the same mass for left- and right-handed sleptons and a massless photino. We obtain, for the Z-intermediate state contribution:

$$W_T = (g_L^2 + g_R^2)\frac{1}{x^2}\left(1 - x - m_s^2/xQ^2\right)\frac{m_s^2}{Q^2}$$
$$W_L = \left(g_L^2 + g_R^2\right)\frac{1}{4x}\left[\frac{1}{2} - \left(3 - \frac{1}{x}\right)\frac{m_s^2}{Q^2} - \frac{1}{x^2}\frac{m_s^4}{Q^4}\right] \qquad (9)$$

where m_s is the mass of the slepton.

Plots of the slepton contribution to the structure function W_L at Q = 93 GeV are given in Fig. 12. Since xW_L dominates over xW_T for scalar particle production, it is inconvenient to use the latter as normalization (as we did for the fermion case). The top two curves are, thus, normalized with respect to $x(2W_T + W_L)$, the cross-section integrated over the observed lepton angle for a given x. To indicate the absolute magnitude of the effect due to slepton production, the two bottom curves depict xW_L normalized with respect to our 'calibration process' $\sigma_{\tau \to \ell}$ (as in Fig. 7).

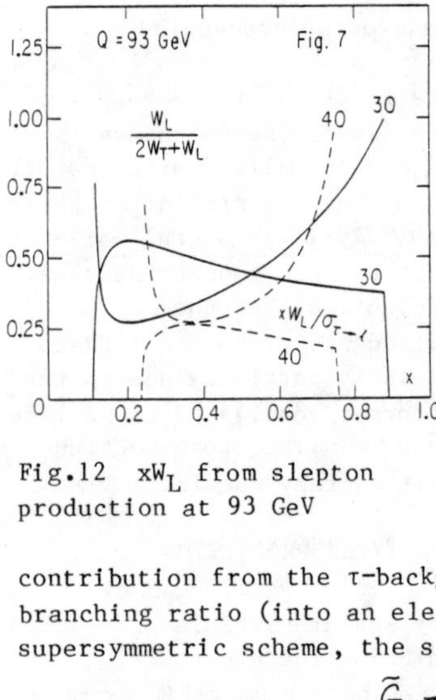

Fig.12 xW_L from slepton production at 93 GeV

The signal is quite large, since (i) there is no mass-suppression factor for scalar particle production; and (ii) the contribution from the τ-background process is multiplied by a branching ratio (into an electron) of 20% whereas, at least in the supersymmetric scheme, the slepton turns into its lepton partner (and a $\tilde{\gamma}$) 100% of the time. In addition to the magnitude of the effect, a striking feature of the scalar particle contribution to the longitudinal structure function is the distinctive shape of the x-distribution. This result is due to the combined effect of scalar coupling and 2-body decay kinematics. The contrast between the two categories of new physics

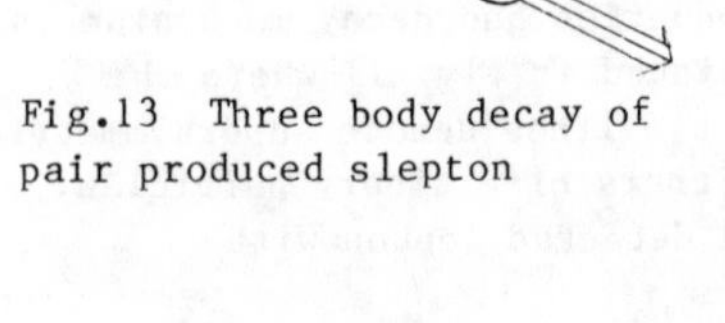

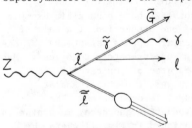

Fig.13 Three body decay of pair produced slepton

considered, heavy fermion production and scalar particle production, could not have been sharper.

In order to separate the effects due to the spin of the slepton and the 2-particle decay phase-space, we also performed a calculation assuming that the slepton decays into a 3-body channel. As a concrete example, we may imagine that the pair-produced supersymmetric electron decays into an electron (observed) and a photino as before, but the latter in turn decays into an ordinary photon plus a 'goldstino' (Fig. 13).

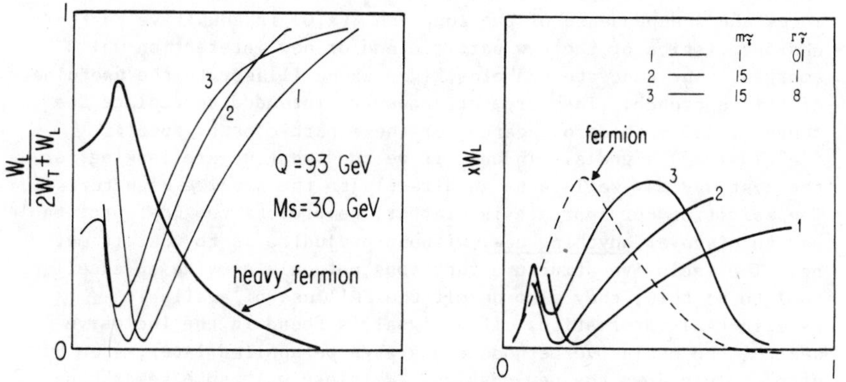

(a) $W_L/(2W_T + W_L)$ (b) $xW_L/\sigma_{T+\gamma}$

Fig. 14 xW_L from slepton production and 3-body decay. Dotted line in each case indicates qualitative feature of corresponding curve for heavy fermion production.

The results in this case clearly depend on the assumed mass and width of the supersymmetric particles. We assume that m_G is negligibly small, and present representative results corresponding to 3 choices of $M_{\tilde{\gamma}}$ and $\Gamma_{\tilde{\gamma}}$. Fig. 14a plots $W_L/(2W_T + W_L)$ from pair produced sleptons, and Fig. 14b depicts the shape of xW_L (in arbitrary units). In each of these graphs, we also included a dotted curve representing the general feature of heavy fermion production (cf. Sec. 4) for easy comparison. We see that the two cases are really quite distinctive.

To apply similar analysis off the Z-resonance, we again need to include other virtual particle intermediate states. In the supersymmetric scheme, there exists extra contributions on top of the familiar s-channel virtual photon contribution--such as t-channel 'photino' exchange. In contrast to the one-particle s-channel diagrams, the t-channel exchange contributions cannot be written in terms of the common structure functions. The situation is therefore more complicated than previously described. Ideally, the experimental signature for the presence of such new contributions would be the observation of the breakdown of Eq. (5) for the angular distribution of the observed electron. In practice, quantitative analysis will be possible only with very high statistics.

5. FURTHER DISCUSSION

We have seen that the non-vanishing of the 1-lepton longitudinal structure function in e^+e^- annihilation signals the onset of new physics beyond that of current standard model processes. Furthermore, the x-dependence of the function R(x,Q) is sensitive to the characteristics of the new particle and/or new interaction which emerges. The concrete examples given above illustrate the usefulness of this approach. They are not, however, intended to replace the conventional methods of search for these particles by specific ('exclusive') signals. Indeed, if we know what we are looking for, the best way may well be to go directly to the precise signatures. The spirit of our approach is, rather, to propose the most profitable way to discover anything new, without prejudice as to what it may be. The inclusive structure functions not only provide an excellent tool to do this, they also permit the full use of available experimental information. If a signal is found in the inclusive search, the exclusive methods can always be applied later, step by step, to pin down the new physics. We close with some remarks on possible extensions of this approach.

(i) We have chosen the 1-lepton inclusive structure functions for analysis because they are both the cleanest to measure and the easiest to calculate. However, the same general principles also apply if the leptons are replaced by elementary hadrons, i.e. quarks. At very high energies, it may be possible to consider 1-jet inclusive cross-section and the associated structure functions[6]. Because of the color factor, hadronic events are more copious, offering an advantage in statistics. If this proves practical, we will gain new windows on new particles and their interaction.

(ii) We can also extend similar analyses to 2-particle inclusive cross-sections and structure functions. Two particle correlations (either lepton-lepton or lepton-quark) clearly offer a great deal more information on the underlying dynamics. Furthermore, generalizations of W_L, to which conventional processes give vanishing contributions, also exist[7,8]. Measurements of such cross-sections should be possible, at least on the Z-resonance. This study is underway.

(iii) It is of interest to apply this approach to $\bar{p}p$ scattering. The partons inside the proton and anti-proton replace the initial state e^+e^- particles in the above analysis. The $\bar{p}p$ colliders allow one to reach a much higher energy range than the e^+e^-

machines. However, uncertainty on the CM frame of the partons which initiate the interaction smears out some features of the final state. A detailed study is needed to determine the usefulness of the inclusive approach in the $\bar{p}p$ process.

(iv) As part of the standard model scheme, the Higgs meson and the top quark have been the subject of many experimental searches and theoretical calculations. Our analysis of l-lepton inclusive structure function due to heavy fermion production provides a new method to find the top quark. The Higgs boson is more elusive. Whether it can be observable in the l-lepton (light) longitudinal structure function, assuming it exists, is currently being studied.

ACKNOWLEDGEMENT

We thank Steven Gottlieb and Jon Rosner for useful discussions. WKT also wishes to thank Jon Rosner and the Enrico Fermi Institute for their hospitality during the initial stages of this work, and likewise to the High Energy Physics Division at Argonne during the completion of the work.

REFERENCES

1. C. Callan and D. Gross, Phys. Rev. Lett. 22, 156 (1969).
2. Y. Kitazawa and Wu-Ki Tung, Phys. Lett., to be published.
3. Y. Kitazawa and Wu-Ki Tung, to be published.
4. S. Gottlieb and T. Weiler, Phys. Rev. D29, 2005 (1984).
5. For a review, see P. Fayet and S. Ferrara, Phys. Rep. 32C, 249 (1977); H. Baer et al. CERN Preprint (1984) discuss the possibility of detecting sleptons in $p\bar{p}$ colliders.
6. J. L. Rosner, CERN Preprint (1985) discusses the angular distribution of 'monojets'. The spirit is close to our approach.
7. C. S. Lam and Wu-Ki Tung, Phys. Lett. 80B, 228 (1979); and Phys. Rev. D21, 2711 (1980).
8. A. P. Contogouris and H. Tanaka (1985), Phys. Rev. D31, 1638.

NEW PHYSICS FROM THE SUPERSTRING

D. V. Nanopoulos

CERN, Geneva
and
Physics Department, University of Wisconsin, Madison, WI 53706

ABSTRACT

In this talk I discuss, in a simplified way, recent attempts to extract observable, four-dimensional physics from ten-dimensional superstrings. Under certain rather natural assumptions a class of models emerges, of the no-scale type and encompassing a new $U(1)_E$ gauge interaction at low energies. The main characteristics of these models are, a rather "heavy" supersymmetric spectrum ($m_{\tilde{q}}, m_{\tilde{\ell}} > O(150$ Gev)) extendable up to the few Tev region, while the mass of the new neutral gauge boson Z_E may be anywhere from O(150 Gev) to O(1 Tev). Hopefully the SuperString Collider (SSC) will comfortably cover this range, while pleasant surprises at the Tevatron, SLC and LEP are not excluded.

A consistent and calculable theory of quantum gravity demands extended fundamental objects, as point field theories of gravity are incurably unrenormalizable. The simplest extended objects, strings, contain among their massless modes, spin-2 particles identifiable with the graviton. String field theory leads in general, to several inconsistencies except if and only if we consider strings containing both fermions and bosons in their particle spectrum, superstrings[1], in D=10 space-time dimensions with $E_8 \times E_8'$ or O(32) as the appropriate gauge group[2]. Such types of superstrings seem to satisfy all the usual rules of field theory, contain no anomalies[2] (Yang-Mills, gravitational or mixed) and seem to be, at least, one-loop finite[3]. The best candidate for a superstring theory is the heterotic string[4], where not only the number of space-time dimensions (D=10), and N=1 supergravity are derived dynamically, but also the existence of the gauge group $E_8 \times E_8'$ or O(32) is explained[4]. Interestingly enough, the point field theory limit of the superstring is N=1 supergravity in D=10 dimensions with $E_8 \times E_8'$ or O(32) as a gauge group plus some extra extremely useful terms. These include the addition of a Lorentz-Chern-Simons term (ω_L) for the field strength $H_{\mu\nu\rho}$ of the antisymmetric tensor field $B_{\mu\nu}$ contained in the N=1, D=10 supermultiplet. The existence of ω_L is essential for cancelling anomalies of all types[2] plus the fact that its supersymmetrization[5] leads automatically to the right combination of R^2 terms, ghost free[6], and essential in compactifying from 10 to 4 dimensions.

The survival of N=1 supersymmetry (SUSY) in four dimensions, essential for the solution of the gauge hierarchy problem, picks out uniquely (?) Calabi-Yau manifolds[8] as the prime candidates for the compactification manifolds (K_6) of the six surplus dimensions. Calabi-Yau manifolds[8], are manifolds with SU(3) holonomy and thus Kähler and Ricci flat. Choosing the ground state of the superstring to be of the admittable type $M_4 \times K_6$ with M_4 a four-dimensional maximally symmetric space and K_6 a 6-dimensional Calabi-Yau manifold, we automatically oblige[7] M_4 to be the highly desirable Minkowski space, i.e., the cosmological constant (Λ_c) is automatically zero at the classical (tree) level. A result rather intriguing and full of far reaching consequences.

The continuation of the absence of anomalies in 4-dimensions imply[9,7] $F \times F \propto R \times R$ which, in the case of the heterotic string, is only satisfied if we identify[7] the SU(3) holonomy group with an SU(3) $\subset E_8 \times E_8'$ or O(32). Thus during compactification $E_8 \times E_8' \to E_6 \times E_8'$ or $O(32) \to O(26) \times U(1)$. While E_6 is an acceptable gauge group[10], O(26) is not since it eventually leads to a vector-like fermion spectrum[7]. Remarkably enough the particle spectrum that emerges in 4-dimensions looks very similar to the one observed experimentally. Beyond the gauge sector that contains the $E_6 \times E_8'$ gauge bosons and gauginos, there are N_g chiral muliplets transforming as the $\underline{27}$ of E_6 (see table 1), some possible relics from $\delta(27+\overline{27})$ as well as some gauge-singlets. The numbers N_g and δ are given by certain topological numbers characterizing the Calabi-Yau manifold K_6. Namely, $N_g = |\chi(K_6)|/2$, with $\chi(K_6)$ the Euler Characteristic number of K_6 and $\delta = b_{1,1}$, the Hodge number of K_6. Simply connected Calabi-Yau manifolds always have $b_{1,1} \geq 1$ and $|\chi(K_6)| >> 1$. That's not very exciting since it gives $N_g >> 1$, while most of us believe that the number of generations (N_g) cannot be a large number. Actually, this fact is one of many reasons to consider not-simply connected Calabi-Yau manifolds[7,11,12]. In this case N_g is reduced considerably (N_g=3 or 4 is possible), δ can be effectively set to zero, $E_6 \times E_8'$ may automatically be broken to smaller groups by the Wilson-loop mechanism[13] with the exciting possibility[11] that the gauge couplings obey the successful grand unification relations while the Yukawa couplings need not. Clearly, primordial E_6-invariance entails the allowable Yukawa terms, but the Yukawa coupling constants need not obey the E_6 Clebsch-Gordan relations.

All these facts are of invaluable importance in constructing realistic models. A large number of generations, especially coming $\underline{27}$ - a piece, will destroy the smooth run of the renormalization group equations (RGE) and will demand unification at unacceptably low values ($\leq O(10^{12}$ Gev)) thus ruining the succesful predictions of $\sin^2\theta_w$ and the longevity of the proton[14,15]. Ditto for large δ's. The Wilson-loop mechanism[13,7,11,12] is the only one, presently known, for breaking $E_6 \times E_8'$ without a simultaneous unacceptable breaking of the highly desirable residual N=1 SUSY in four dimensions.

The need for E_6 breaking at low energies is obvious, while the breaking of $E_8' \to SU(N)$ is demanded[16] by the formation of a gluino-condensate needed[17,16] for local SUGY (supergravity) breaking. Futhermore, the absence of E_6 - Clebsch-Gordan relations between Yukawa couplings, may resolve automatically, long-standing problems related to proton decay or neutrino masses. The possible breakings of E_6 through the Wilson-loop mechanism have been meticulously classified[15], though a small theorem, recently derived[19], suggests a unique solution to the E_6 breaking pattern[20,21]: $E_6 \to G_0 \equiv SU(3)_c \times SU(2)_L \times U(1)_Y \times U(1)_E$. The decomposition of the $\underline{27}$ under $SU(3)_c \times SU(2)_L \times U(1)_Y \times U(1)_E$ is given in table 1. A crucial assumption in deriving such an exciting result is the absence of an intermediate energy scale, which follows from the no-scale philosophy[22] developed during the last few years[23] and naturally emerging from the superstring, as we will discuss later.

Notice the existence of a new gauge interaction $U(1)_E$ at low energies, with a well-defined "charge" relative (see table 1) and absolute (as determined[19-22] by grand unification)

$$\alpha_E\ (M_W) \equiv \frac{g_E^2}{4\pi} = 0.0164 \qquad (1)$$

The interesting phenomenology of a new, low mass (100-1000 Gev), neutral gauge boson Z_E has been discussed in considerable detail elsewhere[19,20,21,24]. Here it suffices to remind the reader that presently available experimental data put a lower bound

$$M_{Z_E} \gtrsim O(200\ \text{Gev}) \qquad (2)$$

which may push down to the infinum[21] of 110 Gev or stretch up[25] to O(300) Gev if we try to satisfy cosmological constraints coming from primordial nucleosynthesis[25].

Concerning the chiral multiplet content, for models with no intermediate scale RGE analysis demands[24,15] the existence of only Ng=3 generations of $\underline{27}$ (as presented in table 1), while δ may be taken naturally and safely to be effectively zero. These facts are of great importance when one tries to build up the Yukawa sector, or more general the super potential, f. Given the particle content of table 1, primordial E_6-invariance allows only the following sets of couplings[19,26]:

$$f_1 = hQHu^c + h_DQ\bar{H}d^c + h_LL\bar{H}e^c \tag{3}$$

$$f_2 = \lambda H\bar{H}N + KDD^cN \tag{4}$$

$$f_3 = \lambda_1D^cLQ + \lambda_2De^cu^c + \lambda_3Dd^c\nu^c \tag{5}$$

$$f_4 = \lambda_4DQQ + \lambda_5D^cu^cd^c \tag{6}$$

and

$$f_5 = \lambda_6HL\nu^c \tag{7}$$

While f_1 contains the usual Yukawa couplings of the "old" standard model ($SU(3)_c\times SU(2)_L\times U(1)_Y$), $f_{2,3,4,5}$ contain new "good" and "bad" couplings. The couplings contained in f_2 are of prime importance[19]:

a) They allow the N-field to get a vacuum expectation value (v.e.v) $\langle N\rangle\equiv X$ and thus break the $U(1)_E$ and provide mass to the Z_E and to the D-quark. Notice that both m_{Z_E}, $m_D \propto\langle N\rangle(\equiv x)$ and in principle, if experimentally desired, they may be pushed higher than the usual $M_{W^\pm,Z^0} \propto \langle \overset{(-)}{H} \rangle(\equiv \overset{(-)}{v})$.

b) They give masses to all Higgs or Higginos, Neons and Ninos, avoiding embarassments with unwanted massless scalars (axion-types) or massless fermions.

c) They help in an essential way the realization of the "no-scale" scenario, as we will shortly see. For example, the presence of the $H\bar{H}N$ coupling is unavoidable[20,21] if $m_{1/2}$, the primordial gauginos mass, is to be different from zero, as the only source left of SUSY breaking in the observable sector[19,20,21]

In sharp contrast with the above remarkable consequences of f_2 (4), one has to be careful[19,26] with $f_{3,4,5}$ (5,6,7). The longevity of protons demands the exclusive presence (for making the D quark unstable) of f_3 (5) or f_4 (6). The absence of large neutrino masses entails the banishing of f_5(7). It is highly remarkable that we have to rely on the existence of discrete symmetries and/or "topological" arguments [27,28] to forbid only f_3(5) or f_4(6) and f_5(7). This is not a formidable task, since for example[19,26] a simple $Z_2\times Z_2'$ symmetry: Z_2: $(L,e^c,\nu^c) \to (-1)(L,e^c,\nu^c)$ and Z_2': $\nu^c\to(-1)\nu^c$, allows $f_{1,2,4}$ but forbids $f_{3,5}$, etc. It is also encouraging that examples have been provided[28], where out of 8436 allowable Yukawa couplings, only 111 survive for "topological" reasons[28]. It should not escape our notice the extra bonus[19,26] coming from the form of the superpotential $f=f_1+f_2+f_{3\text{ or }4}$. There are automatically no problems with flavor changing neutral currents (FCNC) at the tree-level, as well as no weak universality problems because D and d do not mix[19,26]. This later property shows the deep connection between the dis-allowance of dangerous B-violating couplings and the exact validity of weak universality.

Up to now, starting from the superstring we have derived[19] a unique low energy gauge group $Go \equiv SU(3)_c \times SU(2)_L \times U(1)_Y \times U(1)_E$, a unique particle spectrum (table 1), and an (essentially) unique superpotential $f = f_1 + f_2 + f_{3 or 4}$, which suitably bypasses unscathed all the usual traps (FCNC, weak universality, proton stability, (almost?) massless neutrinos, etc.) of the well-established low-energy phenomenology. Next we have to determine dynamically the values of the different gauge coupling constants and mass scales in our model. This formidable task has been realized only recently[19] and here I give a brief resume of the present situation[19].

Compactification on Calabi-Yau manifolds does not only preserve[7] N=1 supergravity in 4-dimensions but it also determines[29,30], at the classical level, the precise form of the two, otherwise arbitrary functions[31], $G(z_i z_i^*)$ and $f_{ab}(z_i)$. The Kähler potential G, a real function of its arguments, determines through its second derivatives G_{ij} (≡metric) the Kähler manifold where the fields z_i "live", or in other words the form of the kinetic terms of the scalar fields z_i. The analytic function $f_{ab}(z_i)$ determines the kinetic terms of the gauge boson and correspondong gauginos.

Using an "educated" dimensional reduction, highly imitative of compactification on a Calabi-Yau manifold, Witten succeded[29] to derive the following forms for G and f_{ab}, at the tree level

$$G = -\ell n(S+S^*) - 3\ell n\,(T+T^*-2\phi_i\phi_i^*) + \ell n|W(S)+W(\phi_i)|^2 \quad (8)$$

and

$$f_{ab} = S\delta_{ab} \quad (9)$$

In writing down (8) and (9) we have used the 4-dimensional gauge-singlet chiral superfields[29]

$$S = e^{3\sigma}\phi^{-3/4} + 3i\sqrt{2}\,D \quad (10)$$

$$T = e^{\sigma}\phi^{3/4} - i\sqrt{2}\,a \quad (11)$$

where the two scalar fields ϕ and σ are the dilation (ϕ) of the ten-dimensional supergravity theory and the "compacton" (σ) which scales the compactification manifold $g_{ij} = e^{\sigma} g^{o}_{ij}$, and the two pseudoscalar fields (D,a) correspond to components of the antisymmetric tensor field $B_{\mu\nu}$ (with ϕ_i denoting collectively the E_6 - $\underline{27}$ chiral multiplet). It should be stressed that, since the d=4 unified gauge coupling is given by

$$g_u^2 = \frac{1}{Re(f_{ab})} : g_u^2 = \frac{1}{S_R} \text{ in leading order} \tag{12}$$

and loop diagrams renormalize g_u, the simple expression (9) will be renormalized by the matter fields in the observable sector. The D and a fields have been identified with the axion (D) and plation[32] (a) fields responsible for solving the strong CP-problem, and also, maybe, the cosmological constant problem[32] (to all orders!) respectively. Notice that recent claims[33] that the effective ten-dimensional point field theory Wess-Zumino terms destroy the simple structure of (8) and (9) have been found to be not valid[34].

The amazing thing about the form of G as given by (8), with its remarkable SU(1,1)/U(1) × SU(n,1)/SU(n)×U(1) structure, is identical to the one that defines the so-called no-scale supergravity[35,22,36-38,23]. No-scale supergravity has been invented[35] by the demand of a naturally vanishing cosmological constant, at least, at the classical level and has been subsequently used[22,36-38] for a dynamical determination of all the mass scales in terms of a primordial mass scale, the Planck Mass $M \left(\equiv M_{P\ell}/\sqrt{8\pi} \equiv 1/\sqrt{8\pi \; G_{Newton}}\right) \equiv 1$. Indeed it is a simple exercise[35] to show that inserting the (T) part of (8) into the scalar potential[31] $\left(G_i = \frac{\partial G}{\partial z_i^*}, \ldots\right)$

$$V = e^G [G i (G^{-1})^i_j G^j - 3] \tag{13}$$

we get $V \equiv 0$, for every value of the field T, while the gravitino mass $m_{3/2} = e^{G/2}$ is different from zero. In other words, we get a theory[35] with a naturally vanishing cosmological constant at the tree-level, local SUSY breaking ($m_{3/2} \neq 0$) but of undetermined magnitude. But that's exactly what we are after[22]. We are going to let quantum corrections dynamically determine the magnitude of local SUSY breaking ($\propto m_{3/2}$) as well as the magnitude of gauge symmetry and global SUSY breaking[22,36-38]. This is the no-scale scenario, which has been implemented successfully in the past[23].

The highly desirable form of G as given by (8) should not come as such a big surprise. As we discussed in the beginning, Calabi-Yau compactification leads to a vanishing cosmological constant for a large class of Calabi-Yau manifolds[7]. Since the observed four-dimensional chiral fields are the zero-modes of the compactification manifold, parametrizing its distortions, it follows that we get $\Lambda_c=0$ for a continuous range of the chiral field-values. It has been proven in the past[36] for one chiral-field, and very recently[30] for any-number of chiral fields, that the (T,ϕ_i) part of (8) is the unique G-function that leads to V=0, for a continuous range of small field-values[30]. Actually, we have been able recently to derive[30], more or less, the whole form of (8) directly from the superstring, without invoking the highly questionable[39] intermediate step of N=1, D=10, $E_8\times E_8'$ point field theory. So several criticisms of Witten's intuitive approach[29] are highly undeserved and unfounded.

Let us show now the importance of every and each term in G eq. (8) and f_{ab} eq. (9) in getting a realisitic model with well-determined and calculable supersymmetry and gauge breaking[19-21]. Let us start with the S-part which is the part directly connected with local SUSY breaking. Indeed, non-perturbative effects of some unbroken subgroup $Q \subset E_8'$ may lead to gaugino condensation[16] which in compliance[18] with the non-trivial form of f_{ab}(9) trigger local SUSY breaking[17,16]. Integrating out the E_8'-gauge-sector degrees of freedom, the seeds of local SUSY breaking are "transmitted" to the derived effective superpotential[17]

$$W(S) \ni h \exp\left(\frac{-3S}{2b_Q}\right) \tag{14}$$

where h is an unknown constant O(1), and b_Q is the first non-trivial term in the beta-function for the subgroup Q of E_8' which condenses (e.g., for SU(N): $b_Q= 3N/16\pi^2$). The effective scalar potential for the S-field is then (using (8), (13) and (14))

$$V = \frac{1}{(16S_RT_R^3)} |W(S)|^2 \; G_S G_{\bar{S}} G_{S\bar{S}} \tag{15}$$

which gives a minimum V=0 at $S_R=\infty$! Thus the dynamical determination of S and hence through (12) of g_u^2 would require an additional contribution $\delta W(S)$ to W(S). The simplest possibility[17] is a constant $C \propto \langle H_{ijk} \rangle \neq 0$, but any other feature of string dynamics generating a $\delta W(S)$ would be equally satisfactory. Assuming this to be the case, generically there will be a value of $S \neq 0$ or ∞ for which $G_S=0$ and hence V=0 (15). This would give a vanishing cosmological constant $\Lambda_c=0$ at the tree level in four dimensions, while local SUSY is broken

$$m_{3/2}=e^{G/2} = \left(\frac{1}{16S_R T_R^3}\right)^{1/2} |W(S)| \qquad (16)$$

Since $1/S_R = g_u^2 = O(1)$ is also fixed, the only uncertainty in (16) is the value of T_R, which will be determined dynamically later. Incidentally, it has been shown recently[40] that contrary to opposite, persistent claims[39], $g_u=0$ cannot be reached by the classical field equations, but for a considerable range in initial field values, these drive S_R towards the region $S_R=O(1)$ where the minimum is located. It is important to notice that (15) implies $M_{S_I} \sim M$, i.e., the axion ($\equiv$D) becomes invisible, not because is superlight but because it is superheavy! Since in this case it decays instantly, no bound on $\langle S_I \rangle$ can be derived from the cosmological relaxation constraint[41].

We now face the problem of dynamical determination of T_R and of supersymmetry breaking in the observable sector, which of course are resolved following the standard no-scale strategy[23]. However, there are certain amusing new facts that deserve some discussion. In the standard SU(n,1) no-scale supergravity[37] the only allowed seed for supersymmetry breaking in the observable sector is non-vanishing primordial gaugino mass, $m_{1/2}$. It has been proven that soft supersymmetry breaking gauge-non-singlet scalar masses are naturally zero not only at the tree-level[37] but at the one-loop level[42,43] as well. Furthermore, this result is easily extended[43] to all higher loop orders if there is no non-trivial superpotential. These results[37,42-43] support our claim[19,21] that the primordial gaugino mass is the largest global SUSY breaking mass parameter in the effective no-scale supergravity model (8) derived from the superstring[29,30]

But who provides us with a non-vanishing $m_{1/2}$? The superstring! We simply observe that in N=1 supergravity[31]

$$m_{1/2} = \frac{1}{4\mathrm{Re}(f_{ab})} e^{G/2} G^i (G^{-1})_i^k f_{ab,k} \qquad (17)$$

which, for sure, demands non-trivial f_{ab} and of the appropriate form. Clearly, (9) is not enough, but as discussed after (12), it is renormalized by non-gravitational interactions[20,21]. The astonishing thing is that if a $H\bar{H}N$ term exists in the superpotential $W(\phi_i)$, as it does in our case $(W(\phi_i)\ f_2(4))$ then[20,21,19]

$$m_{1/2} = \frac{E_6 g_u^2}{4}\left[\frac{1}{16\ S_R T_R^3}\right]^{1/2} \cdot \lambda\ \frac{[|H\bar{H}|^2+|HN|^2+|\bar{H}N|^2]}{H\bar{H}N} \qquad (18)$$

with $E_6 = 27/16\pi^2$. In other words, the superstring-derived no-scale model[19] not only provides us with a nonvanishing $m_{1/2}$, but it predicts a strong correlation between the magnitudes of global SUSY breaking, $m_{W\pm, Z_0}$ and m_{Z_E}. It becomes now apparent why m_{Z_E} cannot be too large (say above Θ(1 Tev)), since then $m_{1/2} > O$(1 Tev) and the gauge hierarchy is jeopardized. That's another reason for excluding intermediate mass scales.

The only remaining problem is who determines T_R? Here we can give a two-fold answer[19,21]. It is conceivable that superstring dynamics (such as loop-effects) determine T_R dynamically[21,42]. In such a case, we have to follow[19] the standard dimensional transmutation strategy[44] in which one keeps $m_{1/2}$ fixed, uses standard non-gravitational interactions to break radiatively $SU(2)_L \times U(1)_Y \times U(1)_E$, thus determines $\langle H\rangle \equiv v$, $\langle \bar{H}\rangle \equiv \bar{v}$, $\langle N\rangle \equiv x$ dynamically and then calculates, self-consistently, $m_{1/2}$ through (18). This we call[19] the Hybrid Dimensional Transmutation model (HYDTRA) in contrast to the pure no-scale scenario[21,19] where T_R (and thus $m_{1/2}$) is left as a dynamical variable to be determined by low energy radiative corrections, in the usual way[22,36-38]. Notice that in both scenarios[19,21], (18) implies $T_R \sim O(1)$ which through (16) gives $m_{3/2} \sim M$. Notice how catastrohpic such a result would be if m_o, the soft-breaking scalar masses were not protected[37,42,43], as discussed above, to be naturally zero (up to non-gravitational radiative corrections) since then $m_0 \sim m_{3/2} \sim M$, i.e., the end of gauge hierachy and beginning of gauge chaos! It should be emphasized that not only $m_{3/2}$ but <u>all</u> other relevant scales like the compactification scale M_c, the superstring scale M_s, the grand unification scale M_u and the gaugino condensation scale Λ_Q, are dynamically determined to be O(M) since they are all expressible as simple functions of the S and T fields[16,20]. Furthermore the dynamical determination of the M_u in conjunction with the dynamical determination (through (12)) of g_u^2, imply the dynamical determination (through the standard lore of RGE's) of <u>all</u> the gauge coupling constants[16,19,20]. An unprecented incident in particle physics.

There are some advantages of the HYDRA over the pure no-scale scenario. In the HYDRA scenario the "hidden sector" (S,T), being there for providing SUSY breaking seeds, decouples completely from the observable sector. So, there is no mystery anymore why globel SUSY breaking and electroweak breaking are of the same magnitude (see (18)). Furthermore, such a complete decoupling of the two sectors resolves automatically the "Polonyi problem"[45]. It has been difficult to see how in the context of conventional cosmology the T_R field, in all models where the SUSY breaking scale is determined through a v.e.v. in the hidden sector, can relax to its minimum and not leave behind an excess of vacuum energy in the form of coherent waves[45]. In the HYDTRA-type models[19] $M_{T_R} \sim M$ and thus T_R, as being superheavy, decays instantly and poses, like the axion $S_I(\equiv D)$ case, no cosmological problems.

A complete and detailed study of the superstring HYDTRA and pure no-scale models has been given in Ref. 19. Here we discuss a few highlights of the HYDTRA-type models, following Ref. 19. Notice that the only free parameters in this type of model are h_u, λ_u and k_u, the values of Yukawa couplings (defined in (3) and (4)) at the unification scale $M_u \sim M$. As it is shown in fig. 1, the arbitrariness in the values of (h_u, λ_u, k_u) is largely reduced if all "known" phenomenological constraints are imposed. Remarkably enough, a non-trivial region in the (h_u, λ_u, k_u) parameter space satisfies all the constraints, for m_t[19] in the (40 to 70) GeV region[19,46]. The previously alluded correlation between $m_{1/2}$ and x/v is shown explicitly in fig. 2, which clearly indicates that the solution to the gauge hierachy problem does not allow[19,46] large values for M_{Z_E}, say bigger than 1 Tev, which makes the experimental Z_E-hunt feasible and exciting. On the other hand, as fig. 3 and table 2 shows, the expected SUSY-spectrum is rather "heavy" (say for the CERN $Sp\bar{p}S$ collider) but hopefully not too heavy for the Tevetron (?) or the Super String Collider (SSC).

Superstring Physics is very exciting, and it is highly remarkable that the first efforts for extracting phenomenology from such complicated theories have not been obviously in vain. Let me now, following the tradition, end by discussing crucial, open problems

Acknowledgements

I would like to thank V. Barger, D. Cline and F. Halzen for their kind invitation and warm hospitality extended to me during the Madison "New Particles 85" conference.

References

1. M.B. Green, Sur. High. Ener. Phys. 3 (1983) 127; J.H. Schwarz, Phys. Rep. 89 (1982) 223.
2. M.B. Green and J.H. Schwarz, Phys. Lett. 149B (1984) 117.
3. M.B. Green and J.H. Schwarz, Phys. Lett. 151B (1985) 21.
4. D. Gross, J. Harvey, E. Martinec and R. Rohm,

Phys. Rev. Lett. 54 (1985) 502; Nuc. Phys. B256 (1985) 253 and Princeton University preprint 85-0694 (1985).
5. S. Cecotti, S. Ferrara, L. Girardello and M. Porrati, Phys. Lett. 164B (1985) 46; L. Romans and N. Warner, Caltech preprint 88-1291 (1985).
6. B. Zwiebach, Phys. Lett. 156B (1985) 315.
7. P. Candelas, G.T. Horowitz, A. Strominger and E. Witten, Nucl. Phys. B258 (1985) 46.
8. E. Calabi, in "Algebraic Geometry and Topology: A Symposium in Honour of S. Lefschetz" (Princeton University Press, 1957), p. 78; S.T. Yau, Proc. Nat. Acad. Sci. 74 (1977) 1798.
9. E. Witten, Phys. Lett. 149B (1984) 351.
10. Y. Achiman and B. Stech, Phys. Lett. 77B (1978) 389; R. Barbieri and D.V. Nanopoulos, Phys. Lett. 91B (1980) 369; R. Barbieri, A. Masiero and D.V. Nanopoulos, Phys. Lett. 104B (1981) 194; R. Barbieri, S. Ferrara and D.V. Nanopoulos, Phys. Lett. 116B (1982) 16.
11. E. Witten, Nucl. Phys. B258 (1985) 75.
12. J.D. Breit, B.A. Ovrut and G. Segre, Phys. Lett. 158B (1985) 33.
13. Y. Hosotani, Phys. Lett. 126B (1983) 309.
14. S. Cecotti, J.P. Derendinger, S. Ferrara, L. Girardello and M. Roncadelli, Phys. Lett. 156B (1985) 318.
15. M. Dine, V. Kaplunovsky, M. Mangano, C. Nappi and N. Seiberg, Nucl. Phys. B259 (1985) 519.
16. E. Cohen, J. Ellis, C. Gomez and D.V. Nanopoulos, Phys. Lett. 160B (1985) 62.
17. M. Dine, R. Rohm, N. Seiberg and E. Witten, Phys. Lett. 156B (1985) 55; J.P. Derendinger, L.Ibenez and H. P. Nilles, Phys. Lett. 155B (1985) 65.
18. S. Ferrara, L. Girardello and H.P. Nilles, Phys. Lett. 125B (1983) 457.
19. J. Ellis, K. Enqvist, D.V. Nanopoulos and F. Zwirner, CERN preprint TH-4323 (1985).
20. E. Cohen, J. Ellis, K. Enqvist and D.V. Nanopoulos, Phys. Lett. 161B (1985) 85.
21. E. Cohen, J. Ellis, K. Enqvist and D.V. Nanopoulos, Phys. Lett. 165B (1985) 76.
22. J. Ellis, A.B. Lahanas, D.V. Nanopoulos and K. Tamvakis, Phys. Lett. 134B (1984) 429.
23. For reviews see: D.V. Nanopoulos, in the Proc. XXII Inter. Conf. on High Energy Physics, Eds. A. Meyer and E. Wieczorek (Leipzig 1984), published by Academie der Wissenschaft der DDR, Vol. II, p. 36.; C. Kounnas, A. Masiero, D.V. Nanopoulos and K. A. Olive, "Grand Unification with and without Supersymmetry and Cosmological Implications" (World Scient. Publ. Company, Singapore (1984)); J. Ellis, Proc. 5th Workshop on Grand Unification, Brown University 1984, ed. K.Kang, H. Fried and P. Frampton (World Scient. Publ. Company, Singapore, 1984), p. 436; A.B. Lahanans and D.V. Nanopoulos, "The road to No-Scale Supergravity" Physics reports to appear.

24. V. Barger, N.G. Deshpande and K. Whisnant, Madison preprint MAD/PH/268 (1985); L.S. Durkin and P. Langacker, Univ. of Pennsylvania Preprint UPR-0287-T (1985); M. Drees, N. Falck and M. Glück, Phys. Rev. Lett 56 (1986) 30.
25. J. Ellis, K. Enqvist, D.V. Nanopoulos and S. Sarkar, CERN preprint TH-4303 (1985).
26. B. Campbell, J. Ellis, K. Enqvist and D.V. Nanopoulos, CERN preprint in preparation.
27. A. Strominger and E. Witten, Princeton preprint (1985).
28. A. Strominger, Inst. of Theor. Phys. Santa Barbara preprint (1985).
29. E. Witten, Phys. Lett. 155B (1985) 151.
30. J. Ellis, C. Gomez and D.V. Nanopoulos, CERN preprint TH-4328 (1985).
31. E. Cremmer et al, Nucl. Phys. B212 (1983) 413.
32. I. Antoniadis, C. Kounnas and D.V. Nanopoulos, Phys. Lett. 162B (1985) 309.
33. K. Choi and J.E. Kim, Seoul Nat. Univ. Preprint SNUHE 85/10 (1985); J.P. Derendinger, L.E. Ibanez and H.P. Nilles, Phys. Lett. 165B (1985) 71, Nucl. Phys. B267 (1986) 365.
34. J. Ellis, C. Gomez and D.V. Nanopoulos, CERN preprint TH-4314 (1985).
35. E. Cremmer, S. Ferrara, C. Kounnas and D.V. Nanopoulos, Phys. Lett. 133B (1983) 61.
36. J. Ellis, C. Kounnas and D.V. Nanopoulos, Nucl. Phys. B241 (1984) 406.
37. J. Ellis, C. Kounnas and D.V. Nanopoulos, Nucl. Phys. B247 (1984) 373 and Phys. Lett. 143B (1984) 410.
38. J. Ellis, K. Enqvist and D.V. Nanopoulos, Phys. Lett. 147B (1984) 94; 151B (1985) 357.
39. M. Dine and N. Seiberg, Phys. Rev. Lett. 55 (1985) 366 and isomorphic Princeton preprints (1985); V. Kaplunovsky, Phys. Rev. Lett. 55 (1985) 1036.
40. J. Ellis, K. Enqvist, D.V. Nanopoulos and M. Quiros, CERN preprint TH-4325 (1985).
41. J. Preskil, M.B. Wise and F. Wilczek, Phys. Lett. 120B (1983) 127; L.E. Abbott and P. Sikivie, Phys. Lett. 120B (1983) 133; M. Dine and W. Fischler, Phys. Lett. 120B (1983) 137.
42. P. Binetruy and M.K. Gaillard, LBL-preprint-19972 (1985); J.D. Breit, B.A. Ovrut and G. Segre, Pennsylvania preprint UPPR-0282-T (1985).
43. G. Diamantis, J. Ellis, A.B. Lahanas and D.V. Nanopoulos, CERN preprint in prepration.
44. J. Ellis, J. Hagelin, D.V. Nanopoulos and K. Tamvakis, Phys. Lett. 125B (1983) 275; C. Kounnas, A.B. Lahanas, D.V. Nanopoulos and M. Quiros, Phys. Lett. 132B (1983) 95 and Nucl. Phys. B236 (1984) 438.
45. G.D. Goughlan et al. Phys. Lett. 131B (1983) 59.
46. J. Ellis, K. Enqvist, D.V. Nanopoulos and F. Zwirner, CERN preprint TH-4350 (1986).

Table 1

Transformation properties under $SU(3)_C \times SU(2)_L \times U(1)_Y \times U(1)_E$ of the chiral matter superfields contained in the 27 of E_6. Group and generation indices are understood. In terms of Y_L and Y_R, associated to the maximal $SU(3)_C \times SU(3)_L \times SU(3)_R$ subgroup of E_6, the conventional weak hypercharge Y and the new hypercharge Y_E are given by $Y = (Y_L + Y_R)/2$, $Y_E = (Y_L - Y_R/4)$. The properly normalized quantities $\hat{Y}$ and $\hat{Y}_E$, such that $\mathrm{Tr}\, T_3^2 = \mathrm{Tr}\, \hat{Y}^2 = \mathrm{Tr}\, \hat{Y}_E^2$, are given by $\hat{Y} = \sqrt{3/5}\, Y$ and $\hat{Y}_E = \sqrt{3/5}\, Y_E$. According to our conventions, ν^c is the $SU(3)_C \times SU(2)_L \times U(1)_Y$-singlet contained in the 16 of $SO(10) \subset E_6$, while N is the SO(10)-singlet.

Chiral superfield	$SU(3)_C \times SU(2)_L \times U(1)_Y \times U(1)_E$ quantum numbers
$Q \equiv \begin{pmatrix} u \\ d \end{pmatrix}$	$(3, 2, 1/6, 1/3)$
u^c	$(\bar{3}, 1, -2/3, 1/3)$
d^c	$(\bar{3}, 1, 1/3, -1/6)$
$L \equiv \begin{pmatrix} \nu \\ e \end{pmatrix}$	$(1, 2, -1/2, -1/6)$
e^c	$(1, 1, 1, 1/3)$
ν^c	$(1, 1, 0, 5/6)$
$H \equiv \begin{pmatrix} H^+ \\ H^0 \end{pmatrix}$	$(1, 2, 1/2, -2/3)$
$\bar{H} \equiv \begin{pmatrix} \bar{H}^0 \\ \bar{H}^- \end{pmatrix}$	$(1, 2, -1/2, -1/6)$
N	$(1, 1, 0, 5/6)$
D	$(3, 1, -1/3, -2/3)$
D^c	$(\bar{3}, 1, 1/3, -1/6)$

Table 2

Particle and s-particle spectrum of the 'hybrid' dimensional-transmutation model for $h_u = 0.025$, (corresponding to $m_{top} = 40$ GeV) and two representative choices of the parameters (k_u, λ_u). Case (a), corresponding to $k_u = 0.055$ and $\lambda_u = 0.070$, gives $x/v = 9.9$ and $\bar{v}/v = 0.4$, and is in agreement both with the particle physics limits and with the cosmological constraints, but is characterized by a rather heavy s-particle spectrum. Case (b), corresponding to $k_u = 0.050$ and $\lambda_u = 0.100$, gives $x/v = 3.9$ and $\bar{v}/v = 0.5$, and is characterized by a lighter s-particle spectrum, but does not satisfy the cosmological limit (3.32)

Particle	Mass (GeV) (a)	Mass (GeV) (b)
Top-quark	42	41
D-quark	530	180
Z_1 ($\sin^2\theta_W = 0.206$)	91.8	90.5
Z_1 ($\sin^2\theta_W = 0.220$)	92.6	91.3
Z_2 ($\sin^2\theta_W = 0.206$)	650	250
Z_2 ($\sin^2\theta_W = 0.220$)	680	260
$H_a^{\pm}$	450, 540	160, 210
$H^{\pm}$	620	260
$\tilde{t}_1, \tilde{t}_2$	1500, 1600	540, 630
$\tilde{u}_R, \tilde{c}_R; \tilde{u}_L, \tilde{c}_L$	1500, 1600	580, 610
$\tilde{b}_R, \tilde{b}_L$	1500, 1600	560, 610
$\tilde{d}_R, \tilde{s}_R; \tilde{d}_L, \tilde{s}_L$	1500, 1600	560, 610
$\tilde{D}_1, \tilde{D}_2$	970, 1900	380, 720
$\tilde{D}_a, \tilde{D}_a^c$	1500, 1500	550, 560
$\tilde{\ell}_R^{\pm}, \tilde{\ell}_L^{\pm}$	430, 540	160, 210
$\tilde{\nu}_R, \tilde{\nu}_L$	520, 540	190, 200
$\tilde{\chi}^0$	140	57
$\tilde{\chi}^{\pm}$	210	110

Figure Captions

Fig. 1. Allowed region in the (k_u,λ_u)-plane for $h_u = 0.025$ (corresponding to $m_{top} \sim 40$ GeV). The region outside the solid lines is excluded if we want to have $m_W \simeq 82$ GeV for a gaugino mass 50 GeV $\leq m_{1/2} < 3$ TeV. The region on the right of the dotted line is excluded because it corresponds to negative or too small squared-masses of the charged scalars $H_a^{\pm}$. In the region below the wavy line $\lambda^2 - g_E^2/12 < 0$, and thus the minimum obtained is not necessarily the global one. The shaded area is the allowed one: in the darker region we can have solutions with $m_{1/2} < 1$ TeV and $x/v > 3$, while in the lighter one these two requirements cannot be met simultaneously.

Fig. 2. Theoretical 'scatter-plot' showing the correlation between the ratio of VEVs x/v and the gaugino mass $m_{1/2}$ for $h_u = 0.025$ (corresponding to $m_{top} \sim 40$ GeV). The 48 points of this figure correspond to the points of a square lattice in the region of the (k_u, λ_u)-plane which allows for $m_{1/2} < 1$ TeV.

Fig. 3. Theoretical 'scatter-plot' showing the dependence of some typical s-particle masses on the ratio x/v, for $h_u = 0.025$ (corresponding to $m_{top} \sim 40$ GeV). As a representative squark mass $m_{\tilde{q}}$ (denoted by crosses) we have taken that of the right-handed up-squarks of the first two generations. The masses of the left-handed and right-handed charged s-leptons $(m_{\tilde{e}_R}, m_{\tilde{e}_L})$ are denoted by full and empty circles, respectively. The masses of the lightest chargino and neutralino $(\tilde{\chi}^{\pm}, \tilde{\chi}^0)$ are denoted by full and empty squares, respectively. The 5×23 points of the figure correspond to the points of a square lattice in the region of the (k_u, λ_u)-plane which allows for $m_{1/2} < 1$ TeV. Interpolating lines have been drawn for the reader's convenience.

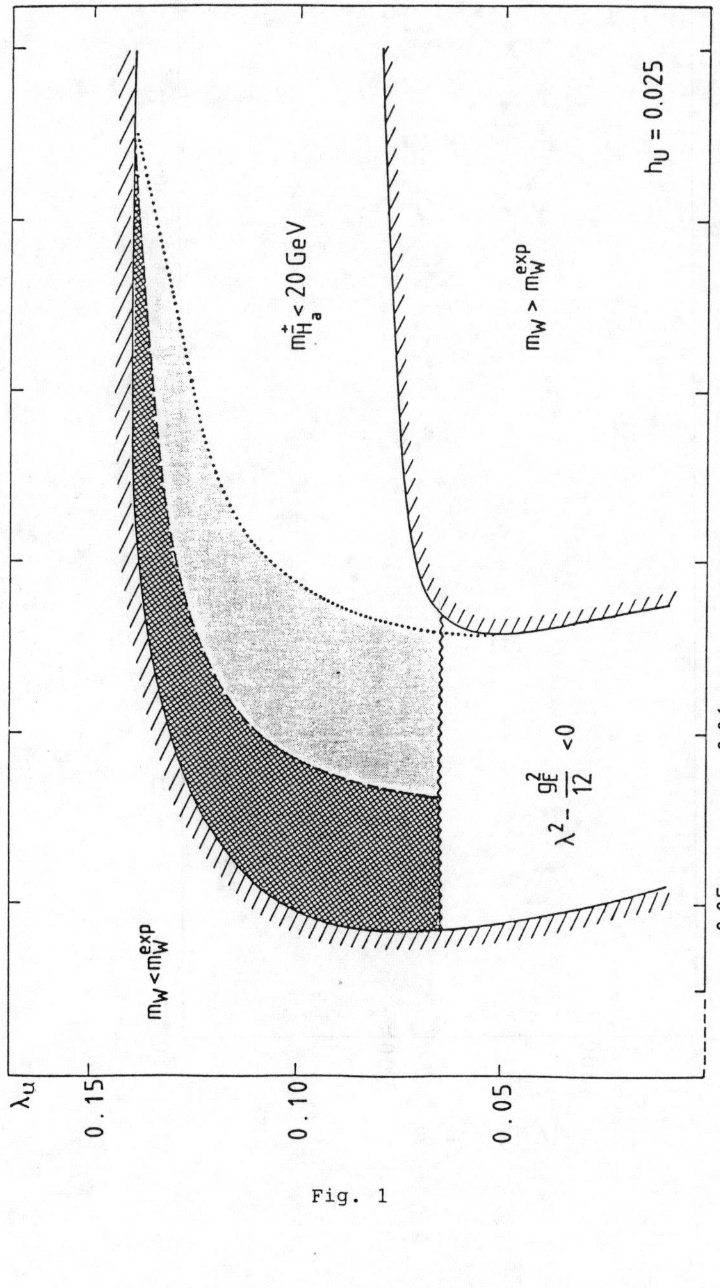
λu
0.15
0.10
0.05
0.05
0.06
0.07
0.08
0.09
0.10
k_U
$h_U = 0.025$
$m_W < m_W^{exp}$
$m_W > m_W^{exp}$
$m^{\pm}_{H_a} < 20$ GeV
$\lambda^2 - \frac{g_E^2}{12} < 0$

Fig. 1

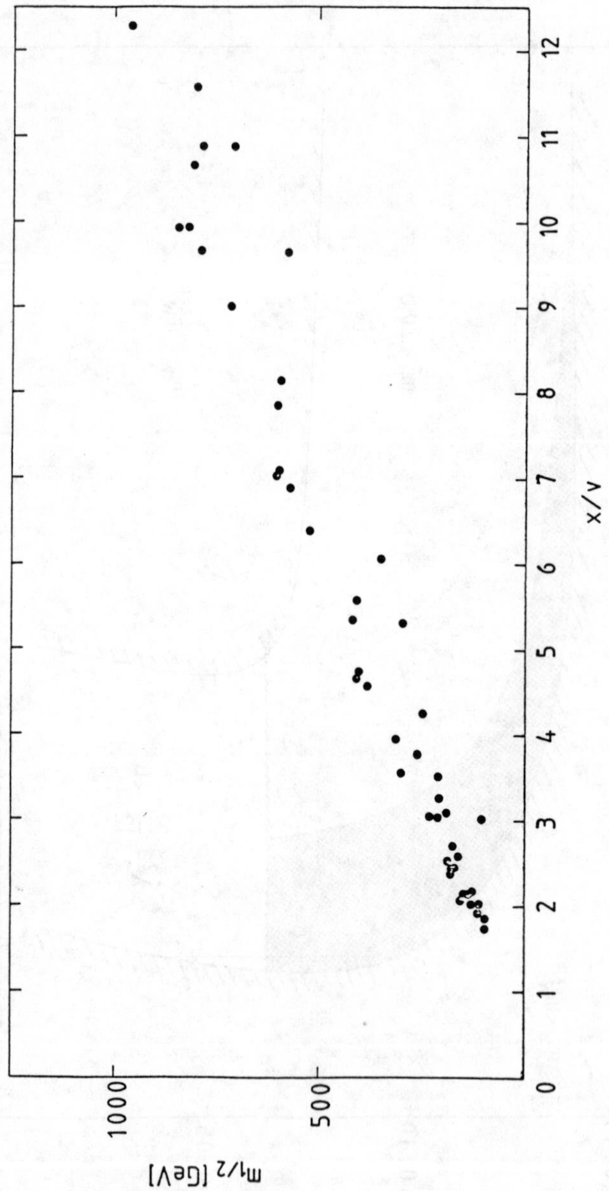

Fig. 2

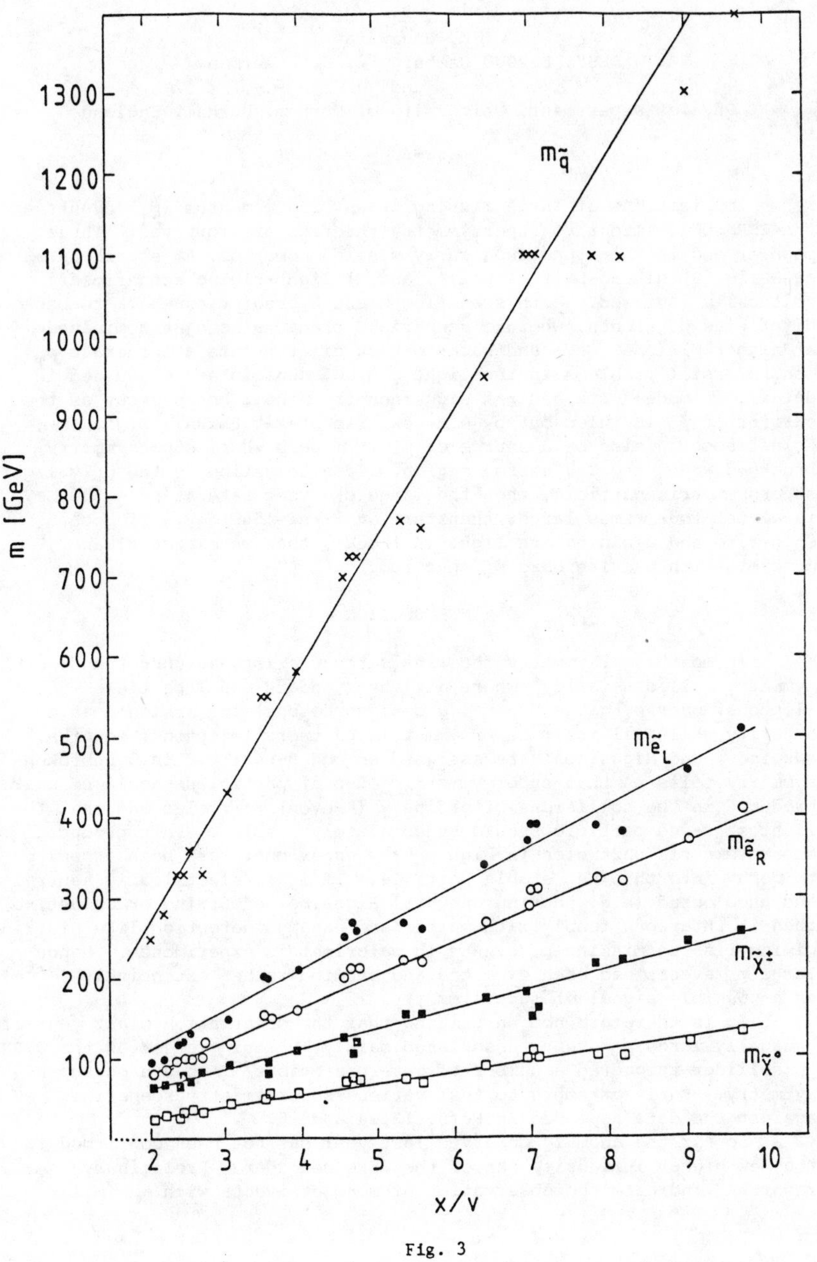
m [GeV]
1300
1200
1100
1000
900
800
700
600
500
400
300
200
100
2
3
4
5
6
7
8
9
10
x/v
$m_{\tilde q}$
$m_{\tilde e_L}$
$m_{\tilde e_R}$
$m_{\tilde\chi^\pm}$
$m_{\tilde\chi^0}$

Fig. 3

LARGE MISSING p_T EVENTS AND SUPERSYMMETRY

K. Hagiwara
DESY, D-2000 Hamburg 52, F.R. Germany
and
Physics Department, University of Durham, Durham, England

ABSTRACT

Implications of large missing transverse momentum ($\not{p}_T$) events at the CERN $p\bar{p}$ collider to supersymmetry theories are surveyed. Three popular models where photinos carry missing momentum, A) squark-pair scenario, B) gluino-pair scenario, and C) light-gluino scenario are critically reviewed. Both scenarios A and B predict non-back-to-back dijet plus $\not{p}_T$ events, whereas scenario C predicts monojet dominance at higher $\not{p}_T$ ($\gtrsim$ 40 GeV) and back-to-back dijet events at moderate $\not{p}_T$. Controversial problems in the light gluino scenario are discussed in detail. A model with neutral Higgsinos from the Z boson decay as the carrier of $\not{p}_T$ is ruled out by e^+e^- experiments at PETRA. Light Goldstinos can also be a source of $\not{p}_T$ in models where supersymmetry breaks down in the TeV energy region. If a Goldstino is the only light supersymmetric particle, then the large $\not{p}_T$ event rate at $\sqrt{s}$ = 630 GeV is almost four times larger than that at $\sqrt{s}$ = 540 GeV. If both a Goldstino and a gluino are light (< 1 GeV), then we expect clean monojets with a quite hard $\not{p}_T$ spectrum.

INTRODUCTION

In most supersymmetry theories[1], there exists an unbroken discrete symmetry called R-parity[2] where all the standard model particles (leptons, quarks, gauge bosons, and Higgs bosons) are assigned an even R-parity while all their supersymmetric partners (sleptons, squarks, gauginos, and Higgsinos) are assigned an odd R-parity. This unbroken symmetry tells us that supersymmetric (R-odd) particles should be pair produced in the collisions of ordinary (R-even) particles and that the lightest R-odd particle should be absolutely stable[3]. Pair produced supersymmetric particles in high energy experiments are both expected to decay into this new stable particle. If it is electrically neutral and uncoloured (e.g. photino, neutral Higgsino, sneutrino or Goldstino) then it interacts feebly with matter and escapes detection like neutrino, giving rise to missing-p_T events in calorimetric experiments. Hence large $\not{p}_T$ events had been expected and studied rather extensively[2,4-6] as a possible signal of supersymmetry.

It is therefore not surprising that the observation[7] last year of unusually large $\not{p}_T$ events associated mainly by a single jet at the CERN $p\bar{p}$ collider triggered a number of interpretations[8-13] based on supersymmetry. Early attempts to test various supersymmetry scenarios against the data appeared in Refs. 12,14 and 15.

So far the anomaly observed last year has not been confirmed in the new higher luminosity run of the CERN collider. Preliminary reports[16] indicate the observation of monojet events with a similar topology to the ones observed last year and also that of dijet plus large $\not{p}_T$ events where most of the dijets have back-to-back configuration in the transverse momentum plane, while acceptance and background estimates have not been presented. It is clearly most important to determine that the observed signals can in no way come from the standard model sources[17,18].

It would be truly exciting if we are observing the first experimental signal of supersymmetry. Even the slightest experimental information on the masses of superpartners will drastically improve our understanding of the supersymmetry breaking mechanism and will give us a clue to determine if the currently most attractive scenario, the N = 1 supergravity grand unification[19] and its eventual embedding into the superstring theory[20] (the first serious candidate of the theory of quantum gravity) is in a correct direction.

In this talk I will review the implications for supersymmetry of the large $\not{p}_T$ events observed at hadron colliders. First, I discuss the most popular scenario where photinos carry the missing momentum; these are A) squark-pair production ($m_{\tilde{q}} \simeq$ 40-50 GeV) followed by the decay $\tilde{q} \to q\tilde{\gamma}$, B) gluino-pair production ($m_{\tilde{g}} \simeq$ 40-50 GeV) with the $\tilde{g} \to q\bar{q}\tilde{\gamma}$ decays, and C) associate production of a light $\tilde{g}$ ($m_{\tilde{g}} \simeq$ 3-5 GeV) and a heavy $\tilde{q}$ ($m_{\tilde{q}} \simeq$ 100 GeV) followed by the decay $\tilde{q} \to q\tilde{\gamma}$. Only this last scenario explains the observed monojet dominance over multijet plus $\not{p}_T$ events and I will discuss in detail the dynamical problems associated with the possibility of having light gluinos. On the second part, I examine the possibility that neutral Higgsinos carry the missing momentum. Neutral Higgsinos are produced via Z boson decay and we find that such a possibility can be ruled out by present e^+e^- annihilation experiments at PETRA/PEP. Finally, I discuss briefly the possibility that light ($\simeq$ massless) Goldstinos carry the missing momentum.

Before we begin, it is worth noting that a candidate for the $\not{p}_T$ carrier at hadron colliders does not necessarily have to be the lightest R-odd particle. Even if it is unstable, an electrically neutral and uncoloured particle can escape detection in calorimetric experiments either if it lives long enough not to decay in the detectors or if it decays mainly into unobservable modes. For example, an unstable photino can be a candidate if it decays into a photon and say, a Goldstino[21], an axiono[22] or a Higgsino[23] very slowly or if it decays mainly into an invisible mode such as a neutrino and a sneutrino[24]. Such possibility should be kept in mind when we examine further implications of the models.

PHOTINOS AS THE CARRIERS OF MISSING MOMENTUM

In order to have a significant cross section at the CERN $p\bar{p}$ collider, a superparticle-pair should be produced either via $O(\alpha_s^2)$ subprocesses $qq \to \tilde{q}\tilde{q}$, $(q\bar{q},gg) \to (\tilde{q}\bar{\tilde{q}},\tilde{g}\tilde{g})$, and $qg \to \tilde{q}\tilde{g}$, or via $O(\alpha)$ subprocesses, i.e., via W and Z boson decays. All other parton subprocesses, e.g. $O(\alpha_s\alpha)$ processes, such as $q\bar{q} \to (\tilde{g}\tilde{\gamma},\tilde{g}\tilde{\omega},\tilde{g}\tilde{z})$ or $qg \to (\tilde{q}\tilde{\gamma},\tilde{q}\tilde{\omega},\tilde{q}\tilde{z})$, are found to give a negligible contribution to the present large $\not{p}_T$ signal. When squarks and/or gluinos are produced via the $O(\alpha_s^2)$ subprocesses they decay, depending on their relative masses, via either $\tilde{q} \to q\tilde{g}$ ($m_{\tilde{q}} > m_{\tilde{g}}$) or $\tilde{g} \to q\bar{\tilde{q}},\bar{q}\tilde{q}$ ($m_{\tilde{g}} > m_{\tilde{q}}$), followed by either

$\tilde{g} \to q\bar{q}\tilde{\gamma}$ or $\tilde{q} \to q\tilde{\gamma}$ with the photino giving missing p_T. Thus depending on the relative masses of the squark and the gluino, we may have three very different scenarios.

$$\text{SCENARIO A: } m_{\tilde{q}} \simeq 40 \text{ GeV} < m_{\tilde{g}} \quad {}^{12,13}$$

If the squark is lighter than the gluino, then the produced squark-pair each decay into a quark and a photino giving rise to a $q\bar{q}\tilde{\gamma}\tilde{\gamma}$ final state. Most of the time, the squark pair is produced near the threshold and each squark has small p_T. For squarks of mass 40 GeV, the photinos each have typically $p_T \simeq 20$ GeV. In order for the two photino momenta to add up vectorally to give large $\not{p}_T$ of typically 30-40 GeV, the two momentum vectors should have rather a small opening angle in the transverse momentum plane. This would then lead to a configuration where the two quark momenta are also aligned and the event has a high probability to be identified as a monojet event in the UA1 jet defining algorithm[7,25]. If squark masses are higher, then no particular final state configuration is required to pass the large $\not{p}_T$ experimental cut and the event would typically be registered as a dijet event.

This is clearly seen in Fig. 1 where the sum (a) and the ratio (b) of monojet and dijet cross sections with $\not{p}_T > 40$ GeV at $\sqrt{s} = 630$ GeV are shown as a function of the squark and gluino masses. Here we assumed that the squark masses are degenerate for 5 flavours and 2 chiralities[12,13] in accordance with the low energy constraints from the absence of large flavour changing neutral currents[26] and of anomalous parity violation in nuclei[27]. When a gluino mass is not much larger than a squark mass, the contribution from $\tilde{g}\tilde{q}$ and $\tilde{g}\tilde{g}$

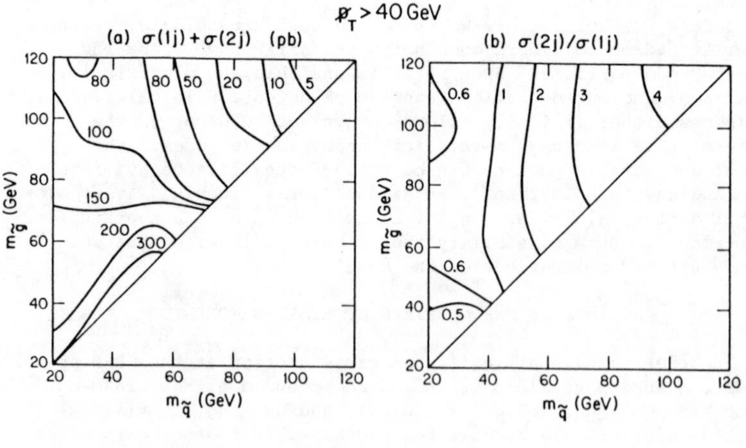

Fig. 1 Predictions for the sum (a) and the ratio (b) of monojet and dijet cross sections with $\not{p}_T > 40$ GeV at $\sqrt{s} = 630$ GeV, taken from Ref. 28.

production processes followed by the cascade decay $\tilde{g} \to q\bar{\tilde{q}}$ or $\bar{q}\tilde{q}$; $\tilde{q} \to q\tilde{\gamma}$ is found to be significant. If we request monojet dominance, then $m_{\tilde{q}} \lesssim 40$ GeV follows from Fig. 1b. Further requirement of $\sigma(\not{p}_T > 40 \text{ GeV}) < 100$ pb would rule out almost all the region in the $m_{\tilde{g}} > m_{\tilde{q}}$ plane except for a tiny window with $m_{\tilde{q}} = 20\text{-}40$ GeV and $m_{\tilde{g}} > 100$ GeV. In the N = 1 supergravity grand unified models[19], there is a severe constraint on the average mass squared of squarks and sleptons

$$\overline{m_{\tilde{q}}^2} - 0.77\, m_{\tilde{g}}^2 = \overline{m_{\tilde{\ell}}^2}$$

which follows[29,30] from the constraint $m_{\tilde{q}} = m_{\tilde{\ell}}$ at the grand unification (GUT) scale via the renormalization group running of the mass parameters[31,30] assuming three generations. This together with the bound $\overline{m_{\tilde{\ell}}^2} > (20 \text{ GeV})^2$ from e^+e^- annihilation experiments[32] forbids a gluino mass much larger than a squark mass. The aforementioned tiny window is hence closed for the N = 1 supergravity GUT models. With four generations of quark-lepton flavours, the numerical coefficient 0.77 in the above equation is replaced[30] by 1.65 and the whole region of scenario A disappears.

SCENARIO B: $m_{\tilde{g}} \simeq 40 \text{ GeV} < m_{\tilde{q}}$ [6,8]

If a gluino is lighter than a squark, then it would decay into a photino and a quark-pair. Even though the final state from $\tilde{g}\tilde{g}$ production now consists of four quarks and two photinos, sufficiently light gluinos ($\lesssim$ 40 GeV) are again found to give mainly monojet plus $\not{p}_T$ events due to the same trigger bias as explained for the scenario A. Fig. 2 shows the one- and multi-jet cross sections with $\not{p}_T > 4\sigma$ at $\sqrt{s} = 540$ GeV, σ being the $\not{p}_T$ resolution[7].

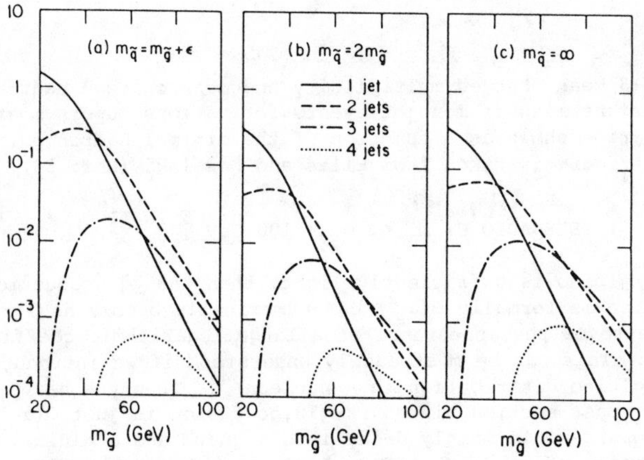

Fig. 2 One- and multi-jet cross sections with $\not{p}_T > 4\sigma$ versus gluino mass at $\sqrt{s} = 540$ GeV taken from Ref. 14.

Although we can again expect monojet dominance with an appropriate cross section at $m_{\tilde{g}} \simeq 40$ GeV, it is very unlikely that this scenario can explain the narrowness of all the observed monojet events with $\not{p}_T > 35$ GeV[7]. Naively, the monojet in this scenario should be broad since it is a combination of up to four quark jets[12,14]. A detailed study of the expected jet structure in various scenarios was performed by Ellis and Kowalski[33], who used a standard jet fragmentation model in e^+e^- jet studies. Fig. 3 shows the expected mean charged multiplicity and the charged particle invariant mass as a function of the minimum hadron p_T in a jet[33]. The solid, dashed, and dash-dotted lines are the predictions of scenarios A, B and C respectively. Unless we start observing multi-prong monojets as well as dijets in association with large $\not{p}_T$, this scenario should also be discarded.

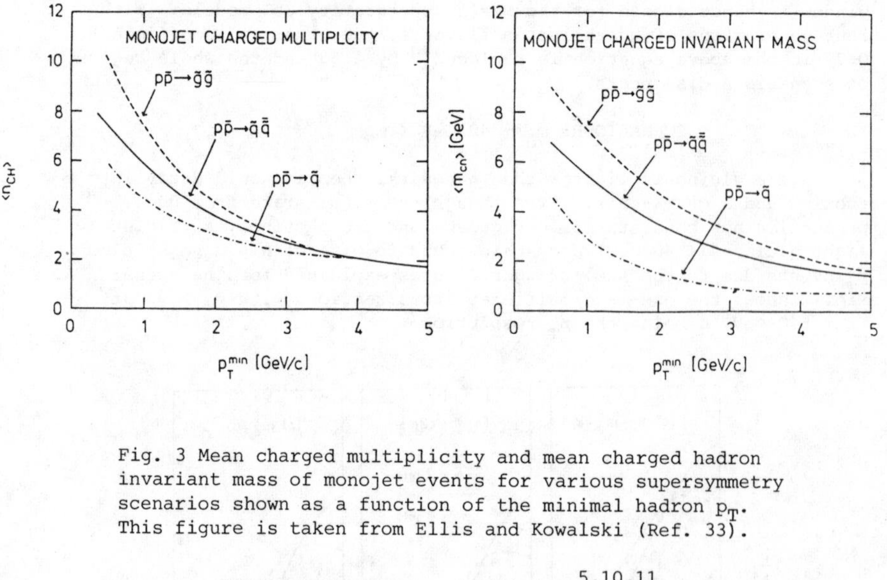

Fig. 3 Mean charged multiplicity and mean charged hadron invariant mass of monojet events for various supersymmetry scenarios shown as a function of the minimal hadron p_T. This figure is taken from Ellis and Kowalski (Ref. 33).

SCENARIO C: $m_{\tilde{g}} << m_{\tilde{q}} \simeq 100$ GeV [5,10,11]

If a gluino is sufficiently light, then the $\tilde{g}\tilde{q}$ production cross section which is formally $O(\alpha_s^2)$ can numerically become as large as $O(\alpha_s)$ because of the appearance of a large logarithmic coefficient $\ln(m_{\tilde{q}}^2/m_{\tilde{g}}^2)$. This can be most easily understood if we introduce an effective gluino distribution in a nucleon[34]. Then the heavy squark production cross section via quark-gluino fusion is just $O(\alpha_s)$. The produced squark would mainly decay into a quark and a gluino, but it can also decay into a quark and a photino which gives a clean monojet event. Even after multiplying by the small branching fraction of $O(\alpha/\alpha_s)$, the monojet cross section is still $O(\alpha)$ and could be significant.

This scenario was first proposed by Herrero et al.[5] as a possible clean signal of supersymmetry at hadron colliders. After the observation[7] of monojet events was reported, we[10] were forced to reinvent the same scenario by systematically searching for a supersymmetry scenario which can give rise to a clean monojet. With $m_{\tilde{q}} \simeq 100$ GeV, we can expect a clustering of the events near $\not{p}_T \simeq 50$ GeV due to the Jacobian peak. With $m_{\tilde{g}} \simeq 3$ GeV, the monojet cross section is expected to be about 20 pb at the CERN collider[10,14]. Furthermore, we do expect in this scenario that the resulting monojets should be narrower than a typical high p_T jet observed at the collider. There are two reasons for this. First our monojet is always a quark jet whereas a typical high p_T jet at the collider is a statistical mixture of a quark and a gluon jet. Second, the lowest order QCD contribution to the mass of the jet system turns out to be significantly smaller[14,35] than that for the mass of the e^+e^- annihilation jets. Hence we should expect that the monojet in this scenario is narrower than a typical hadron jet at similar p_T observed at the collider or extrapolated from e^+e^- jets. Quantitative significance of such an effect at the hadron level, however, remains unclear.

Implications of this scenario for a light gluino of mass about 3-5 GeV are quite involved. Gluinos, being colour octets, can be quite copiously pair-produced in hadron collisions[36]. Once they decay into photinos, a high p_T gluino-pair should lead to large $\not{p}_T$ at hadron colliders[6], while small p_T gluinos can lead to anomalous signals[36] in the beam dump experiments[37]. Early studies at the collider[7] seemed to rule out the gluino mass all the way up to 40 GeV, whereas the mass values of interest ($m_{\tilde{g}} \simeq 3$ GeV, $m_{\tilde{q}} \simeq 100$ GeV) lie near the boundary of the sensitivity of the present beam dump experiments[37]. It is therefore of great interest to study carefully if a light gluino can still be a viable possibility in the face of the collider data. There are three relevant issues to consider: the gluino distribution in a nucleon[5,11,34,38-40], the gluino fragmentation[35,39-41], and the contribution from the QCD $2 \to 3$ processes[42].

It is usually assumed that when one probes a nucleon at much higher momentum scale (Q) than a heavy parton mass ($m > 1$ GeV) there appears to be an effective parton distribution and that the distribution can be calculated via the Altarelli-Parisi equation[43] with an appropriate decoupling condition[44]. In our case, the probe scale $Q \simeq m_{\tilde{q}} \simeq 100$ GeV seems, at first sight, to be sufficiently larger than the parton mass $m_{\tilde{g}} = 3$-5 GeV to ensure the validity of the approximation where one calculates the squark production cross section via the fusion $q\tilde{g} \to \tilde{q}$ by using an Altarelli-Parisi generated effective gluino distribution in a nucleon. We examined[40] this by generating an effective gluino distribution with the simplest decoupling condition

$$\tilde{g}(x, Q < 2m_{\tilde{g}}) = 0 \ .$$

Surprisingly, this overestimates the squark production cross section by more than a factor of three compared to the lowest-order calculation with the exact $qg \to \tilde{q}\tilde{g}$ kinematics. Because of this discrepancy, a squark production cross section of 2 nb (which is needed to explain the monojet rate) is obtained with $m_{\tilde{g}}$ as large as 20 GeV[11,39] in the $q\tilde{g} \to \tilde{q}$ approximation whereas the same cross section requires $m_{\tilde{g}} < 5$ GeV[10] in the exact lowest order calculation.

To clarify the problem, we examined[40] the $q\tilde{g} \to \tilde{q}$ calculation in the single logarithmic approximation (or equivalent gluino approximation, $E_{\tilde{g}}A$). The total squark production cross section is expressed in the lowest order as

$$\sigma(p\bar{p} \to \tilde{q}X) = \int d\sqrt{\hat{s}} \left(\frac{dL}{d\sqrt{\hat{s}}}\right)_{qg} \hat{\sigma}(qg \to \tilde{q}\tilde{g})$$

where $\hat{\sigma}$ denotes the subprocess cross section and $(dL/d\sqrt{\hat{s}})_{qg}$ is the parton luminosity density[45], $\sqrt{\hat{s}}$ being the invariant mass of the colliding quark-glue system. In the $E_{\tilde{g}}A$, $\hat{\sigma}$ can be expressed as a convolution of the $g \to \tilde{g}$ splitting function[43,34] and the $q\tilde{g} \to \tilde{q}$ fusion cross section. We find that this convolution integral gives a resonable approximation to the exact $\hat{\sigma}$ except in the tiny region near the threshold

$$m_{\tilde{q}} < \sqrt{\hat{s}} \lesssim m_{\tilde{q}} + m_{\tilde{g}}$$

where it gives a non-vanishing contribution whereas the exact cross section vanishes. This small discrepancy in the subprocess cross section $\hat{\sigma}$, however, leads to the major discrepancy in the total cross section σ because the quark-gluon luminosity is rapidly falling with increasing $\sqrt{\hat{s}}$ in the relevant region. We show in Fig. 4a and b the product of the luminosity function and $\hat{\sigma}$ as a function of $\sqrt{\hat{s}}$ with the exact kinematics and with the $E_{\tilde{g}}A$, respectively. The area under the curves show the total cross section. We see that the area under the $m_{\tilde{g}} = 20$ GeV curve (dotted line in Fig. 2b) with the $E_{\tilde{g}}A$ is almost as large as the area under the $m_{\tilde{g}} = 5$ GeV curve (solid line in Fig. 2a) with exact kinematics.

The threshold suppression effect for the heavy quark leptoproduction was carefully studied by Glück et al.[46], whose modified gluon-to-heavy quark kernel was subsequently used in the parametrization of Ref. 45. Shown in Fig. 4c are the $E_{\tilde{g}}A$ results obtained by using their modified kernel. The discrepancy with the exact results is still large. This simply reflects the fact that the kinematics in the leptoproduction (space-like probe) and in hadroproduction (time-like probe) is different. Although further modification of the splitting function might lead to resonable distributions, we find it safe to use the exact lowest order results whenever possible. Similar care should be taken generally when one makes use of the Altarelli-Parisi generated heavy quark distributions.

When the gluino mass is below 10 GeV, the large $\not{p}_T$ (of typically > 25 GeV) comes only from p_T much greater than $m_{\tilde{g}}$, and hence the $\tilde{g} \to \tilde{g}_h$ ($\tilde{g}$-containing-hadron) fragmentation effects should become important. We may use[35,40] the parametrization of Peterson et al.[47] for the heavy quark fragmentation which interpolates well between the charm and bottom fragmentation functions[48]. A notable difference between the gluino and heavy quark fragmentation function is that the former, being colour octet, radiates off more gluons than the latter and receives larger scaling violation effects. Convenient parametrizations of the gluino fragmentation functions are given in Ref. 40.

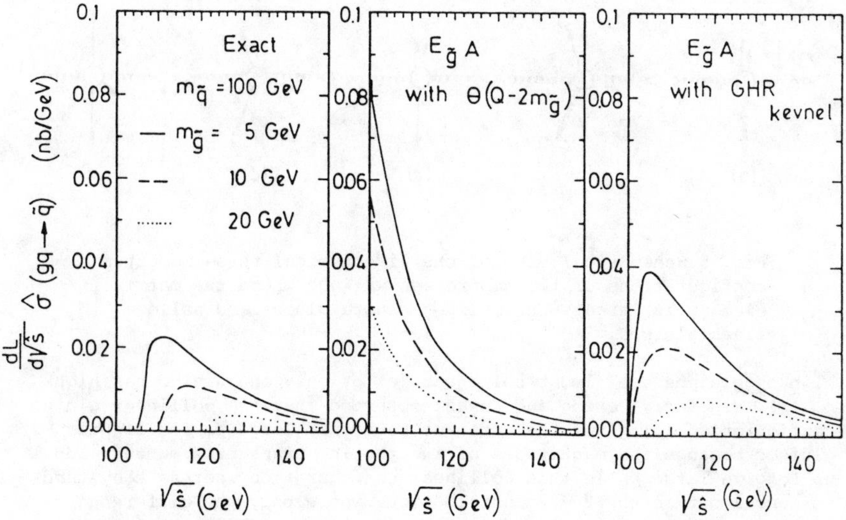

Fig. 4 Product of the subprocess cross section $\hat{\sigma}(ug \to \tilde{u}\tilde{g})$ and the ug luminosity function for $m_{\tilde{u}}$ = 100 GeV plotted against $\sqrt{\hat{s}}$ with (a) exact kinematics, (b) $E_{\tilde{g}}A$ with the theta function threshold, and (c) $E_{\tilde{g}}A$ with the kernel of Glück et al.[46]. Figs. 4a and b are taken from Ref. 40.

Recently, Herzog and Kunszt[42] pointed out that the QCD 2 → 3 processes, $gg \to g\tilde{g}\tilde{g}$ and $qg \to q\tilde{g}\tilde{g}$, give rise to large $\not{p}_T$ events with the rate larger than the 2 → 2 ($gg \to \tilde{g}\tilde{g}$ and $q\bar{q} \to \tilde{g}\tilde{g}$) contributions for smaller gluino masses ($m_{\tilde{g}}$ = 10 to 20 GeV). Fig. 5 shows schematically the five typical momentum configurations for the process $gg \to g\tilde{g}\tilde{g}$ where the matrix element becomes large. Curly lines and solid lines denote gluon and gluino three-momentum vectors, respectively, in the colliding gluon c.m. frame. The configurations may be labelled as (a) $g \to \tilde{g}$ splitting, (b) $\tilde{g}$-excitation, (c) gluon emission colinear to a gluino, (d) gluon emission colinear to initial gluons, and (e) soft gluon emission. Among these, the latter three configurations give similar final states to the leading order one, i.e. a back-to-back gluino pair. Hence the loop correction in the same order is required to know the actual magnitude of the correction. In particular, the leading logarithmic terms in the configuration (c) and (d) are already taken into account by the scaling violations of the $\tilde{g}$ fragmentation function and the g distribution in a nucleon, respectively. On the other hand, the configurations (a) and (b) appear only in the α_s^3 or higher orders. The 2 → 3 processes hence give the leading contributions to these two configurations where two

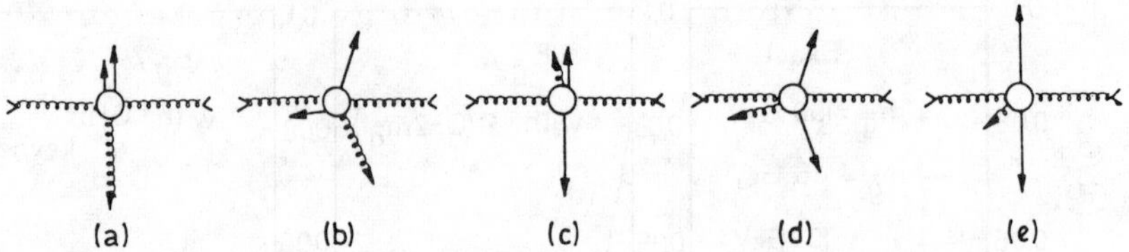

Fig. 5 Schematical view of the five typical three momentum configurations of the subprocess $gg \to g\tilde{g}\tilde{g}$ where the matrix element is large. Curly lines denote gluons and solid lines gluinos.

high p_T gluinos are almost collinear (a) or only one gluino has high p_T (b). In fact, Herzog and Kunszt observed that the collinear gluino configuration (a) gives the dominant source of large $\not{p}_T$ for lighter gluinos because the magnitudes of two photino transverse momenta add up to give large $\not{p}_T$ in this collinear configuration whereas they tend to cancel out in the $2 \to 2$ contributions where only the difference of the two photino transverse momenta gives $\not{p}_T$ in the back-to-back configuration.

To examine this $g \to \tilde{g}$ splitting effect quantitatively for the relevant mass range, $m_{\tilde{g}}$ = 3-5 GeV, we introduce a scalar source for two gluons[49] and a spinor source for a quark and gluon system. By attaching a gluino-pair to a gluon leg emitted from these sources (see Fig. 6), we obtain very simple amplitudes with the correct leading $g \to \tilde{g}$ splitting behaviour in the collinear configuration with exact $2 \to 3$ kinematics. Normalizing the magnitude and aligning the jet axis to those in the dominant $2 \to 2$ subprocesses ($gg \to gg$ and $qg \to qg$), we obtain a simple $2 \to 3$ cross section which simulates the exact $2 \to 3$ cross section in the collinear gluino-pair configuration (Fig. 4a) but gives negligible contributions in all the other configurations shown in Fig. 4.

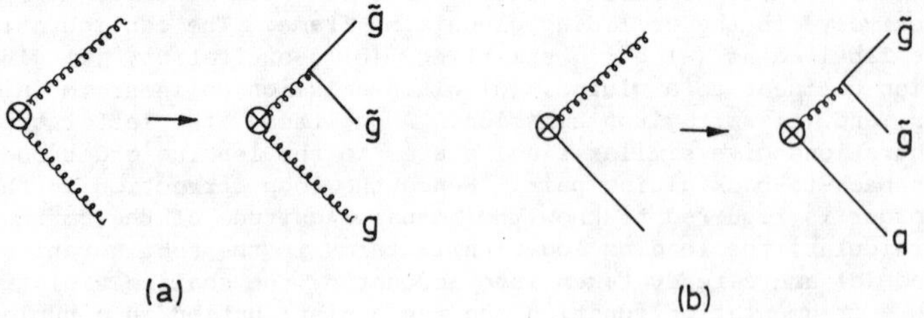

Fig. 6 Feynman diagrams for a scalar source to gg and $g\tilde{g}\tilde{g}$ (a) and for a spinor source to qg and $q\tilde{g}\tilde{g}$ (b).

Shown in Figs. 7a and b are, respectively, the $2 \to 2$ and the $2 \to 3$ subprocess contributions to the $\not{p}_T$ distributions[50] for $m_{\tilde{g}} = 3$ GeV at $\sqrt{s} = 630$ GeV. The importance of the $\tilde{g} \to \tilde{g}_h$ fragmentation effect is clearly seen in these figures. Also shown is the significance of the experimental $\not{p}_T$ resolution (4σ) cut; dotted, dashed and solid lines denote the results obtained with no 4σ cut, by imposing the 4σ cut with spectator $E_T = 25$ GeV, and with the 4σ cut

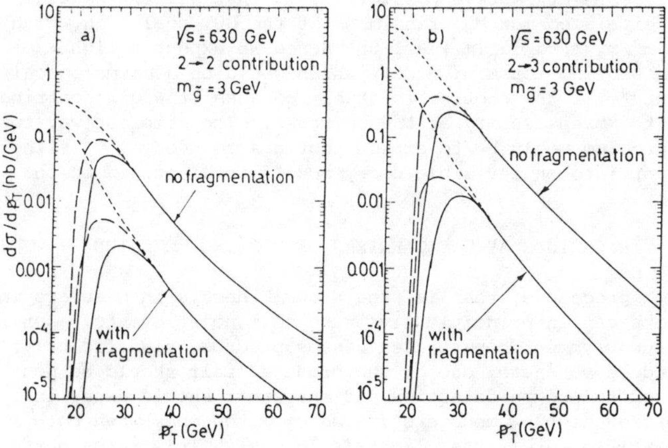

Fig. 7 Missing p_T distributions from the $2 \to 2$ (a) and the $2 \to 3$ (b) processes for $m_{\tilde{g}} = 3$ GeV in $p\bar{p}$ collisions at $\sqrt{s} = 630$ GeV taken from Ref. 50. See text for more details.

with spectator $E_T = 40$ GeV, respectively. The significance of the possible large spectator E_T effect on the experimental $\not{p}_T$ resolution cut was emphasized by Haber[51]. Hence the solid curves with fragmentation and large spectator E_T give the most conservative estimates. It is clear that the $2 \to 3$ processes give by far the dominant contribution to the large $\not{p}_T$ events. We find for $m_{\tilde{g}} = 3$ GeV (5 GeV) the integrated rates 25 pb (60 pb) from the $2 \to 2$ processes whereas 120 pb (210 pb) arise from the $2 \to 3$ processes. Roughly half of the events are classified as monojet and the remaining half as dijet events according to the UA1 jet selection criteria[7,25]. These contributions are peaked around $\not{p}_T \simeq 30$ GeV and have almost always back-to-back jet activities, that is, $\not{p}_T$ vector is not isolated in the transverse plane. Almost no events survive above $\not{p}_T \simeq 40$ GeV. Hence a typical event may look just like a jet-jet fluctuation of an ordinary dijet event where one of the jet p_T is overestimated and the other underestimated. The large expected rates (50-100 events with the accumulated integrated luminosity of 0.4 pb^{-1}), however, means that either this signal should be observed or a light gluino should be ruled out in the near future. The estimates for the rate are very conservative since (i) we use smaller values for the QCD coupling, (ii) we have not taken into account the 'excitation' type configuration (Fig. 4b), and (iii) we expect a significant enhancement[52] from higher order corrections.

Needless to say the above arguments do not apply if gluinos are rather stable[10,53]. However, for gluinos of mass around 3-5 GeV, we should requite a near degeneracy of gluino and photino masses[10]. A theoretically more attractive possibility is that all the neutral gauginos receive a common Majorana mass at the GUT scale, in which case we expect[31] $m_{\tilde{\gamma}}/m_{\tilde{g}} \simeq 8\alpha/3\alpha_s(m_{\tilde{g}}) \simeq 1/6$. Hence we expect a light ($\sim$ 0.5 GeV) photino for $m_{\tilde{g}} \lesssim 3$ GeV, which would be a cosmological embarrassment[54] if the photino is stable because it would contribute too much to the mass density of the universe. The simplest way to avoid this problem would be to expect photinos to decay[21-23] either very slowly or into an invisible decay mode[24] as discussed in the introduction.

HIGGSINOS AS THE CARRIERS OF MISSING MOMENTUM

Order α processes, that is, the W and Z decays into supersymmetric particle pairs can in principle give rise to monojet events. Here the colourless supersymmetric particles can be produced significantly. In order to produce monojets, one of the produced pair should be neutral and long-lived to escape detection and the other should decay into hadrons. Narrow high p_T monojets can be expected only when both supersymmetric particles are rather light (< 10 GeV). This rules out[32] charged superparticles as candidates and leaves the neutral decay modes of the Z boson. Since the Z boson does not couple to photino and zino at the tree level, the unique possibility is the Z decay into Higgsinos.

The decay mode $Z \to \tilde{h}_1\tilde{h}_2$ was, to my knowledge, first studied by Ellis et al.[4] and its implication to the collider monojets studied in Ref. 55. We assume $\tilde{h}_1$ to be light (< a few GeV) and stable or long-lived to escape detection. The heavier Higgsino $\tilde{h}_2$ would then decay into $\tilde{h}_1$ and a quark-pair via a virtual Z exchange. We show in Fig. 8 the Feynman diagram of the subprocess. The monojet production rate at the CERN collider has been estimated[55] to be about 20 pb for $\not{p}_T > 35$ GeV with the maximum $Z\tilde{h}_1\tilde{h}_2$ coupling[56].

Once such light Higgsinos exist, they can be produced in the present e^+e^- annihilation experiments at PEP/PETRA via a virtual Z exchange. Depending on the mass difference between $\tilde{h}_1$ and $\tilde{h}_2$, a typical event would look like a monojet plus $\not{p}_T$ or a dijet plus $\not{p}_T$ in lower energy e^+e^- experiments. Also shown in Fig. 8 is the experimental bound for the mass and coupling of the Higgsinos obtained by the JADE collaboration[57] from the nonobservation of such events. This bound is so stringent that virtually no monojet event can be expected from this source at the CERN collider.

Perhaps the most important lesson learned from this example is that it exemplifies the fruitful interactions between the hadron collider experiments at higher energies and the e^+e^- collider experiments at lower energies but in a cleaner environment. The aforementioned JADE bound was obtained in the region where the Higgsino pair production cross section should be more than two order of magnitude smaller than that at the

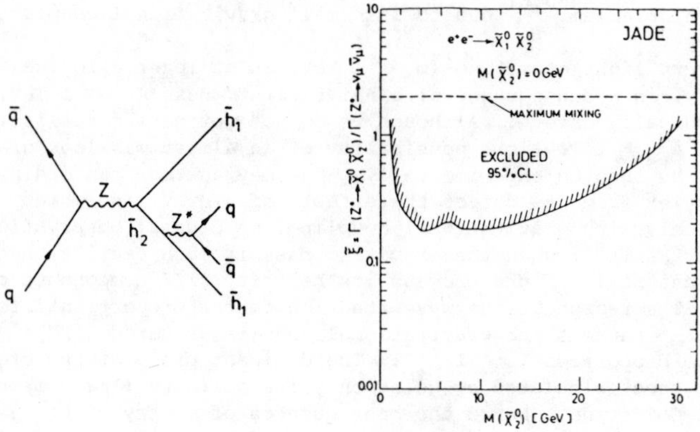

Fig. 8 Feynman diagram and the experimental bound from Ref. 57 for the Higgsino-pair production and decay. $\tilde{\chi}_1^o$ and $\tilde{\chi}_2^o$ of Ref. 57 correspond, respectively, to $\tilde{h}_1$ and $\tilde{h}_2$.

CERN collider. In e^+e^- collider experiments it would also have been easy[55] to distinguish between monojets originating from a fermion (Higgsino) pair or from a scalar (Higgs boson) pair[58] by using the monojet angular distribution whereas such a determination would be extremely hard at hadron colliders. The slightest hint from hadron collider experiments could be examined in detail in e^+e^- collider experiments.

GOLDSTINOS AS THE CARRIERS OF MISSING MOMENTUM

In a locally supersymmetric theory (supergravity theory)[19], a massless Goldstino is absorbed into the helicity ±1/2 component of a massive gravitino (denoted by $\tilde{G}$ like the Goldstino for reasons given below) due to the super-Higgs effect[59]. A general relation[59] between the gravitino mass ($m_{\tilde{G}}$) and the supersymmetry breaking scale (Λ_{ss}) reads

$$m_{\tilde{G}} \sim \Lambda_{ss}^2/M_P$$

where $M_P \sim 10^{19}$ GeV is the Planck mass. In the most attractive class of supergravity models[19] $m_{\tilde{G}} \sim m_W$ and we do not expect the gravitino to play a role in high energy physics apart from the possible cosmological problem[60,61] due to its small gravitational coupling ($\sim 1/M_P$).

A very light gravitino ($m_{\tilde{G}} < 1$ keV) is at least a logical possibility, as a consequence of a naive first guess $\Lambda_{ss} \sim 1$ TeV, and is cosmologically safe[60]. Although we do not yet have a realistic model with $\Lambda_{ss} \sim 1$ TeV, the possibility of an almost massless gravitino should not be ignored experimentally. One may ask how can ordinary high energy experiments detect the effects of a gravitino which interacts only with gravitational coupling. A crucial observation was made by Fayet[62]: when the gravitino mass is much smaller than the physical mass scale of the process its helicity ±1/2 components couple strongly. The reason is the wave function factor proportional to $1/m_{\tilde{G}}^2 \sim M_P^2/\Lambda_{ss}^2$ cancels the gravitational coupling squared $(1/M_P)^2$ to give a rate proportional to $1/\Lambda_{ss}^2$. The dominant piece of the cross section is then calculated by using only its helicity ±1/2 components, namely the Goldstino. Hence the consequences of a very light gravitino in high energy experiments follow from massless Goldstino dynamics[62,63]. This is analogous to the well known fact[64] that the high energy interaction of massive gauge bosons in spontaneously broken gauge theories is dominated by their helicity zero components and the dominant piece is determined by the Higgs sector without the gauge interaction. Hence we examine the possibility that massless Goldstinos ($\tilde{G}$) carry away the missing momentum.

The simplest case, at least as far as the number of participating new particles is concerned, is when all the superpartners, except the Goldstino, are heavy at the energy scale probed by the CERN collider. In this case supersymmetry is realized nonlinearly[65] simply because no supermultiplets appear at our energy and high dimensional interactions including two Goldstinos appear. One of such lowest-dimension operators gives the four fermion coupling between a quark-pair and a Goldstino-pair as depicted in Fig. 9. One gluon is attached to the vertex because the process $q\bar{q} \to \tilde{G}\tilde{G}$ does not lead to any $\not{p}_T$; the process $q\bar{q} \to g\tilde{G}\tilde{G}$ leads to an observable monojet plus $\not{p}_T$ events. Consequences of this type of interaction at hadron colliders were studied by Nachtman et al.[66]. Fig. 9 shows only a partial contribution from the $q\bar{q} \to g\tilde{G}\tilde{G}$ subprocess to the $\not{p}_T$ spectrum at two collider c.m. energies, 540 GeV and 630 GeV, for illustration. First of all the $\not{p}_T$ spectrum is soft, which is essentially determined by the one gluon emission from the initial parton legs. Secondly, there is a striking energy dependence which gives an almost 4 times larger rate for the events with $\not{p}_T > 40$ GeV at $\sqrt{s} = 630$ GeV as compared to the one at $\sqrt{s} = 540$ GeV. This striking energy dependence is a consequence of the high dimensionality of the four-fermion operator; since each Goldstino couples derivatively according to the low energy theorem[67], the $\bar{q}q\tilde{G}\tilde{G}$ operator has dimensionality eight and the coupling is proportional to $1/\Lambda_{ss}^4$ as compared to the Fermi coupling which is proportional to $1/m_W^2$. Because of this, the subprocess cross sections scale as

$$d\hat{\sigma}(q\bar{q} \to g\tilde{G}\tilde{G}) \propto \hat{s}^3/\Lambda_{ss}^8$$

which gives rise to the strong energy dependence. The qualitative behaviour of the monojet cross section shown in Fig. 9 should not change on including all the leading order ($\sim \alpha_s/\Lambda_{ss}^8$) contributions. Since there is no hint[16] of an increase in the monojet cross section

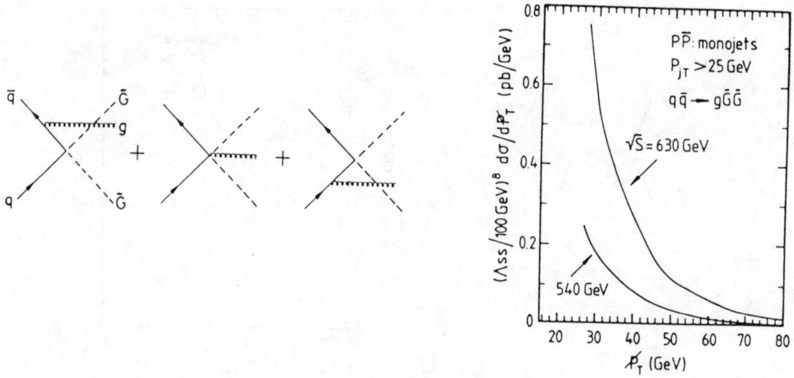

Fig. 9 Feynman diagrams of the process $q\bar{q} \to g\tilde{G}\tilde{G}$ and its contribution to the monojet p_T distribution at $\sqrt{s}$ = 540 GeV and 630 GeV.

at $\sqrt{s}$ = 630 GeV, this interesting possibility cannot be regarded as an explanation of the observed[7] monojet events.

Finally, there is a possibility of clean monojets if in addition to the massless Goldstino a very light ($\lesssim$ 100 MeV) gluino exists. In this case the Goldstino couples to the derivative of the gluon-gluino supercurrent and its low energy interaction can be expressed by the effective Lagrangian[68],

$$L_{eff} = -\frac{1}{2\Lambda_{ss}^{2}}\, \overline{\psi_{\tilde{G}}}\, \sigma_{\mu\nu}\, \not{\partial}\, (F^{a\mu\nu}\psi^{a}_{\tilde{g}})$$

where $F^{a\mu\nu}$ denotes the usual gauge covariant gluon field strength. It is then straightforward to calculate the cross sections for the processes $q\bar{q} \to \tilde{g}\tilde{G}$ and $gg \to \tilde{g}\tilde{G}$ which scale as $\hat{s}/\Lambda_{ss}^4$. The produced gluino would then hadronize into a colour singlet bound state[69] $(\tilde{g}g)$ or $(\tilde{g}q\bar{q})$ and the $\tilde{g}$-jets would just look like ordinary jets. We show in Fig. 10 the Feynman diagram and the partial contribution[70] to the monojet cross section at $\sqrt{s}$ = 540 GeV and 630 GeV from the subprocess $q\bar{q} \to \tilde{g}\tilde{G}$. The $\not{p}_T$ spectrum is very hard and the energy dependence is moderate. The supersymmetry breaking scale Λ_{ss} of about a few hundred GeV would then lead to a desirable monojet rate. In this scenario, it would be natural to expect the photino to also be light and the analogous process $q\bar{q} \to \tilde{\gamma}\tilde{G}$ and $e^+e^- \to \tilde{\gamma}\tilde{G}$ would occur with the relative rate α/α_s. The subsequent $\tilde{\gamma} \to \gamma\tilde{G}$ decay[21] leads to a single photon plus $\not{p}_T$ event. Probably the best place to test this scenario is again in e^+e^- annihilation processes at high energies.

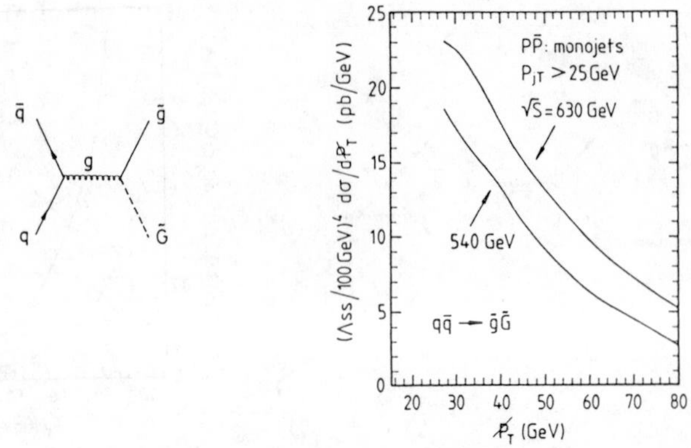

Fig. 10 Feynman diagram of the process $q\bar{q} \to \tilde{g}\tilde{G}$ and its contribution to the monojet p_T distribution at $\sqrt{s}$ = 540 GeV and 630 GeV.

CONCLUSIONS

I have reviewed various attempts to understand the observed monojet events at the CERN collider[7,16] as the first experimental signal of supersymmetry. The monojet dominance and the narrowness of the jets seem to rule out the two most attractive scenarios where either squarks[12,13] or gluinos[6,8] of mass around 40-50 GeV are pair produced and decay into photinos. A scenario where the Z boson decays into two Higgsinos[4,55] is ruled out by e^+e^- experiments at PETRA. An interesting scenario[66] where the Goldstino is the only light superparticle fails to account for the apparently non growing rate with energy. Among the possibilities I examined, only two scenarios do not immediately contradict the observations. These are the scenario[5,10] where a heavy ($\sim$ 100 GeV) squark and a relatively light ($\sim$ 5 GeV) gluino are produced and the scenario[70] where a massless Goldstino and a very light ($\lesssim$ 100 MeV) gluino are produced. These scenarios should be able to be tested in the near future by examining carefully the $\not{p}_T$ events with back-to-back dijet configuration for the former one and by searching a single photon and $\not{p}_T$ events at e^+e^- annihilation for the latter.

I concentrated only on the $\not{p}_T$ plus hadronic jet signal of various supersymmetry scenarios because it is the only observed anomaly which hints at new physics. Recently a number of authors studied the implications of supersymmetry at hadron colliders by examining signals with lepton plus jet[71] and also the purely leptonic[72] signals. Detailed study of these pure- and semi-leptonic signals of supersymmetry is worth pursuing since the relative cleanness of such events could beat the lower rate and also because we can expect a substantial improvement in luminosity in the near future.

ACKNOWLEDGEMENTS

I would like to thank H. Baer, V. Barger, S. Jacobs, S. Komamiya, W.-Y. Keung, R.J.N. Phillips and J. Woodside for fruitful collaborations which led to the work on which this talk is based. I am grateful to S. Komamiya, A.D. Martin, J. Woodside and especially S. Jacobs for valuable help in preparing the talk. I also thank M. Della Negra, S.D. Ellis, C. Fukunaga, M. Glück, K. Hikasa, H. Kowalski, M. Mohamadi, E. Reya and D.P. Roy for informative conversations.

REFERENCES

1. For reviews see e.g., P. Fayet and S. Ferrara, Phys. Rep. 32, 249 (1977);
H.E. Haber and G.L. Kane, Phys. Rep. 117, 75 (1985);
S. Dawson, E. Eichten and C. Quigg, Phys. Rev. D31, 1581 (1985).
2. P. Fayet, Phys. Lett. 69B, 489 (1977);
G.R. Farrar and P. Fayet, Phys. Lett. 76B, 575 (1978).
3. Possible breakdown of R-parity was studied e.g. in L. Hall and M. Suzuki, Nucl. Phys. B231, 419 (1984);
J. Ellis, G. Gelmini, C. Jarlskog, G.G. Ross and J.W.F. Valle, Phys. Lett. 150B, 142 (1985).
4. J. Ellis, J. Hagelin, D.V. Nanopoulos and M. Srednicki, Phys. Lett. 127B, 233 (1983).
5. M.J. Herrero, L.E. Ibáñez, C. López and F.J. Yndurain, Phys. Lett. 132B, 199 (1983).
6. E. Reya and D.P. Roy, Phys. Lett. 141B, 442 (1984);
J. Ellis and H. Kowalski, Phys. Lett. 142B, 441 (1984).
7. G. Arnison et al. (UA1 collaboration), Phys. Lett. 139B, 115 (1984).
8. E. Reya and D.P. Roy, Phys. Rev. Lett. 52, 881 (1984).
9. H.E. Haber and G.L. Kane, Phys. Lett. 142B, 212 (1984).
10. V. Barger, K. Hagiwara, W.-Y. Keung and J. Woodside, Phys. Rev. Lett. 53, 641 (1984).
11. M.J. Herrero, L.E. Ibáñez, C. López and F.J. Yndurain, Phys. Lett. 145B, 430 (1984).
12. J. Ellis and H. Kowalski, Nucl. Phys. B246, 189 (1984).
13. V. Barger, K. Hagiwara and W.-Y. Keung, Phys. Lett. 145B, 147 (1984);
A.R. Allan, E.W.N. Glover and A.D. Martin, Phys. Lett. 146B, 247 (1984).
14. V. Barger, K. Hagiwara, W.-Y. Keung and J. Woodside, Phys. Rev. D31, 528 (1985).
15. E. Reya and D.P. Roy, Phys. Rev. D32, 645 (1985).
16. UA1 collaboration: M. Mohammadi, talk at the Conference on Collider Physics at Ultra-High Energies, Aspen, USA, January 1985;
C. Rubbia, talk at the 5th Topical Workshop on Proton-Antiproton Collider Physics, Saint Vincent, Italy, February 1985;
F. Pauss, talk at the German Physical Society Meeting, München, W. Germany, March 1985.
17. G. Altarelli, R.K. Ellis and G. Martinelli, Z. Phys. C27, 617 (1985);
J.-R. Cudell, F. Halzen and K. Hikasa, Phys. Lett. 157B, 447 (1985);
S.D. Ellis, R. Kleiss and W.J. Stirling, Phys. Lett. 158B, 341 (1985).

18. P. Aurenche and R. Kinnunen, Z. Phys. C28, 261 (1985);
R. Odorico, University of Bologna preprint, IFUB 85/1 (1985);
E.W.N. Glover and A.D. Martin, Z. Phys. C (in press).
19. For reviews and references to the original literature, see
H.P. Nills, Phys. Rep. 110, 1 (1984);
P. Van Niuwenhuizen, Phys. Rep. 68, 189 (1981).
20. M.B. Green and J.H. Schwarz, Phys. Lett. 149B, 117 (1984);
Nucl. Phys. B255, 93 (1985);
D.J. Gross, J.A. Harvey, E. Martinec and R. Rohm, Phys. Rev. Lett. 54, 502 (1985); Nucl. Phys. B256, 253 (1985);
P. Candellas, G. Horowitz, A. Strominger and E. Witten, Nucl. Phys. B258, 46 (1985).
21. N. Cabibbo, G. Farrar and L. Maiani, Phys. Lett. 105B, 155 (1981).
22. J.E. Kim, A. Masiero and D.V. Nanopoulos, Phys. Lett. 139B, 346 (1984).
23. H. Komatsu and J. Kubo, Phys. Lett. 157B, 90 (1985).
24. R.M. Barnett, H.E. Haber and K. Lackner, Phys. Lett. 126B, 64 (1983);
H.E. Haber, in these proceedings.
25. G. Arnison et al. (UA1 collaboration), Phys. Lett. 123B, 115 (1983); 136B, 294 (1984).
26. J. Ellis and D. Nanopoulos, Phys. Lett. 110B, 44 (1982);
R. Barbieri and R. Gatto, Phys. Lett. 110B, 211 (1982);
T. Inami and C.S. Lim, Nucl. Phys. B207, 533 (1982);
M. Suzuki, University of California-Berkeley Report No.UCB-PTH/82/8 (1982);
J.F. Donoghue, H.P. Nills and D. Wyler, Phys. Lett. 128B, 55 (1983);
A.B. Lahanas and D.V. Nanopoulos, Phys. Lett. 129B, 46 (1983);
L. Baulieu, J. Kaplan and P. Fayet, Phys. Lett. 141B, 198 (1984).
27. M. Suzuki, Phys. Lett. 115B, 40 (1982);
M.J. Duncan, Nucl. Phys. B214, 21 (1983).
28. V. Barger, K. Hagiwara, W.-Y. Keung, R.J.N. Phillips and J. Woodside, Phys. Rev. D32, 806 (1985).
29. J. Ellis and M. Sher, Phys. Lett. 148B, 309 (1984);
L. Hall and J. Polchinski, Phys. Lett. 152B, 335 (1985);
K. Enqvist, D.V. Nanopoulos and A.B. Lahanas, Phys. Lett. 155B, 83 (1985);
M. Glück, E. Reya and D.P. Roy, Phys. Lett. 155B, 284 (1985).
30. L.E. Ibáñez, C. López and C. Muñoz, Nucl. Phys. B256, 218 (1985).
31. K. Inoue, A. Kakuto, H. Komatsu and S. Takeshita, Prog. Theor. Phys. 67, 1889 (1982);
C. Kounnas, A.B. Lahanas, D.V. Nanopoulos and M. Quires, Phys. Lett. 132B, 135 (1983); Nucl. Phys. B236, 438 (1984);
L.E. Ibáñez and C. López, Nucl. Phys. B233, 511 (1984);
A. Bouquet, J. Kaplan and C.A. Savoy, University of Louis Pasteur preprint, CRN/HE 85-02 (1985).
32. See e.g. S. Yamada, Proceedings of the 1983 International Symposium on Lepton and Photon Interactions at High Energies, Ithaca, New York, edited by D.G. Cassel and D.L. Kreinick (Cornell University, Ithaca, 1984), p.525.
33. J. Ellis and H. Kowalski, CERN report, TH.4072 (1984).
34. B.A. Cambell, J. Ellis and S. Rudaz, Nucl. Phys. B198, 1 (1982);
I. Antoniadis, C. Kounnas and R. Lacaze, Nucl. Phys. B211, 216 (1983);
C. Kounnas and D.A. Ross, Nucl. Phys. B214, 317 (1983);
S.K. Jones and C.H. Llewellyn Smith, Nucl. Phys. B217, 145 (1983).

35. A. De Rújula and R. Petronzio, CERN report TH-4070 (1984).
36. G.L. Kane and J. Leveille, Phys. Lett. 112B, 227 (1982);
R.R. Harrison and C.H. Llewellyn Smith, Nucl. Phys. B213, 223 (1983); B223, 542(E) (1983).
37. F. Bergsma et al. (CHARM collaboration), Phys. Lett. 121B, 429 (1983);
R.C. Ball et al., Phys. Rev. Lett. 53, 1314 (1984).
38. N.D. Tracas and S.D.P. Vlassopulos, Phys. Lett. 149B, 253 (1984);
X.N. Maintas and S.D.P. Vlassopulos, Phys. Rev. D32, 604 (1985).
39. R.M. Barnett, H.E. Haber and G.L. Kane, Phys. Rev. Lett. 54, 1983 (1985).
40. V. Barger, S. Jacobs, J. Woodside and K. Hagiwara, DESY report 85-032 (1985).
41. J. Ellis and H. Kowalski, Phys. Lett. 157B, 437 (1985).
42. F. Herzog and Z. Kunszt, Phys. Lett. 157B, 430 (1985).
43. G. Altarelli and G. Parisi, Nucl. Phys. B126, 298 (1977).
44. E. Witten, Nucl. Phys. B104, 445 (1976);
L.F. Abbott and M.B. Wise, Nucl. Phys., B176, 373 (1980).
45. E. Eichten, I. Hinchliffe, K. Lane and C. Quigg, Rev. Mod. Phys. 56, 579 (1984).
46. M. Glück, E. Hoffman and E. Reya, Z. Phys. C13, 119 (1982).
47. C. Peterson, D. Schlatter, I. Schmitt and P.M. Zerwas, Phys. Rev. D27, 105 (1983).
48. See e.g., J.M. Izen, DESY report 84-104 (1984).
49. K. Shizuya and S.-H.H. Tye, Phys. Rev. Lett. 41, 787 (1978).
50. K. Hagiwara and S. Jacobs, in preparation.
51. H.E. Haber, talk at the Conference on Collider Physics at Ultra-High Energies, Aspen, USA, January 1985; see also Ref. 39.
52. A.H. Mueller and P. Nason, Phys. Lett. 157B, 226 (1985).
53. G.R. Farrar, Phys. Rev. Lett. 53, 1029 (1984).
54. H. Goldberg, Phys. Rev. Lett. 50, 1419 (1983);
J. Ellis, J.S. Hagelin, D.V. Nanopoulos, K. Olive and M. Srednicki, Nucl. Phys. B238, 453 (1984).
55. H. Baer, K. Hagiwara and S. Komamiya, Phys. Lett. 156B, 177 (1985); ibid, 452E (1985).
56. J.-M. Frère and G.L. Kane, Nucl. Phys. B223, 331 (1983);
J. Ellis, J.-M. Frère, J.S. Hagelin, G.L. Kane and S.T. Petcov, Phys. Lett. 132B, 436 (1983).
57. W. Bartel et al. (JADE collaboration), Phys. Lett. 155B, 288 (1985).
58. S.L. Glashow and A. Manohar, Phys. Rev. Lett. 54, 526 (1985);
S.F. King, Phys. Rev. Lett. 54, 528 (1985);
H. Georgi, Phys. Lett. 153B, 294 (1985).
59. S. Deser and B. Zumino, Phys. Rev. Lett. 38, 1433 (1977);
E. Cremmer, B. Julia, J. Scherk, P. van Nieuwenhuizen, S. Ferrara and L. Girardello, Phys. Lett. 79B, 231 (1978).
60. H. Pagels and J. Primack, Phys. Rev. Lett. 48, 223 (1982);
A. Bouquet and C.E. Vayonakis, Phys. Lett. 116B, 219 (1982).
61. S. Weinberg, Phys. Rev. Lett. 48, 1303 (1982);
L.M. Krauss, Nucl. Phys. B227, 556 (1983).
62. P. Fayet, Phys. Lett. 70B, 461 (1977).
63. P. Fayet, Phys. Lett. 84B, 421 (1979); Phys. Lett. 86B, 272 (1979); Phys. Lett. 117B, 460 (1982).
64. J.M. Cornwall, D.N. Levin and G. Tiktopoulos, Phys. Rev. D10, 1145 (1974);
B.W. Lee, C. Quigg and H.B. Thacker, Phys. Rev. D16, 1519 (1977).

65. S. Samuel and J. Wess, Nucl. Phys. B221, 153 (1983).
66. O. Nachtman, A. Reiter and M. Wirbel, Z. Phys. C27, 577 (1985).
67. B. de Wit and D.Z. Freedman, Phys. Rev. Lett. 35, 827 (1975); Phys. Rev. D12, 2286 (1975);
W.A. Bardeen, unpublished.
68. M. Fukugita and N. Sakai, Phys. Lett. 114B, 23 (1982).
69. M. Chanowitz and S. Sharpe, Phys. Lett. 126B, 225 (1983).
70. K. Hagiwara, K. Hikasa, S. Jacobs and D. Zeppenfeld, in preparation.
71. I.I. Bigi and S. Rudaz, Phys. Lett. 153B, 335 (1985);
L. Hall and S. Raby, Phys. Lett. 153B, 433 (1985);
V. Barger, W.-Y. Keung and R.J.N. Phillips, Phys. Rev. D32, 320 (1985);
H. Baer and X. Tata, CERN report TH.4147 (1985); TH.4158 (1985);
A.P. Contogouris, H. Tanaka and S.D.P. Vlassopulos, McGill University preprint (1985).
72. H. Baer, J. Ellis, D. Nanopoulos and X. Tata, Phys. Lett. 153B, 265 (1985);
H. Baer and X. Tata, Phys. Lett. 155B, 278 (1985); and the references therein.

THE FOURTH GENERATION AND N=1 SUPERGRAVITY

H Goldberg
Northeastern University, Boston, Mass 02115

ABSTRACT

The inclusion of a 4th generation of matter fields in the RG equations of N=1 supergravity yields the following results: small 4th generation Yukawas at GUT energies ($h_X \sim 0.2$) allow electroweak breaking at the weak scale, with $m_t = 40$, $m_{\tilde{e}_R} \gtrsim 30$ GeV. Larger values ($h_X \gtrsim 0.5$) require the breaking to occur via dimensional transmutation. When $h_X \gtrsim 1$, the dimensional transmutation is forced to take place at energies $\gtrsim$ weak scale, yielding upper bounds on 4th generation masses: $m_{t',b'} < 140$ GeV, $m_{\tau'} < 70$ GeV. There are many solutions in which the b', t' are unstable to superpartner decay ($b' \to \tilde{b}' \tilde{\gamma}$, etc) and the lightest Higgs can be as massive as 35 GeV.

INTRODUCTION

In this talk I will explore the interplay of a 4th generation of matter fields and N=1 supergravity. If SU(2)xU(1) is obtained through the evolution of coupling constants via the RG equations[1-9], the existence of only 3 generations and a top quark with mass ~ 40 GeV[10] demands that the values of coupling constants at GUT energies (M_X) are very close to those which produce a dimensional transmutation[11,5,6] at the weak scale. Even so, there are problems with phenomenology, especially concerning the limits on the scalar electron mass.

With 4 generations, I will show that these problems are circumvented. However, it will be seen that as the 4th generation Yukawas at GUT energies $\{h_X\}$ become $\gtrsim 0.5$, electroweak breaking is forced once more to occur via dimensional transmutation (DT) for any values of the soft-breaking parameters $\{A_X, \mu_X, \xi\}$. This then implies rather low upper bounds for the masses of the 4th generation fermions $m_{t', b'} \lesssim 140$ GeV, $m_{\tau'} \lesssim 70$ GeV.

GENERAL 4th GENERATION RESULTS

Independently of the soft breaking details, I find the following (undoubtedly known to others)

(i) For $\alpha_3(m_w) = \frac{1}{8}$, the unification mass $M_X = 2.9 \times 10^{16}$ GeV and $\alpha_G = 1/13.3$

(ii) with $\xi \equiv M_o/m_{3/2}$ (M_o = common gaugino mass at M_X), I find
$m_{\tilde{Q}}^2 = m_{3/2}^2 (1 + 7.4\xi^2)$ + D term

$$m_{\tilde{e}_R}^2 = m_{3/2}^2 (1 + 0.13\xi^2) + \text{D term} \qquad (1)$$

$$\begin{aligned} m_{\tilde{g}}(m_W) &= (\alpha_3(m_W)/\alpha_G)M_o \\ &= 1.67\ \xi m_{3/2}\ \text{(4 generations)} \\ &= 3.0\ \xi m_{3/2}\ \text{(3 generations)} \end{aligned} \qquad (2)$$

$$\begin{aligned} \therefore (m_{\tilde{Q}}^2)_{min} &= m_{3/2}^2 + 2.66\ m_{\tilde{g}}^2 \\ (m_{\tilde{e}_R}^2)_{min} &= m_{3/2}^2 + 0.05\ m_{\tilde{g}}^2 \end{aligned} \qquad (3)$$

∴ unlike the case for 3 generations, 4 generation SUSY-GUTs implies a large gap between $m_{\tilde{Q}}^2$ and $m_{\tilde{g}}^2$. E.g., for $m_{\tilde{e}_R} > 30$ GeV, $m_{\tilde{g}} \geq 40$ GeV, $m_{\tilde{Q}} \geq 71$ GeV. Also, $m_{\tilde{e}_R}$ may be kept comfortably above 30 GeV by an appropriate choice of $m_{3/2}$.

RG ANALYSIS WITH 4 GENERATIONS

As usual, the tree effective potential is given by

$$V_{tree} = m_1^2 |H_1|^2 + m_2^2 |H_2|^2 - m_3^2 (H_1H_2 + c.c.) + \text{D term}$$

SU(2)xU(1) breaking can occur at energy scales Q for which

$$m_1^2 m_2^2 - (m_3^2)^2 < 0.$$

The tree vacuum becomes unstable at scales Q where

$$\Sigma(Q) \equiv m_1^2(Q) + m_2^2(Q) - 2|m_3^2(Q)| < 0. \qquad (4)$$

The tree vacuum is then decided by the one-loop correction, i.e. <u>dimensional transmutation</u> [11,5,6] (DT) occurs at an energy Q_o where $\Sigma(Q_o) = 0$. More operationally, Q_o is determined by requiring one-loop dominance

$$\sigma(Q_o) \equiv Q \left.\frac{d\xi}{dQ}\right|_{Q_o} >> 1 \qquad (5)$$

The theory fails to describe nature if $\sigma(Q_o) >> 1$ for $Q_o > 300$ GeV. Transition to DT occurs when $\sigma(Q) > 1$.

The RG equations for a heavy 4th generation (with light or massless neutrino) plus the 3rd generation top quark are easily derived as extensions of existing work on 3 generations. Labelling the 4th generation quark doublet and charged lepton as

$(t',b'\tau')$, we choose the boundary conditions for the soft-breaking parameters to be conventional:

$$A_t(M_X) = A_{t'}(M_X) = A_{b'}(M_X) = A_{\tau'}(M_X) = A_X$$

$$m_{\tilde{Q}}^2(M_X) = \ldots = m_{\tilde{\tau}'}^2(M_X) = m_{3/2}^2$$

$$\mu(M_X) \equiv \hat{\mu}_X m_{3/2}$$

$$m_1^2(M_X) = m_2^2(M_X) = m_{3/2}^2(1 + \hat{\mu}_X^2)$$

$$m_3^2(M_X) = - \hat{\mu}_X B_X m_{3/2}$$

$$B_X = A_X - 1$$

$$M_1(M_X) = M_2(M_X) = M_3(M_X) = \text{gaugino mass}$$

$$\equiv \xi\, m_{3/2} \tag{6}$$

For Yukawa couplings, I maintain the SU(5) relation:

$$h_{b'}(M_X) = h_{\tau'}(M_X) \tag{7}$$

I also choose

$$h_{t'}(M_X) \geq h_{b'}(M_X) \tag{8}$$

For reasons described elsewhere, I restrict

$$0.3 \leq |\xi| \leq 1 \tag{9}$$

The upper bound comes from Eqs. (2), (3) and the bounds $m_{\tilde{g}} \leq 50$ GeV, $m_{\tilde{e}_R} \geq 30$ GeV. The lower bound is not essential, but constrains $\langle H_1\rangle \simeq \langle H_2\rangle$ when DT occurs. The bounds on ξ translate, via Eq. (3), into bounds on $m_{3/2}$. Choosing $m_{\tilde{g}} = 40$ GeV,[12,13] we have $24 \text{ GeV} \leq m_{3/2} \leq 80$ GeV.

The procedure now consists of choosing a set of parameters $(h_t, h_{t'}, h_{b'}, h_{\tau'})_X \equiv \{h_X\}$, ξ, and $\hat{\mu}_X$, and then searching for values of A_X which will lead to phenomenologically acceptable SU(2)xU(1) breaking. By this I mean (a) correct value of $v/\sqrt{2} = \sqrt{|\langle H_1\rangle|^2 + |\langle H_2\rangle|^2} = 175$ GeV. (b) $m_{\tilde{t}'}, m_{\tilde{b}'}, m_{\tilde{\tau}'}, m_{\tilde{W}^\pm} \geq 30$ GeV.

(c) $m_{\tilde{t}} > 35$ GeV. (d) SU(3)xU(1) vacuum stability[2,6]. (e) $m_{h^o} >$ 9 GeV, where h^o is lightest Higgs scalar, in order to ensure $T \not\to h^o\gamma$

I state first the following result: small values of 4th generation Yukawas $0.2 \leq h_X \leq 0.3$ allow large regions of parameter space (including m_t = 40 GeV) to generate completely acceptable low energy phenomenology without requiring DT. Thus, the original pressure[5,8,9] to adopt this scheme with 3 generations in order to accommodate a low value of m_t is relieved, because now $(h_{t'}, h_{b'}, h_{\tau'})$ will drive SU(2)xU(1) breaking. In showing my results, however, I will choose to single out the part of parameter space which will give acceptable low energy phenomenology and SU(2)xU(1) breaking via DT. The reasons are (1) the transition to the case of larger h's will be clearly seen (2) this mode of breaking can occur for low (t', b', τ') masses and (3) the values of A_X, μ_X will fall into an interesting range.

As shown in Refs. 5 and 6, DT should occur at $Q_o = \sqrt{e/2}\, v = 290$ GeV. The low energy fermion masses are given by

$$m_f(Q_o) = h_f(Q_o) \cdot (Q_o/\sqrt{2e})$$
$$\simeq h_f(Q_o) \cdot 125 \text{ GeV}. \qquad (10)$$

Figures (1), (3) and (5) show the regions of $\hat{\mu}_X, \xi)$ space permitting SU(2)xU(1) breaking via DT for representative and progressively increasing sets of $\{h_X\}$. Table I gives corresponding ranges of particle spectra. Figures (2) displays the $(A_X, \xi > 0)$ support for SU(2)xU(1) breaking via DT for the parameters of Fig. 1. Figure (4), which is crucial for the development of this paper, will be discussed shortly.

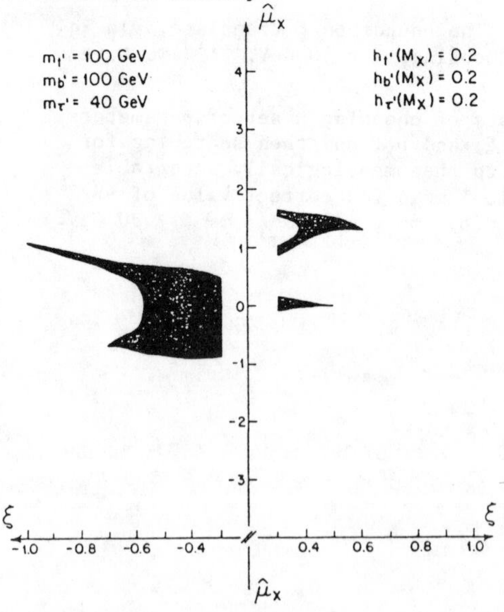

Fig.1. Regions of $(\hat{\mu}_X, \xi)$ space (shaded) which allow SU(2)xU(1) breaking via dimension transmutation for a representative set of small 4th generation Yukawa couplings $\{h(M_X)\}$ shown on graph. There are larger areas which allow SU(2)xU(1) breaking without dimensional transmutation.

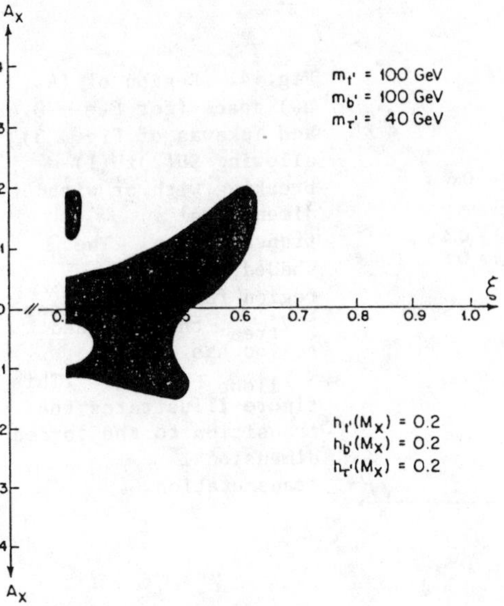

Fig. 2. Region of (A_X, $\xi > 0$) space (shaded) corresponding to Fig. 1.

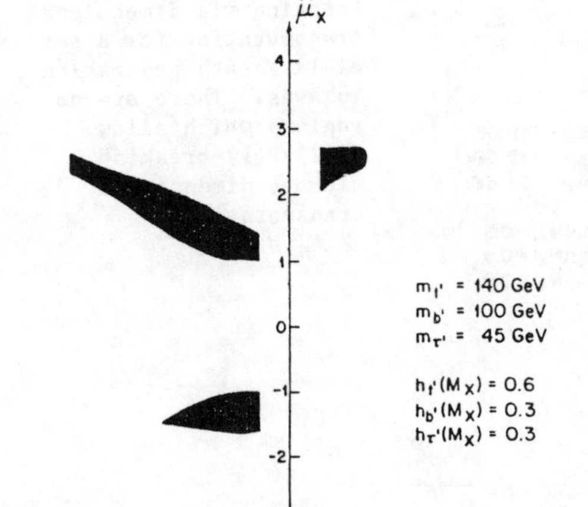

Fig. 3. Regions of ($\hat{\mu}_X$,ξ) space (shaded) which allow SU(2)xU(1) breaking via dimensional transmutation for a representative set of small 4th generation Yukawa couplings $\{h(M_X)\}$ shown on graph. There are larger areas which allow SU(2)xU(1) breaking without dimensional transmutation.

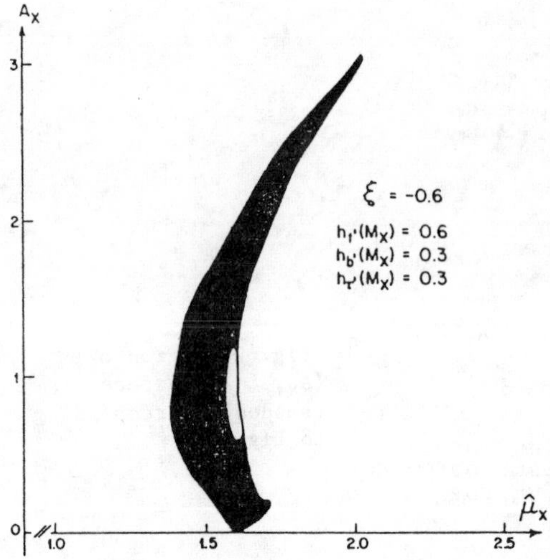

Fig. 4. Region of (A_X, $\hat{\mu}_X$) space (for $\xi = -0.6$ and Yukawas of Fig. 3) allowing SU(2)xU(1) breaking with or without dimensional transmutation. The shaded area is the region for which $V_{1loop} > V_{tree}$, the unshaded region has $0.6\ V_{tree} < V_{1loop} < V_{tree}$. This figure illustrates the transition to the forced dimensional transmutation.

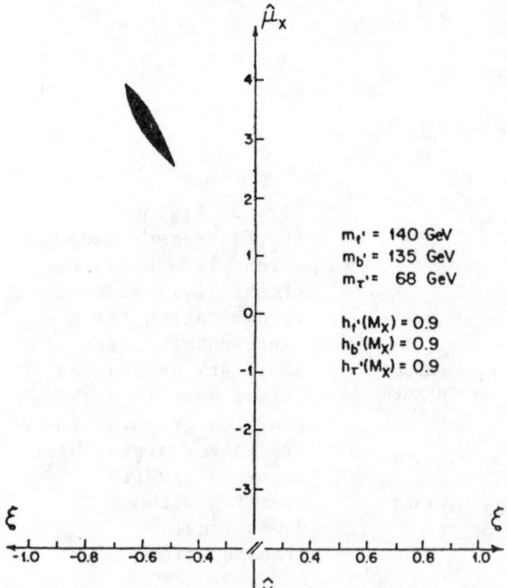

Fig. 5. Region of ($\hat{\mu}_X$, ξ) space (shaded) which allows SU(2)xU(1) breaking via dimensional transmutation for a set of large 4th generation Yukawas. There are no regions which allow SU(2)xU(1) breaking without dimensional transmutation.

Table 1. Parameters and Spectra. All masses are in GeV units; $\tilde{b},\tilde{\tau},\tilde{W}$ refer to the lightest b',τ',W superpartners; h^o is the radiative Higgs scalar; $\tilde{h}$ is the "axino" with mass μ.

	Fig. 1	Fig. 3	Fig. 5
$m_t, m_{t'}, m_{b'}, m_{\tau'}$	40,100,100,35	40,140,110,45	40,140,135,68
$(h_t, h_{t'}, h_{b'}, h_{\tau'})_{M_X}$	0.066,0.2,0.2,0.2	0.10,0.6,0.3,0.3	0.11,0.9,0.9,0.9
A_X: $\xi > 0$	(-2,2)	(-1,1)	----
$\xi < 0$	(-0.7,2)	(-0.5,3)	(1,4)
$m_{\tilde{b}'}$	50-100	33-95	91-97
$m_{\tilde{\tau}'}$	30-75	30-75	36-65
$m_{\tilde{W}}$	30-83	40-72	65-70
m_{h^o}	12-33	10-35	20-21
$m_{\tilde{h}}$	0-150	40-90	50-60

I note the following:

(a) Figures (1) and (2) show that some very small values of both A_X and μ_X will allow SU(2)xU(1) breaking via DT.

(b) Figure (4) shows, for $\xi = -0.6$ and the Yukawas of Fig. 3, the region of the $(A_X, \hat{\mu}_X)$ plane supporting phenomenologically acceptable SU(2)xU(1) breaking with or without DT. Examination of the figure reveals a key point: only in a very small region is the potential even approximately tree-dominated (see Figure caption). This is not the case for the smaller Yukawas of Fig. 1. As the GUT Yukawas become larger (Fig. 5), the tree dominated region shrinks to zero, and the region of support becomes a line. That is, the electroweak breaking is forced to proceed via DT. The reason is simple and will be noted shortly.

(c) As the Yukawas $\{h_X\}$ increase toward 1 (which is still very perturbative: $(h_x/4\pi)^2 < 0.007$), the possibility of phenomenologically acceptable SU(2)xU(1) breaking (which by now must occur via DT) begins to disappear (Fig. 5): m_1^2 and m_2^2 are being driven negative too quickly. From the Yukawa RG equations and Eq. 14, this leads to upper bounds on the 4th generation fermion masses:

$$m_{t'} < 140 \text{ GeV}$$

$$m_{b'} < 135 \text{ GeV}$$

$$m_{\tau'} < 70 \text{ GeV} \qquad (11)$$

The first two correspond to approximate fixed point values for the 4th generation Yukawas[14-17] the last does not.

Before proceeding to a discussion of spectra and other points of interest, I would like to explain briefly why the SU(2)xU(1) breaking is forced to proceed via DT as the Yukawas become large. In the (m_1^2, m_2^2) plane, the, large Yukawas are driving both m_1^2 and m_2^2 downward from their common value $m_{3/2}^2(1 + \hat{\mu}_X^2)$ at M_X. Very soon after the RG trajectory crosses the hyperbola $m_1^2 m_2^2 = (m_3^2)^2$ delimiting electroweak breaking, it also is at or very close to the straight line $m_1^2 + m_2^2 = 2|m_3^2|$ delimiting tree vacuum stability, and signalling onset of DT. (This discussion can be easily followed with reference to Fig. 5 of Ref. 4). Thus for large Yukawas, SU(2)xU(1) breaking implies DT.

COMMENTS, SPECTRA, AND CONCLUSIONS

1) With small 4th generation Yukawas ($h_X \sim 0.2 - 0.3$) one may comfortably achieve electroweak breaking (with or without DT), with no strain on keeping $m_t = 40$ GeV or $m_{\tilde{e}_R} \geqslant 25$ GeV.

Larger Yukawas ($h_X > 0.5$) force the breaking to occur via DT, while $h_X \simeq 1$ represent a maximum allowable Yukawa. This enforces strong upper bounds on 4th generation masses in N = 1 supergravity: $m_{t', b'} < 140$ GeV, $m_{\tau'} < 70$ GeV.

(2) A glance at Table + shows that the lightest Higgs ("radiative Higgs") may be as heavy as 35 GeV. This is much larger than the usual bound $m_{h^o} < 7$-8 GeV found in 3 generations[6], and thus the presence or absence of the decay $T \to h^o\gamma$ is of no particular significance in 4 generation supergravity.

(3) Again, from Table I, we see that there are a range of situations where the decays $\tau' \to \tilde{\tau}' \tilde{\gamma}$, $b' \to \tilde{b}'\tilde{\gamma}$, of the τ', b' to superpartners are permitted. ($m_{\tilde{\gamma}} \simeq 7$ GeV corresponding to $m_{\tilde{g}} = 40$ GeV). This complicates b' and τ' detection, but does create a large new source of events with missing energy in collider physics. The phenomenology of production and decay of $(b', \tilde{b}')$ and $(\tau', \tilde{\tau}')$ constitutes a considerable future study[18].

(4) The results presented in this paper are essentially unchanged if the boundary conditions on A_X, B_X at M_X are those suggested by some other models. In particular, $A_X = B_X = 0$[19] produces the same bounds on Yukawas and 4th generation masses.

ACKNOWLEDGEMENT

I would like to thank Dick Arnowitt, Mark Claudson, Lawrence Hall, Pran Nath, and Mike Vaughn for helpful discussions during the course of this work. My thanks also to Howie Baer for pointing out some errors in the references in the original preprint.

This research was supported in part by NSF Grant PHY-841643. The results in this talk have been submitted for publication in Physics Letters.

REFERENCES

[1] K. Inoue, A. Kakuto, H. Komatsu, and S. Takeshita, Prog. Theor Phys. 68 (1982) 927.
[2] L. Alvarez-Gaume, J. Polchinski, and M.B. Wise, Nucl. Phys. B221 (1983) 499.
[3] L.E. Ibanez and C. Lopez, Nucl. Phys. B233, 511 (1984).
[4] M. Claudson, L.J. Hall, and I. Hinchliffe, Nucl. Phys. B228 (1983) 501.
[5] J. Ellis, J. Hagelin, D.V. Nanopoulos, and K. Tamvakis, Phys. Lett. 125B (1983) 275.
[6] C. Kounnas, A. Lahanas, D.V.Nanopoulos, and M. Quiros, Nucl. Phys. B236 (1984) 438.
[7] L.E. Ibanez, C. Lopez, and C. Munoz, CERN preprint CERN-TH. 4071/84.
[8] J. Ellis and M. Sher, Phys. Lett. 148B (1984) 309.
[9] K. Enqvist, D.V.Nanopoulos, and A.B. Lahanas, Phys. Lett. 155B (1985) 83.
[10] G. Arnison et al. (UA1 Collaboration). Phys. Lett. 147B (1984) 493.
[11] S. Coleman and E. Weinberg, Phys. Rev. D7 (1973) 1888.
[12] J. Ellis and H. Kowalski, Nucl. Phys. B246 (1984) 441.
[13] M. Gluck, E. Reya, and D.P. Roy, Universitat Dortmund preprint DO-TH 85/1, January, 1985. See also E. Reya and D.P. Roy, Phys. Rev. Lett. 53 (1984) 881.
[14] C. Hill, Phys. Rev. D24 (1981) 691.
[15] M. Machacek and M.T. Vaughn, Phys. Lett. 103B (1981) 427.
[16] J. Bagger, S. Dimopoulos, and E. Masso, Nucl. Phys. B253 (1985) 397.
[17] J.W. Halley, E.A. Paschos, and H. Usler, Phys. Lett. 155B (1985) 107.
[18] For some recent phenomenological studies of detecting 4th generation leptons at colliding beam facilities, see V. Barger, H. Baer, A.D. Martin, E.W.N. Glover, and R.J.N. Phillips, Phys. Rev. Lett. 133B, 449 (1983); Phys. Rev. D29, 2020 (1984); S. Gottlieb and T. Weiler, Phys. Rev. D29, 205 (1984); UCSD-10P10-244 (1985); D. Cline and C. Rubbia, Phys. Lett. 127B, 277 (1983); H. Baer, V. Barger, and R.J.N. Phillips, MAD/PH/213, Feb. 1985; H. Baer, invited talk at the New Particles '85 Conference, University of Wisconsin-Madison, May 1985; S. Willenbrock and D. Dicus, Texas preprint UTTG-03-85 (1985).
[19] I. Affleck, M. Dine and N. Seiberg, IAS preprint August 1984; E. Witten, Princeton preprint (1985); M. Mangano, Princeton preprint (April 1985); J. Ellis, C. Kounnas, and D.V. Nanopoulos, Nucl. Phys. B247, 373 (1984). I would like to thank D.V. Nanpoulos for bringing this possibility to my attention.

SIGNATURES OF SUPERSYMMETRY AT THE CERN COLLIDER

Howard E. Haber*
Santa Cruz Institute for Particle Physics
University of California, Santa Cruz, CA 95064

ABSTRACT

This paper describes work performed in collaboration with R. Michael Barnett and Gordon L. Kane. A comprehensive analysis of missing energy events at the CERN Collider which would arise from a supersymmetric theory is presented. The rates for gluino and scalar-quark production are computed subject to the new 1984 UA1 cuts, triggers and resolutions. Using the newly reported 1984 data, it is argued that the monojet events are highly unlikely to come from supersymmetric particle production. This implies very restrictive limits: $M_{\tilde{g}}$ and $M_{\tilde{q}}$ must be larger than 60–70 GeV. These conclusions have been obtained assuming that photinos are stable and are the lightest supersymmetric particle. The implications of a Higgsino which is lighter than the photino are briefly discussed.

1. INTRODUCTION

In the last few years the CERN $p\bar{p}$ Collider has provided a first look into a new energy regime. The W and Z bosons have been discovered, and their observed properties[1)] have provided another resounding success for the Standard Model. Nevertheless, many theorists believe that this model is only an approximation to a more fundamental theory which will supercede the Standard Model at an energy scale between 100 GeV and 1 TeV. The basic problem concerns the mechanism of electroweak symmetry breaking. At present, there is no experimental evidence for the existence of an elementary Higgs boson which is required by the Standard Model. More important, the Higgs mechanism is unnatural—the origin of the electroweak symmetry breaking scale remains totally unexplained (unless one tolerates a very precise fine tuning of parameters in the high energy theory as one must do in grand unified models). All attempts to overcome this problem lead one to expect that new physics beyond the Standard Model should appear around the energy scale characterized by electroweak symmetry breaking. This observation provides the motivation for expecting that new phenomena might be observable at the CERN Collider (or at new accelerators which will turn on in the next few years). The

* Partially supported by a grant from the Department of Energy.

most popular approach which attempts to solve the problem discussed above is supersymmetry. In supersymmetric theories, elementary Higgs fields are still present. But, their masses can be kept naturally small because they are related by supersymmetry to fermion masses which can be protected by chiral symmetries.

Our goal is to test the hypothesis that supersymmetry is related to the scale of electroweak symmetry breaking.[2,3] If this hypothesis is true, then there must exist supersymmetric partners to all presently known particles with masses of the order of the electroweak symmetry breaking scale. (It is also possible that some supersymmetric partners are massless to first approximation and obtain their masses radiatively. One could then have superpartners with masses of order $g^2 m_W$, i.e. considerably smaller than a few hundred GeV.) Hence, to test the hypothesis advanced above, we propose to search for new supersymmetric particles which can be produced at colliders provided that their masses are not too large. Clearly, the CERN Collider only covers the lower portion of the energy regime of interest. At best, one might hope for some early evidence for departures from the Standard Model at CERN. More realistically, one can use the results of the CERN Collider to put limits on hypothetical supersymmetric particle masses. A more thorough search for new physics beyond the Standard Model will undoubtedly require higher energy machines.

In early 1984, the analysis of data from the 1983 run at the CERN $p\bar{p}$ Collider (at $\sqrt{s} = 540$ GeV) by the UA1 collaboration resulted in the report of candidates for events which seemed to be unexplainable by the Standard Model.[4] These events were precisely of the type expected from supersymmetry: events with jets and missing transverse momentum. This led to a plethora of papers attempting to explain these events as being evidence for new physics. By far the most popular explanation[5–15] was in terms of the production of supersymmetric particles—either scalar-quarks or gluinos. In the fall of 1984, more data was taken at a slightly higher energy, $\sqrt{s} = 630$ GeV. More than twice the luminosity (as compared to the 1983 run) was collected. It seems clear from the reports of the 1984 data[16] that the missing-energy events which are seen are (for the most part) less dramatic and possibly entirely explained by Standard Model backgrounds. Thus, the enthusiasm for the possible discovery of supersymmetry at CERN has certainly been dampened. On the other hand, the analysis of the 1984 data will provide much more stringent limits on supersymmetric particle masses.

In order to search for evidence for supersymmetry at hadronic colliders, we make use of one key property of nearly all supersymmetric models. One can define a multiplicative quantum number called R-parity[17] which is defined as $R = (-1)^{3B+L+2J}$ for a particle with baryon number B, lepton number L

and spin J. A quick computation shows that $R = 1$ for all presently known particles and $R = -1$ for their supersymmetric partners. This implies that there exists a lightest supersymmetric particle (LSP) which is stable, weakly interacting and the end result of the decay chain of any heavier supersymmetric particle. From cosmological arguments, we take the LSP to be color and electrically neutral.[18] In the standard scenario, one takes the photino ($\tilde{\gamma}$) to be the LSP. We will assume this to be the case in most of the subsequent analysis. (Near the end, we will consider how our conclusions change if the Higgsino were the lightest supersymmetric particle.) The LSP behaves like a neutrino; therefore any event in which the LSP is emitted will exhibit missing transverse momentum. Hence, the basic signal for supersymmetry consists of events with missing transverse momentum which is not due to neutrinos.

At hadron colliders, the supersymmetric particles with the largest cross-sections are those with color, i.e. the gluino ($\tilde{g}$) and the scalar-quark ($\tilde{q}$).[19–22] Therefore, we shall concentrate in this analysis on the production and decay of scalar-quarks and gluinos in $p\bar{p}$ collisions. Although the masses of the $\tilde{g}$ and $\tilde{q}$ are *a priori* unknown, their couplings to quarks and gluons are precisely related by the supersymmetry to the strong interaction gauge coupling constant. Furthermore, gluinos are color octets, so that enhanced color factors lead to gluino pair production cross-sections which are about an order of magnitude larger than that of a heavy quark of the same mass. The production cross-sections for scalar-quarks are somewhat smaller. Here, we shall assume that there are five degenerate flavors of scalar-quarks; in addition, for each flavor one can produce either $\tilde{q}_L$ or $\tilde{q}_R$. (In many models, the $\tilde{t}_L$ and $\tilde{t}_R$ are substantially split in mass from the other scalar-quarks, so we will omit them from further consideration.) Thus the total $\tilde{q}\tilde{q}$ production cross-section can end up being almost comparable to the $\tilde{g}\tilde{g}$ cross-section.

In the calculations described below, we shall compute the production cross sections for scalar-quarks and gluinos as a function of $M_{\tilde{q}}$ (the common mass of the scalar-quarks) and $M_{\tilde{g}}$. The mass of the photino is generally expected to be quite small compared to the gluino mass. In some supersymmetric models, the $\tilde{\gamma}$ and $\tilde{g}$ masses are related via:[23]

$$\frac{M_{\tilde{\gamma}}}{M_{\tilde{g}}} = \frac{8}{3}\frac{\alpha}{\alpha_s} \approx \frac{1}{6}\,. \tag{1}$$

In all the calculations presented in this paper, we have taken $M_{\tilde{\gamma}} = 0$. Although this is not likely to be true, our results are not especially sensitive to the precise value of the photino mass, assuming that $M_{\tilde{\gamma}} \lesssim 10$ GeV.

2. PARTON MODEL PRELIMINARIES

The QCD-improved parton model is used to compute the production cross-sections for pairs of supersymmetric particles. This requires the calculation of all $2 \to 2$ scattering subprocesses involving quark and gluon initial states and scalar-quark and/or gluino final states. The processes to consider are listed below:

$$gg \to \tilde{g}\tilde{g} \tag{2a}$$

$$q\bar{q} \to \tilde{g}\tilde{g} \tag{2b}$$

$$qg \to \tilde{q}\tilde{g} \tag{3}$$

$$gg \to \tilde{q}\bar{\tilde{q}} \tag{4a}$$

$$q\bar{q} \to \tilde{q}\bar{\tilde{q}} \tag{4b}$$

$$qq \to \tilde{q}\tilde{q} \tag{4c}$$

$$\bar{q}\bar{q} \to \bar{\tilde{q}}\bar{\tilde{q}}\ . \tag{4d}$$

In process (4b) we include the contribution due to the production of an on-shell W and Z (if it is kinematically allowed). In general, it will be necessary to include:

$$q\bar{q} \to \tilde{g}\tilde{\gamma} \tag{5a}$$

$$qg \to \tilde{q}\tilde{\gamma}\ . \tag{5b}$$

However, despite the favorable signature (lots of missing energy due to the direct production of the photino), processes (5a–b) are suppressed by a factor α/α_s as compared to processes (2)–(4) due to the appearance of the electromagnetic vertex $q\tilde{q}\tilde{\gamma}$. It turns out that after applications of the UA1 cuts and triggers, processes (5a–b) do not lead to an observable cross-section at this time.

Once supersymmetric particles are produced, they will quickly decay. If $M_{\tilde{g}} > M_{\tilde{q}}$, then $\tilde{g} \to q\bar{\tilde{q}}$ and $\tilde{q} \to q\tilde{\gamma}$. On the other hand, if $M_{\tilde{g}} < M_{\tilde{q}}$, then $\tilde{g} \to q\bar{q}\tilde{\gamma}$ via the exchange of a virtual scalar-quark, whereas,

$$\tilde{q} \to \begin{cases} q\tilde{g}, & BR = \frac{r}{1+r} \\ q\tilde{\gamma}, & BR = \frac{1}{1+r} \end{cases} \tag{6}$$

where $r = \frac{4}{3}\frac{\alpha_s}{\alpha e_q^2}$ and e_q is the electric charge of the quark (in units of e).

These two cases must be considered separately. For example, in the latter case, $\tilde{q}\tilde{q}$ final states will be very difficult to observe since the dominant decay chain for the scalar quark will be $\tilde{q} \to q\tilde{g} \to qq\bar{q}\tilde{\gamma}$. The end result starting from $\tilde{q}\tilde{q}$ will be six quarks and two photinos. This implies that the event will not contain much missing transverse energy, and therefore it will be difficult to use such events as evidence for new physics.

So far we have only considered $2 \to 2$ subprocesses, which contribute to $O(\alpha_s^2)$. Recently, Herzog and Kunszt[24)] have argued that the $2 \to 3$ subprocesses:

$$gg \to \tilde{g}\tilde{g}g \tag{7a}$$

$$q\bar{q} \to \tilde{g}\tilde{g}g \tag{7b}$$

$$gq \to \tilde{g}\tilde{g}q \tag{7c}$$

will be important if the gluino is light. If the perturbative series is trustworthy, then the $O(\alpha_s^3)$ processes (7a–c) along with one-loop corrections to processes (2a–b) should add up to a result which is smaller than the leading $O(\alpha_s^2)$ Born terms. However, it turns out that when various triggers and cuts are applied to the total cross-section, it is possible that the $O(\alpha_s^3)$ contributions are enhanced significantly. In fact, we find that this indeed occurs for certain ranges of values of the gluino mass when the UA1 triggers and cuts are imposed. To explain the effect, consider the effect of requiring that the missing transverse energy be larger than some fixed number: $E_T^{miss} > E_0$. Suppose that the gluino is light (say, $M_{\tilde{g}} = 5$ GeV). Then in general, $\tilde{g}\tilde{g}$ events will not survive the E_T^{miss} cut. The reason is that light gluinos are typically quite energetic. Since to a very good approximation, the transverse momenta of the gluinos is entirely generated by the hard subprocess, their transverse momenta are nearly back-to-back. When the gluinos decay ($\tilde{g} \to q\bar{q}\tilde{\gamma}$), the two photinos will be nearly back-to-back. Since $E_T^{miss} = |\vec{p}\,^{\tilde{\gamma}}_{T1} + \vec{p}\,^{\tilde{\gamma}}_{T2}|$, we see that E_T^{miss} will in general be small and these events will fail to pass the E_T^{miss} cut. How then do any $\tilde{g}\tilde{g}$ events survive the cut? If the decays of the two gluinos are sufficiently asymmetric, it is possible to generate $E_T^{miss} \geq E_0$. As $M_{\tilde{g}}$ becomes *smaller*, $\sigma_{TOTAL}(p\bar{p} \to \tilde{g}\tilde{g} + X)$ increases rapidly, while at the same time, the probability that a $\tilde{g}\tilde{g}$ event passes the cut decreases rapidly. When fragmentation effects are taken into account, one finds that $\sigma(p\bar{p} \to \tilde{g}\tilde{g} + X)$ subject to the E_T^{miss} cut actually decreases as $M_{\tilde{g}}$ becomes smaller for $M_{\tilde{g}} \lesssim 20$ GeV.

Now, consider the effect of the E_T^{miss} cut on the processes (7a–c). If the final state gluon (or quark) is hard, then one possible configuration is one where the $\tilde{g}\tilde{g}$ pair is emitted in the same hemisphere recoiling against the gluon (or quark). In this configuration, the photinos resulting from the gluino decays are often emitted in the same hemisphere so that it is much easier to have $E_T^{miss} = |\vec{p}_{T1}^{\,\tilde{\gamma}} + \vec{p}_{T2}^{\,\tilde{\gamma}}| \geq E_0$. Of course, the cross-section for the $2 \to 3$ processes decreases as the transverse momentum of the hard gluon (or quark) increases. Nevertheless, it is clear that the $2 \to 3$ processes of eqs. (7a–c) are more likely to survive the E_T^{miss} cut than the $2 \to 2$ processes. So, it is conceivable that when the E_T^{miss} cut is applied, the $2 \to 3$ processes will result in *more* events passing the cuts as compared with the $2 \to 2$ processes. This indeed occurs as first shown by Herzog and Kunszt.[24)] We find that these processes play an important role in ruling out the possiblity of a light gluino.

If gluinos are light, another effect must be considered. Since a gluon can split into a $\tilde{g}\tilde{g}$ pair, a nonzero gluino distribution function will be generated inside the parton.[25)] Due to the fact that gluinos are color octets, for $M_{\tilde{g}} = m_b$, one finds that $f_{\tilde{g}}(x, Q^2) = 6 f_b(x, Q^2)$, i.e. gluinos are six times more plentiful than b-quarks inside the proton in the case of $M_{\tilde{g}} = 5$ GeV. Given a gluino distribution function,[25)] new subprocesses must be considered. One can resonantly produce scalar-quarks via:[12)]

$$\tilde{g} + q \to \tilde{q} \tag{8}$$

followed by scalar-quark decay according to eq. (6). This process can be very efficient for producing events with large E_T^{miss} if scalar quarks are heavy (say, $M_{\tilde{q}} \approx 100$ GeV). For example, if $\tilde{q} \to q + \tilde{\gamma}$, then $E_T^{miss} \approx \frac{1}{2} M_{\tilde{q}}$. This is just a consequence of the Jacobian peak analogous to that which is seen in the missing energy spectrum which results from $W \to e\nu$ decay. Of course, this decay mode is disfavored as shown in eq. (6) since the scalar-quark is assumed to be heavier than the gluino. The dominant decay is $\tilde{q} \to q\tilde{g} \to qq\bar{q}\tilde{\gamma}$. However, by including fragmentation effects, the resulting E_T^{miss} spectrum due to the photino is softened to such an extent, it turns out that the rarer $\tilde{q} \to q + \tilde{\gamma}$ decay dominates the missing energy events for the light gluino scenario when UA1 cuts and triggers are applied.

Barger et al.[8)] have recently argued that using subprocess (8) along with the gluino structure function as computed by using the Altarelli-Parisi evolution equations substantially overestimates the inclusive scalar-quark cross section. The crux of their argument is that in the above application,

$$\alpha_s log(M_{\tilde{q}}^2 / M_{\tilde{g}}^2) \sim 1,$$

whereas this quantity must be much larger than unity in order to justify the leading log approach which is being employed in obtaining the gluino structure function. In fact, the non-leading log terms turn out to be very important in the present case. Barger et al. argue that it is appropriate in this case simply to ignore the gluinos in the proton.

One could include the effect of the gluino structure function, if one were careful not to neglect terms of similar importance to the ones being kept. Ironically, the UA1 collaboration has introduced a new trigger in their analysis of 1984 data[26,16] which significantly enhances the $\tilde{g}\tilde{g}$ production processes (eq. (2)) with respect to process (8). As a result, the uncertainty in the relevance of the gluino structure function is not crucial for interpreting the 1984 data. Henceforth, we shall neglect the gluino content of the proton.

3. ASPECTS OF THE MONTE CARLO EVENT GENERATION

Using the formalism described in section 2, we may obtain formulas for the process $p\bar{p} \to \textit{final state partons}$, where the final state consists of quarks, gluons and photinos which are the decay products of supersymmetric particles. Of course, the calorimeters of the UA1 and UA2 detectors measure energy deposition of final state hadrons and *not* the partons themselves. Furthermore, a given event consists of far more activity than a hard $2 \to 2$ (or even $2 \to 3$) scattering. One can make a list of many effects which clearly take place which are not included in the simple parton formalism. Such a list would include: initial-state gluon radiation, final-state gluon radiation, fragmentation and hadronization of final state partons, interaction of spectators; recombination of spectators into color singlet final states, etc. In comparing a theoretical prediction with actual experimental results, all these effects must be accounted for in some way. In constructing our Monte Carlo event generator, we have decided to incorporate these effects only in the crudest way. To do anything more sophisticated would be pointless—without making use of a full detector simulation appropriate for analyzing the results of a given experiment. This is clearly the job of the experimental groups themselves. We shall argue, however, that even with our crude implementation of "real world" effects beyond the parton model, we will be in the position to obtain reasonable estimates of the magnitude of various differential cross-sections of interest as well as the effects of triggers and cuts imposed by the UA1 collaboration on their data. This will enable us to estimate which ranges of parameters of supersymmetric particle masses are allowed or ruled out given the current data. Precise limits must await a more complete analysis of the various experimental groups at the CERN $Sp\bar{p}S$ Collider.

The procedure of our Monte Carlo Event Generator is as follows. The parton model formulas lead to a phase space integral to be performed over all final state four-momenta. This integration is performed by Monte Carlo techniques; the result is a series of four-momenta for all final state particles (which result after the decay of all intermediate states). Each set of four-momenta is called an "event". For example, if the hard scattering was $gg \to \tilde{g}\tilde{g}$ followed by $\tilde{g} \to q\bar{q}\tilde{\gamma}$, an event would consist of six four-momenta of two quarks, two antiquarks and two photinos respectively. The parton four-momenta obtained in this way are interpreted as jets as observed in the calorimeters of real experiments. This is a major approximation—hadronization of final state quarks and gluons is omitted. (Such a method has been recently dubbed the "parton Monte Carlo".) Such an approximation will preclude us from studying many aspects of the data. For example, it is difficult to discuss jet multiplicities and single jet masses in this context. Nevertheless, such a procedure probably does not do so badly in estimating the gross features of the data: e.g., p_T-distribution of jets, two-jet invariant mases, etc. An important feature which must not be neglected is fragmentation, since this effect can play an important role in determining the E_T^{miss} spectrum of a given event.

For example, a gluino which is produced by some hard process must first "fragment" into a supersymmetric hadron (say a $g\tilde{g}$ or $\tilde{g}q\bar{q}$ bound state) which then decays weakly, emitting a photino which escapes the detector. If the momentum of the gluino inside the supersymmetric hadron is less than that of the original gluino, then the photino spectrum will be degraded compared to the spectrum which would have resulted had fragmentation been ignored. Let us define z to be the momentum fraction of the gluino (or scalar-quark) inside the supersymmetric hadron. We have implemented fragmentation effects by generating a z at random (on an event-by-event basis) according to a distribution $D(z, \hat{s})$ suggested by De Rujula and Petronzio[27)] (where $\hat{s}$ is the partonic center-of-mass energy squared). The function D peaks near $z = 1$; the peak becomes more pronounced as the gluino mass increases. For smaller values of $M_{\tilde{g}}$, values of $z < 1$ are more probable, resulting in a softer energy spectrum for the photino which is emitted by the decaying gluino.

The result of the fragmentation procedure is to modify somewhat the four-momenta of the final state partons which make up our events. (For a more detailed discussion, see Ref. 11.) Since we have claimed that the final state quarks and gluons will be interpreted as jets (and the final state photinos result in the missing energy), we must decide precisely how to define a jet which corresponds to the jet definition of the UA1 collaboration.[28)] If the jets are "close" to each other, they must be identified as a single entity. Here, we follow a simplified version of the UA1 jet-finding algorithm to determine how to identify the jets in the final state. Namely, two jets which satisfy $(\Delta\phi)^2 + (\Delta\eta)^2 < 1$ are coalesced into a single jet, where $\Delta\phi$ and $\Delta\eta$ are the relative azimuthal angles and pseudorapidities of the two jets. In fact, all jets obtained in this way would not necessarily be observed as jets in the actual experiments. The UA1 collaboration requires that in order that a cluster of particles be called a jet, it must have transverse energy above a certain threshold ($E_T{}^{jet} \geq 12$ GeV). Those clusters which do not pass the jet threshold requirements are relegated to the "background" of the event.

At this point, an event consists of final state jets, missing transverse energy and a background due to jets which were not energetic enough to be defined as actual jets. However, this final state has resulted entirely from the Born approximation to the hard constituent scattering. We have so far neglected a number of important effects: the possibility of gluon radiation, the effects of the remnants of the orignal p and $\bar{p}$ which are colliding, the generation of hadrons in the color neutralization of the final state jets, etc. Consider first the effects on the final state jets themselves. First, a jet can lose energy—in the fragmentation process or in gluon bremsstrahlung. Second, the jet can gain energy—particles from processes which contribute to the event "background" processes (independent of the jet formation) which can stray into the cone which defines the jet in question. All in all, the prediction of jet distributions (e.g., the p_T distribution) should be only marginally changed by including such effects. Only the single jet multiplicities and invariant masses are totally unreliable (where hadronization plays a crucial role).

An important quantity to consider is the total scalar transverse energy of an event denoted by E_T. This is defined experimentally by adding in a scalar fashion the transverse energy deposited in all calorimeter cells. This quantity is used by the UA1 collaboration in defining one of the triggers and one of the missing energy cuts so we must consider it here in detail. It is clear that such a quantity is incalculable in the framework of perturbative QCD. In particular, the sum of the scalar transverse energies of the final-state partons is a severe underestimate of the value of E_T. To get an idea of the magnitude of such an underestimation, consider $p\bar{p}$ collisions with *no* large p_T jets in the final state (the so-called "minimum bias" events). The UA1 collaboration has obtained the E_T distribution of such events,[29)] which we reproduce here in Fig. 1. Consider the top graph which corresponds to $|\eta| < 3$. A noteworthy feature of this distribution is the long non-gaussian tail at both low and high E_T. Quantitatively, the mean of the distribuiton (24 GeV) is quite different from the median (18 GeV) and both numbers are larger than the value of E_T where the distribution peaks (12 GeV). To obtain E_T in our parton-Monte Carlo, one is tempted to superimpose this minimum-bias background on top of the underlying hard scattering. In fact, this is known to be incorrect. If one studies the UA1 sample of large-p_T two-jet events, and subtracts out the two jets in each event, the E_T distribution of the remaining background is substantially harder, roughly twice that of minimum bias.[28)] Qualitatively, such an effect is due in part to initial-state radiation which occurs when the constituents which participate in the hard scattering are pulled out the p and $\bar{p}$, and in part to the final-state radiation of the final state quarks and gluons in the process of fragmentation and hadronization.

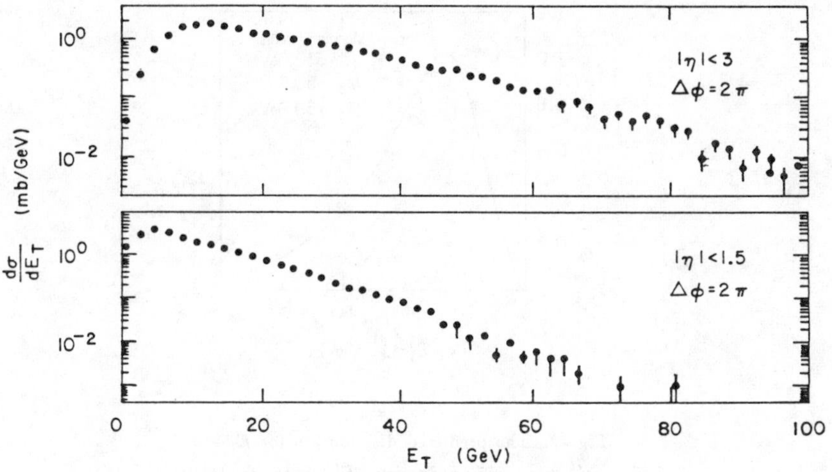

Fig. 1. E_T distribution of minimum bias events at $\sqrt{s} = 540$ GeV, as measured by the UA1 collaboration (taken from Ref. 29). The two sets of data correspond to different ranges of pseudorapidity, η.

In order to proceed with the analysis of supersymmetric particle production, one needs to know what the background distributions are in order to compute the E_T of a given Monte Carlo event. On the basis of the discussion above, it is likely that the background E_T distribution in events with supersymmetric particle production will be roughly twice that of minimum bias (in analogy to the two jet events). However, it is at present unknown how to theoretically compute the background E_T distributions given a hypothetical hard scattering subprocess. Given this current state of ignorance, we have decided on the following course of action. We have rescaled the E_T distribution shown in the top graph of Fig. 1 such that the mean is 40 GeV. (In fact, the median of our rescaled distribution is 32 GeV). The total scalar transverse energy for a given event is then given by

$$E_T = \sum_{\text{jets}} E_{Ti} + \sum_{\substack{\text{low energy}\\ \text{clusters}}} E_{Ti} + E_r \tag{9}$$

where by "low energy clusters" we mean clusters which were not called jets since their transverse energy was below the jet threshold of 12 GeV.

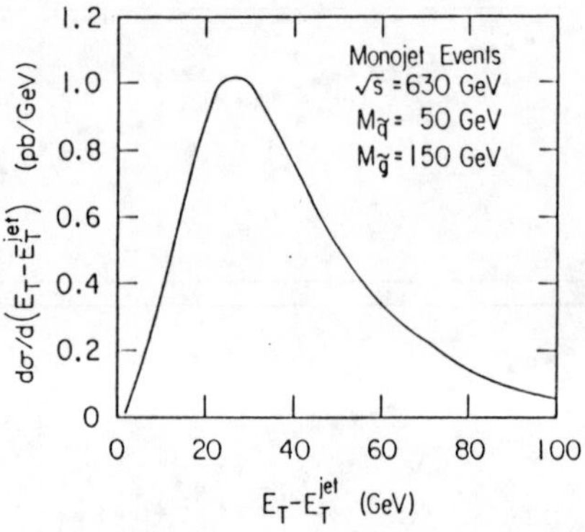

Fig. 2. The predicted distribution for $E_T - E_T^{jet}$ due to $\tilde{q}\tilde{q}$ production. E_T^{jet} is the transverse energy of the leading jet and E_T is the total scalar transverse energy which has been obtained according to eq. (9).

Let us denote:

$$E_T^{extra} = E_T - \sum_{jets} E_{Ti}. \tag{10}$$

From our discussions above, we expect the distribution of E_T^{extra} to be roughly twice that of minimum bias (i.e. with a mean of about 50 GeV). We have found in our computations that

$$\sum_{\substack{\text{low energy}\\ \text{clusters}}} E_{Ti} \approx 10\ GeV \tag{11}$$

in the mean. (This value is closely tied to the assumed value for the jet threshold.) This explains why we chose E_r based on a distribution whose mean was 40 GeV. In studying the monojet events, we can study the published 1983 data[4)] and compute $E_T^{extra} = E_T - E_T^{monojet}$ on an event-by-event basis. Although the statistics are quite low, we find that E_T^{extra} (of events with $E_T^{miss} > 40$ GeV) is distributed with a mean of about 50 GeV, in agreement with the discussion above. The precise distribution of E_T^{extra} is quite sensitive to the UA1 cuts and triggers which are discussed below. As an example, we display in Fig. 2 the distribution we have obtained (after cuts) based on our choice of E_r described above for the case of $M_{\tilde{q}} = 50$ GeV and $M_{\tilde{g}} = 150$ GeV.

The choice for E_r can have a tremendous impact on the number of events which pass the UA1 missing energy trigger. This is unfortunate given that E_r

is not so well understood theoretically. This uncertainty will be an important factor in determining the reliability of our final numbers.

Consider now the UA1 cuts and triggers which were used in their 1984 data analysis.[26,16] First, we state the UA1 cuts. If any of the conditions below are met, the event would be excluded:

$$E_T^{miss} < 15 \text{ GeV} \tag{12}$$
$$E_T^{miss} < 4\sigma \tag{13}$$

where

$$\sigma \equiv 0.7\sqrt{E_T} \tag{14}$$
$$|\text{angle}(\vec{p}_T^{\,miss}, (\text{vertical})\,| < 20^\circ \tag{15}$$
$$|\,\eta^{jet(1)}\,| < 2.5. \tag{16}$$

The purpose of the cuts (12) and (13) is to eliminate missing transverse energy events due to mismeasurement in the calorimetry.

Two further cuts (used first with the 1984 data) are especially helpful for eliminating events resulting from mismeasurement of QCD two-jet events. These cuts, however, use a less restrictive definition of jets (so that they can be applied to monojets as well as dijets):

$$E_T{}^{jet} \geq 8 \text{ GeV} \tag{17}$$

(We caution the reader that eq. (17) is only used in conjunction with the cuts shown below, eqs. (18, 19)). The cuts eliminate events with

$$\text{angle}(\vec{p}_{jet(2)}, -\vec{p}_{\text{jet}(1)}) < 30^\circ \tag{18}$$
$$\text{angle}(\vec{p}_{jet(2)}, \vec{p}_T^{\,miss}) < 30^\circ \tag{19}$$

where jet(1) is the jet with the largest transverse energy, and jet(2) is any jet satisfying eq. (17). Although these cuts are helpful in removing QCD background, they can also be quite effective in removing monojets and dijets of supersymmetric origin.

We now describe the conditions which can trigger an event. There were three UA1 triggers in the 1984 run.[26,16,30] Events were kept if they satisfied *any* of the following conditions (where jet(1) is the jet with the highest E_T):

$$1. \quad E_T(\text{jet } 1) > 25\,\text{GeV} \quad (85\%\text{ run}) \tag{20}$$

$$E_T(\text{jet } 1) > 30\,\text{GeV} \quad (15\%\text{ run}) \tag{21}$$

$$2.\quad E_T(\text{jet } 1) > 15 \text{ GeV} \tag{22a}$$

and

$$|E_T^{miss}(\text{left} - \text{hemisphere}) - E_T^{miss}(\text{right} - \text{hemisphere})| > 17\,\text{GeV} \tag{22b}$$

$$3.\quad E_T(\text{total in } |\eta| < 1.4) > 80\,\text{GeV}. \tag{23}$$

This last trigger uses a more restrictive η region than used for other cuts and triggers ($|\eta| < 2.5$). (To implement this last trigger, a rescaled distribution based on the lower plot shown in Fig. 1 was used.)

The second trigger (22) was not used for the 1983 run. While it apparently allows relatively few extra events from QCD sources, we found that it allowed an order-of-magnitude more events from supersymmetric sources in the case of a light gluino with $M_{\tilde{g}} \lesssim 10$ GeV.

In principle, it is a straightforward task to implement these cuts and triggers inside our Monte Carlo. However, in attempting to mimic the UA1 procedure as closely as possible, we had to take into account a number of effects. First, experimental measurements are not perfect and it is easy to introduce "fake" E_T^{miss} where none exists. This is taken into account by "smearing" our jet four-momenta and the transverse energy of the event background. That is, we alter these quantities which come out of our Monte Carlo according to some appropriate Gaussian resolution function. Second, the hadronic energy as measured by the UA1 calorimeter tends to underestimate the actual value. The magnitude of this effect is not precisely known; it is somewhere between a 0 to 20% effect.[30)] These issues are described in great detail in ref. 11. The end result is somewhat sensitive to how one precisely implements the above effects. We will summarize the sensitivities to our assumptions and approximations after presenting our results.

4. RESULTS FOR MONOJETS AND MISSING-ENERGY EVENTS

The results presented in this section are the consequence of the full analysis described in previous sections. For a given value of $M_{\tilde{q}}$ and $M_{\tilde{g}}$, we compute the number of events from all possible supersymmetric processes which would survive the UA1 cuts and triggers. It is convenient to summarize these results by a set of contour plots which indicate the total number of events of supersymmetric origin which would be observed under 1984 UA1 running conditions per 100 nb^{-1}. These are shown in Figs. 3–6.

Although monojet and dijet cross-sections are well defined in our procedure, we believe that it is best to combine all missing energy events (monojets,

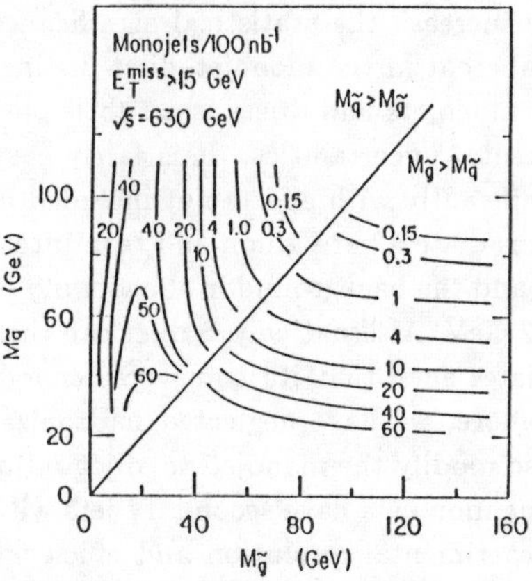

Fig. 3. The number of monojets per 100 nb^{-1} passing the new 1984 UA1 cuts and triggers shown as a contour plot as a function of $M_{\tilde{q}}$ and $M_{\tilde{g}}$. The total 1984 UA1 data sample has an integrated luminiosity of about 270 nb^{-1}.

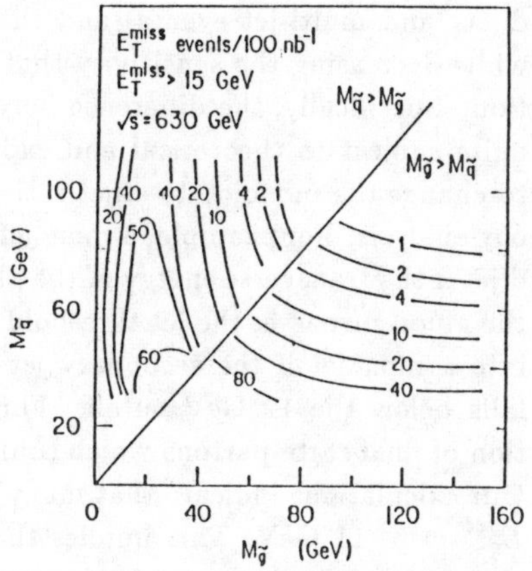

Fig. 4. The number of missing-energy events (monojets, dijets plus multi-jets) per 100 nb^{-1} passing the new 1984 UA1 cuts and triggers. See caption to Fig. 3.

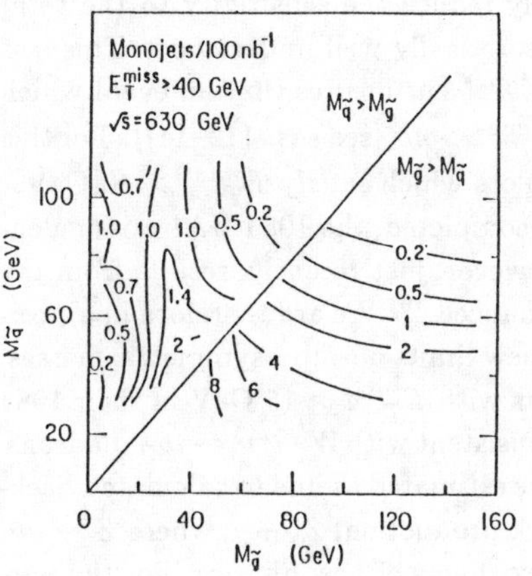

Fig. 5. The number of monojets per 100 nb^{-1} passing the new 1984 UA1 cuts and triggers. In addition, an additional cut, $E_T^{miss} \geq 40$ GeV, is imposed. See caption to Fig. 3.

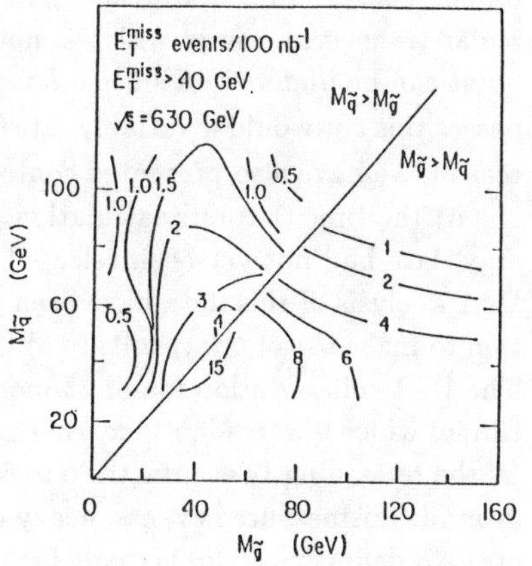

Fig. 6. The number of missing-energy events (monojets, dijets plus multi-jets) per 100 nb^{-1} passing the new 1984 UA1 cuts and triggers. In addition, an additional cut, $E_T^{miss} \geq 40$ GeV, is imposed. See caption to Fig. 3.

dijets and multi-jet events) and thereby increase the statistical significance while decreasing the sensitivity that is inherent in the monojet–dijet distinction. Specifically, the difference between monojets and dijets is a subtle one, quite subject to theoretical and experimental uncertainties. It is fairly easy to change the monojet–to–dijet ratio significantly with effects not included in our analysis. For example, a monojet can radiate a hard gluon and turn into a dijet if the transverse energy of the gluon and the background in the vicinity of the gluon surpasses the jet threshold of 12 GeV. Alternatively, a dijet can turn into a monojet if the secondary jet radiates such that its transverse energy falls below the 12 GeV cutoff. Furthermore, we have neglected hadronization of final state partons which could also modify the monojet–to–dijet ratio. Our calculations indicate that many of the monojets have secondary jets with $E_T^{jet} = 6-11$ GeV. This implies that experimental resolution and efficiency also play important roles in determining the precise number of dijets. Our conclusion is that, although the monojet–dijet distinction is useful in studying gross features of the data, much work is necessary before one can draw detailed conclusions.

Therefore, we have separately presented contour plots for monojet events and for total missing energy events. As we advocated in our previous work,[10)] one can also make a cut on E_T^{miss} which is more severe than the one usually employed by the UA1 collaboration. This is useful, because it can substantially reduce backgrounds and also significantly reduce the sensitivity to the total scalar transverse energy which is not theoretically well understood. This last point can be understood since a *harder* E_T^{miss} cut insures that an event which passes this cut would invariably satisfy $E_T^{miss} > 4\sigma$ (see eqs. (12–14)). For this reason, we have also presented contour plots which satisfy $E_T^{miss} > 40$ GeV.

At the time that our calculation was completed, the 1984 UA1 missing energy data had not yet been released. However, just recently, results from the UA1 analysis of this data have been announced.[16)] We are therefore in a position to make use of our results to obtain new limits on supersymmetric masses. The UA1 collaboration found 23 monojets with $E_T^{miss} \geq 15$ GeV in their 1984 run; of which 9 were identified as being consistent with $W \to \tau\nu, \tau \to \nu +$ hadrons. Of the remaining 14 events, the 6 to 8 were estimated as due to calculated background: mismeasured events, heavy quark production, $Z^\circ + g$, where $Z^\circ \to \nu\bar{\nu}$, etc. No definitive claim is made for the existence of new physics. For the purposes of establishing a conservative set of limits for scalar-quark and gluino masses, we will assume that the upper limit for the number of events which could represent new physics is 13 events at 90% confidence limit. Since the integrated luminosity of the 1984 run was about 270 nb^{-1}, the 90% confidence upper limit for monojet production due to new physics is then 5 events per 100 nb^{-1}.

The UA1 collaboration also reports that after the back-to-back cuts (eqs. (18,19)), two dijets and no multijet-events remain from the 1984 run compared with an estimated background of 2 events. We conclude that at 90% confidence level, the upper limit for dijet events is 2 events per 100 nb^{-1}, whereas the upper limit for *all* missing energy events (with $E_T^{miss} \geq 15$ GeV) is 5 events per 100 nb^{-1}.

If we impose a harder E_T^{miss} cut on top of the UA1 cuts and triggers: $E_T^{miss} \geq 40$ GeV, we note that only 6 events of the reported missing energy events in the 1984 run would survive. Assuming the corresponding background to be about 2 events, we deduce that the 90% confidence upper limit for missing energy events with $E_T^{miss} \geq 40$ GeV is 3 events per 100 nb^{-1}.

The data described above can now be compared directly with our contour plots, and limits on scalar-quark and gluino masses may be deduced. We find that the best limits are obtained from Fig. 4, which shows the total number of missing-energy events with $E_T^{miss} \geq 15$ GeV.

$$M_{\tilde{q}} > 60\text{–}70 \text{ GeV (depending on } M_{\tilde{g}}) \tag{24}$$

$$M_{\tilde{g}} > 50\text{–}60 \text{ GeV (depending on } M_{\tilde{q}}) \ . \tag{25}$$

Note that these limits are better than one obtains from the monojets alone (Fig. 3). The reason is that for large supersymmetric masses, dijet production should dominate over monojet production, even after the back-to-back cuts (eqs. (18,19)) are imposed. Thus, the results obtained by combining all missing energy events are both more restrictive and more reliable (the latter, because we need not distinguish among events with different numbers of jets).

If one wishes instead to assume that we can accurately separate monojets and dijets, then the above results suggest that it will be useful to examine the dijet rate separately. This can be done by subtracting Fig. 3 from Fig. 4. The resulting limits are slightly better than those shown in eqs. (24,25):

$$M_{\tilde{q}} > 65\text{–}75 \text{ GeV} \tag{26}$$

$$M_{\tilde{g}} > 60\text{–}70 \text{ GeV}. \tag{27}$$

If we consider the implications based on the harder cut $E_T^{miss} \geq 40$ GeV, we see that the limits are not as good for scalar-quark masses, and are non-existent for gluino masses. This occurs because the missing-energy events are predicted to be heavily populated in the $E_T^{miss} < 40$ GeV region (and similarly for the Standard Model backgrounds). As $M_{\tilde{g}}$ and $M_{\tilde{q}}$ become large, the E_T^{miss} distributions get quite hard (as can be seen by comparing Figs. 3 and 4 with Figs. 5 and 6), but then phase space considerations cut off the rate.

Could some of the observed monojet events be due to the production of gluinos or scalar quarks? There are two factors which argue against the monojets coming from gluino or scalar quark production. Both are consequences of the fact that the appropriate event rate (2–3 monojets/100 nb^{-1}) only occurs for large $M_{\tilde{q}}$ or $M_{\tilde{g}}$ ($\approx$ 60 GeV). For such masses we would predict 4–6 dijets/100 nb^{-1}, and these certainly have not been observed. Furthermore, at these masses one would expect significant numbers of monojets with $E_T^{miss} >$ 45 GeV, and only one such event was observed in the 1984 run.

The two observed dijet events (surviving the back-to-back cuts) in the 1984 run have $E_T^{miss} \geq 55$ GeV. Although there is a roughly equal background expected, these backgrounds are unlikely to have so much E_T^{miss}. A 70–90 GeV scalar-quark could give dijets with such characteristics and with this rate, and would produce very few monojets. Clearly, such speculation must await until considerably more data has been accumulated.

How precise should we treat the numbers we have obtained (shown on our contour plots, Figs. 3–6)? There are a number of uncertainties which enter into our calculation, both from theoretical sources and experimental sources. First, there are the usual uncertainties inherent in calculations based on perturbative QCD. The calculation depends on a choice of Λ_{QCD} and a set of structure functions. (We have chosen $\Lambda = 0.29$ GeV and use the EHLQ structure functions[31)] coresponding to this choice.) Higher order corrections are neglected (no "K-factor"is used), except that we do investigate the effects of a hard $2 \to 3$ partonic subprocess (see. eq. (7)). As a result, the scale used in the running coupling constant and the structure functions is undetermined and a particular choice (we choose $Q^2 = \hat{s}$) will imply some uncertainty in the final result. Another question as to the reliability of our results concern the effects of ignoring hadronization (our final-state quarks and gluons are not converted into hadronic jets). In general, for the rates and distributions we discuss, we believe this approximate should not have major consequences. Ellis and Kowalski[6)] who have studied the effects of hadronization have reached this conclusion.

Other uncertainties in our calculation are related to the precise definition of jets and the residual "background"of a given event. For example, nonperturbative QCD corrections and spectator activity can add to jet momenta; whereas energy can leave the jet cone (e.g. by gluon bremsstrahlung) and thus not be included in the definition of the jet. We have argued above that these effects can be very important for certain quantities such as monojet–to–dijet ratio, although the sensitivity to our final results can be substantially reduced by combining all missing-energy events. A related uncertainty involves the transverse energy distribution of the "background." This quantity is not theoretically understood, and experimental information on the back-

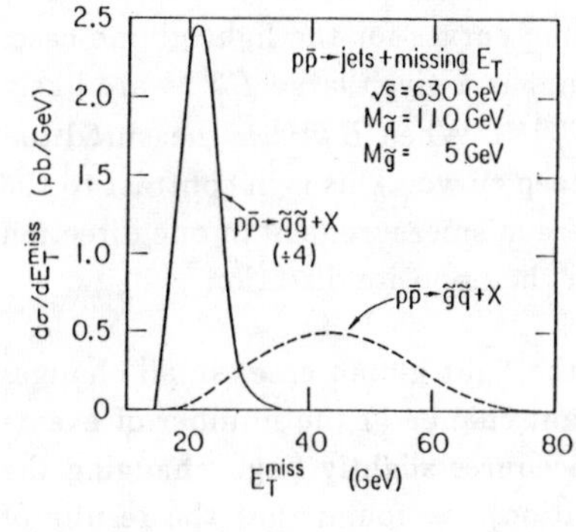

Fig. 7. The E_T^{miss} distribution expected from $\tilde{g}\tilde{g}$ (solid curve) and from $\tilde{q}\tilde{q}$ (dashed curve) production at $\sqrt{s} = 630$ GeV, if $M_{\tilde{g}} = 5$ GeV and $M_{\tilde{q}} = 110$ GeV. In order to fit both curves on the same graph, we have divided the $\tilde{g}\tilde{g}$ curve by 4. The $\tilde{g}\tilde{g}$ distribution is cut off at small E_T^{miss} by experimental cuts. All of the new 1984 UA1 cuts and triggers are included.

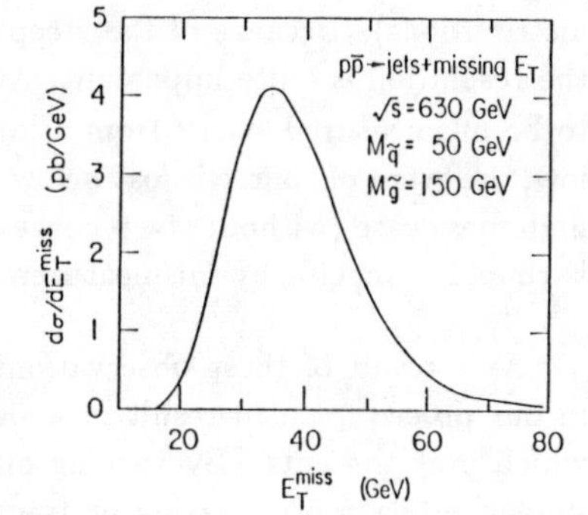

Fig. 8. The E_T^{miss} distribution expected from $\tilde{q}\tilde{q}$ production at $\sqrt{s} = 630$ GeV, if $M_{\tilde{q}} = 50$ GeV and $M_{\tilde{g}} = 150$ GeV. All of the new 1984 UA1 cuts and triggers are included.

ground underneath two jet events is scarce at present. The background is an important part of the total scalar transverse energy, E_T of an event which is used in the missing-energy cuts (eqs. (13,14)).

All in all, we expect that the uncertainties in our numbers is roughly a factor of two. This implies an uncertainty in our quoted mass limits of about 5 GeV. However, in the particular case where the gluino is light (say $M_{\tilde{g}} \lesssim$ 20 GeV), much more care must be given to the estimation of uncertainties. In fact, the numbers shown in our contour plots are much less certain in this regime. We now turn to the case of the light gluino in more detail.

5. IS THERE A LIGHT GLUINO WINDOW?

Taken at face value, the results of Section 4 imply that there is no region where $M_{\tilde{g}}$ is small which does not predict too many missing energy events compared to what is seen by UA1. However, the uncertainties of our calculation become especially severe in this region. To explain, let us examine Figs. 7 and 8. When the gluino is quite light, all events passing cuts and triggers are on the tails of distributions. For the case of the 5 GeV gluino in Fig. 7, the curve would rise far above the figure as E_T^{miss} decreases except that it is cut-off by the imposed cuts. This is unlike the curve for higher masses (Fig. 8) where the cuts have little impact (the events are not on the tail of distributions). Because of the steeply falling curves for the light gluino case, the resolution is quite important. All events at a given large E_T^{miss} are likely to be mismeasured events from a lower E_T^{miss}. When E_T^{miss} is measured too low, the event of course is lost below the steep curve. This is in contrast to the high mass case (without steep curves) where mismeasurement in one direction is roughly canceled by mismeasurement in the opposite direction.

As a result of these observations, in the light gluino case, small changes in our procedures can result in a significant change in the number of events which pass the cuts. By varying our procedures slightly (e.g. changing the implementation of smearing or fragmentation), we found that the results of these variations could reduce the number of events passing the cuts by a factor of two if the gluino were light. For heavier gluinos (and scalar-quarks), the sensitivity to the methods of smearing and fragmentation was negligible. As another example, if the gluino is light, we find that our results are very sensitive to small changes in the parametrization of the transverse energy distribution of the background (E_r in eq. (9)). By changing the mean or the shape of this distribution, one can greatly effect the number of events which pass the missing-energy cuts (eqs. 12–14), if $M_{\tilde{g}}$ is small. On the other hand, if $M_{\tilde{g}}$ and $M_{\tilde{q}}$ are large, then typically E_T^{miss} is large (see Fig. 8). In this case, one usually finds that $E_T^{miss} > 4\sigma$ for all reasonable choices of the background distribution of E_r. Another way to reduce the sensitivity to the background distribution is to impose a harder E_T^{miss} cut. Of course, this leads to fewer events (both predicted and observed) passing the cuts, thereby weakening the mass limits substantially.

For all the reasons mentioned above, we believe that the numbers shown on the contour plot may be uncertain by as much as a factor of 4 or 5 (if all effects are added linearly). In spite of this large uncertainty, our predicted rate of events is large enough to rule out nearly all possible masses for a light gluino. For $M_{\tilde{q}} = 100$ GeV, we would predict 26 events per 100 nb^{-1} for $M_{\tilde{g}} = 5$ GeV and 13 events per 100 nb^{-1} for $M_{\tilde{g}} = 3$ GeV. (The decrease in number of events passing the cuts is due to gluino fragmentation.) Note that all these events would be monojets; the rate for dijet production here is negligible. Indeed, these numbers are much larger than the 5 event per 100 nb^{-1} upper limit we used in the previous section to set our limits. Even with a factor of 4 uncertainty, we can rule out $M_{\tilde{g}} = 5$ GeV, whereas the possibility of $M_{\tilde{g}} = 3$ GeV is very marginal. We conclude that, based on the 1984 UA1 data, what has been referred to as a "window" allowing light gluinos is at best a "peephole"; and most likely, all light gluinos can be ruled out. (For even lighter gluinos, we may quote the results of beam-dump experiments. A recent

BEBC experiment[32] sets the limit $M_{\tilde{g}} > 3$–4 GeV at the 90% confidence level, depending on the value of $M_{\tilde{q}}$.)

There are two basic reasons as to why the production rate for light gluinos is large enough to rule out their existence. First, as mentioned below in eq. (7), the production of $\tilde{g}\tilde{g}g$ (with $p_T(g)) > 10$ GeV) is enhanced by virtue of the fact that the emission of a hard gluon increases the probability that the event passes the missing-energy cuts. In fact, we find that for light gluinos, this process is the *dominant* supersymmetric subprocess; it is a factor of two more likely to pass the cuts than events arising from a $\tilde{g}\tilde{g}$ final state. Second, the 1984 run contained a new trigger (not used in the 1983 run) given by eq. (22). This trigger was particularly efficient at finding missing-energy events, although the increase in efficiency depends on the value of the supersymmetric masses. In general, the efficiency is increased by about a factor of two. However, when gluinos are light, the increased efficiency for finding $\tilde{g}\tilde{g}$ events due to this new trigger becomes a factor of ten! This reflects the large sensitivity of the light gluino events to changes in the cuts and triggers. The $\tilde{g}\tilde{g}$ events which pass the UA1 cuts and triggers lie close to the boundaries of those cuts and triggers when the gluino is light. This is in part due to the fact that the probability that a $\tilde{g}\tilde{g}$ event (with $M_{\tilde{g}}$ small) actually passes the UA1 cuts and triggers is quite small (e.g. for $M_{\tilde{g}} = 5$ GeV, less than 10^{-4} of all $\tilde{g}\tilde{g}$ events survive the analysis).

As an historical aside, we note that light gluinos probably could not be ruled out based on the 1983 UA1 data alone.[15] The absence of the new trigger (eq. (22)) implies that fewer $\tilde{g}\tilde{g}$ events would pass the cuts. Under the 1983 running conditions, we found that for $M_{\tilde{g}} = 5$ GeV, the number of predicted $\tilde{g}\tilde{g}$ events tended to be larger than the number of events seen. (For the most part, these events populated the region $15 \leq E_T^{miss} \leq 32$ GeV and were therefore not considered in our previous analysis given in ref. 10.) However, by taking into account the large uncertainties discussed above, we felt that we could not convincingly rule out the light gluino window based on the 1983 data.

6. THEORETICAL MODELS AND ALTERNATIVES

Throughout this paper, we have treated the gluino and scalar-quark masses as independent parameters. Furthermore, we took the photino to the be lightest supersymmetric particle (LSP) and assumed that its mass could be neglected. If we are willing to adopt a particular approach to low energy supergravity model building,[33] we can constrain certain parts of the $M_{\tilde{q}} - M_{\tilde{g}}$ plane.

For illustration purposes, let us consider a class of models which have been often referred to as being "minimal" supergravity models.[33–36] These models are "minimal" in two respects. First, they consist of the minimal number of particles: the Standard Model particles with two Higgs doublets and their supersymmetric partners. Second, these models depend on a minimal set of parameters: the gravitino mass ($m_{3/2}$), the gluino mass ($M_{\tilde{g}}$), μ (a supersymmetric Higgsino mass), A (a parameter related to the super-Higgs mechanism) and v_2/v_1 (the ratio of vacuum expectation values of the two Higgs fields). These models are obtained in two steps. First, starting with an $SU(3) \times SU(2) \times U(1)$ (or grand unified) gauge theory coupled to supergravity, an effective renormalizible field theory at the Planck mass ($M_{p\ell}$) is obtained in the limit of $M_{p\ell} \to \infty$. Second, renormalization group equations are used in order that we may obtain the effective low energy theory which is valid at the electroweak scale. It turns out that the effective theory at the Planck scale does not break the $SU(2) \times U(1)$ gauge symmetry. Thus, in order that the $SU(2) \times U(1)$ electroweak group be broken in the standard way in the low-energy effective theory (such that the $SU(3)_{color} \times U(1)_{EM}$ remains conserved), it must happen that the parameters of the theory, which evolve via the renormalization group, satisfy an appropriate set of conditions. These conditions are quite restrictive and tend to reduce much of the freedom in the choice of parameters of the model. For example, if we assume that $m_t \approx 40$ GeV (as may be indicated by recent results[37]) of UA1), then it follows that $v_1 \approx v_2$ and $B\mu m_{3/2} \sim O(m_{3/2}^2)$, where $B \equiv A - 1$ at the Planck scale is a number of order unity. The renormalization group analysis of such models has been performed; for simplicity, we exhibit below the results obtained by ref. 36 in the limit of $v_1 = v_2$:

$$M_{\tilde{\gamma}} \approx \frac{1}{6} M_{\tilde{g}} \tag{28}$$

$$M_{\tilde{q}L}^2 = m_{3/2}^2 + C_{\tilde{q}L} M_{\tilde{g}}^2 \tag{29}$$

$$M_{\tilde{q}R}^2 = m_{3/2}^2 + C_{\tilde{q}R} M_{\tilde{g}}^2 \tag{30}$$

$$M_{\tilde{\ell}_L}^2 = m_{3/2}^2 + C_{\tilde{\ell}_L} M_{\tilde{g}}^2 \tag{31}$$

$$M_{\tilde{\ell}_R}^2 = m_{3/2}^2 + C_{\tilde{\ell}_R} M_{\tilde{g}}^2 \tag{32}$$

where the constants are given by: $C_{\tilde{q}L} = 0.85$, $C_{\tilde{q}R} = 0.78$, $C_{\tilde{\ell}_L} = 0.08$, and $C_{\tilde{\ell}_R} = 0.02$. Strictly speaking, eqs. (29) and (30) must be modified somewhat for the third generation; we refer the reader to ref. 36 for the details. Eq. (28) follows from eq. (1); in the models where the gluino mass is not too large, the LSP is approximately a pure photino with the mass shown above. Using eqs. (29)–(32), it follows that

$$M_{\tilde{\ell}_R}^2 \approx M_{\tilde{q}}^2 - 0.8\, M_{\tilde{g}}^2. \tag{33}$$

This is an interesting constraint, in that it implies that the gluino cannot be much heavier than the scalar quark.

Another constraint can be obtained by considering the cosmological implications of a light photino. In particular, since photinos are the LSP (and thus stable), their annihilation rate must be sufficiently efficient to reduce their abundance in the early universe to a cosmologically acceptable level. (Another cosmologically acceptable solution—to have the photino nearly massless like the neutrino—is unacceptable, since by eq. (28), it would imply a nearly massless gluino, which is almost certainly ruled out.) A calculation of Ellis et al.[18] shows that $M_{\tilde{\gamma}} \gtrsim 0.5$ GeV if $M_{\tilde{q}} \gtrsim 20$ GeV and $M_{\tilde{\gamma}} \gtrsim 5$ GeV if $M_{\tilde{q}} \gtrsim 100$ GeV. (The efficiency of photino annihilation decreases as the scalar-quark mass increases.) Using eq. (28), this leads to a lower limit on the gluino mass as a function of the scalar-quark mass.

The two constraints discussed above substantially limit the region of the $M_{\tilde{q}} - M_{\tilde{g}}$ plane which is consistent with the minimal low energy supergravity model described above. These have been discussed in detail in refs. 38 and 39. We have emphasized earlier that many of our conclusions (and our strict mass limits) depend on the assumption that the photino is the LSP. It is of interest to consider whether it is possible to construct models where this assumption is not valid, and what the implications are for another candidate for the LSP.

We consider here the case where the LSP is a light Higgsino. In the minimal model discussed above, the Higgsino mass turns out to be $M_{\tilde{H}} \sim \mu \sim O(m_{3/2})$. Hence, unless the gluino mass is large enough (implying a large value for $M_{\tilde{\gamma}}$ via eq. (28)), the Higgsino will not be the LSP. However, as shown in ref. 40, one can easily generalize the minimal model in such a way that the parameter μ is not constrained to be of $O(m_{3/2})$. In such a model, the Higgsino will be the LSP as long as $\mu < M_{\tilde{\gamma}}$.

Let us summarize some of the phenomenological implications of this alternative scenario. First, the photino will tend to be the second lightest supersymmetric particle, and hence only two decay channels are available: $\tilde{\gamma} \to f\bar{f}\tilde{H}$ or $\tilde{\gamma} \to \gamma\tilde{H}$. The latter decay occurs via a one-loop Feynman diagram (see refs. 40 and 41). The three-body tree level decay of the photino occurs via the exchange of a virtual $\tilde{f}$. Because the $\tilde{H}f\tilde{f}$ Feynman rule is proportional to the mass of the *fermion* (m_f), this decay rate is negligible and the decay $\tilde{\gamma} \to \gamma\tilde{H}$ is the dominant one. A calculation of the lifetime for the two-body decay yields approximately:

$$\tau_{\tilde{\gamma}} \approx 10^{-11} sec. \left(\frac{1 \text{ GeV}}{M_{\tilde{\gamma}}} \right)^3 \tag{34}$$

which indicates that the photino decay is prompt unless $M_{\tilde{\gamma}}$ is sufficiently light. Note that because the photino is now unstable, the cosmological limits for the photino obtained in ref. 18 no longer apply. Instead, one now finds cosmological limits on the Higgsino mass: either $M_{\tilde{H}} \lesssim 100$ eV or $M_{\tilde{H}} \gtrsim m_b$. Since the Higgsino and gluino masses are logically independent, there is no phenomenological reason which rules out a massless Higgsino.

Scalar-quark and gluino decays are, however, unchanged. In principle, one could have $\tilde{g} \to q\bar{q}\tilde{H}$ and $\tilde{q} \to q\tilde{H}$. However, as stated above, the $\tilde{H}f\tilde{f}$ vertex is proportional to m_f and hence these decay rates can be neglected as compared with the standard ones involving the photino. Two cases can be envisioned. If the photino is long lived (see eq. (34)), then it will escape the Collider detectors, and there is no change in any of the results obtained in this paper. However, if the photino decays promptly, $\tilde{\gamma} \to \gamma\tilde{H}$, then the phenomenology changes drastically. First, when supersymmetric particles are produced, the resulting missing-energy spectrum softens considerably. As a result, fewer events pass the UA1 E_T^{miss} cuts, and the limits on supersymmetric masses obtained in Sec. 4 are significantly weakened. Although we have not yet implemented this possibility in our Monte Carlo program, we may quickly obtain an estimate as to the new limits. We do this by noting that Dawson[42)] has investigated the implication of missing-energy events for supersymmetric models which violate R-parity. In these models, the photino is the LSP but is unstable and decays via $\tilde{\gamma} \to \gamma\nu$. Thus, the signature is identical to the case we are considering here, so we may use her results. Dawson finds (see Fig. 11 of ref. 42) that the number of events which pass the UA1 cuts and triggers is roughly a factor of five less than for the case of a stable photino. This suppression factor is roughly independent of $M_{\tilde{q}}$ and $M_{\tilde{g}}$. If we reduce the numbers which appear in Figs. 3–6 by a factor of five, we would obtain the following allowed regions for $M_{\tilde{q}}$ and $M_{\tilde{g}}$:

$$M_{\tilde{g}} \lesssim 5 \text{ GeV} \quad \text{or} \quad M_{\tilde{g}} \gtrsim 40 \text{ GeV} \tag{35}$$

$$M_{\tilde{q}} \gtrsim 45\text{–}60 \text{ GeV} . \tag{36}$$

In eq. (36), the stronger limit is obtained as we take the gluino mass approaching the scalar-quark mass. Note that the "light gluino window" has returned. That is, we no longer feel able to rule out gluino masses of order $M_{\tilde{g}} \lesssim 5$ GeV since the number of predicted events passing the UA1 cuts has been significantly reduced.

Suppose we accept the possibility that a few of the monojets could be due to supersymmetry where the Higgsino is the LSP. As argued above, when scalar-quarks and gluinos are produced, they decay into photinos which subsequently decay into Higgsinos: $\tilde{\gamma} \to \gamma\tilde{H}$. Thus, these events should contain photons! Can this be ruled out? At present, the answer seems to be negative. The photon could not be easily distinguished from a π° in the UA1 detector, so these events would just exhibit extra observed neutral energy.

Note, however, that some limits do exist from e^+e^- physics. The process $e^+e^-\to\tilde{\gamma}\tilde{\gamma}$, $\tilde{\gamma}\to\gamma\tilde{H}$ can take place which yields $e^+e^-\to\gamma\gamma+$ missing energy. Such a process has been searched for at PETRA; no events of this type above background have been seen.[43] This implies that the cross-section for $e^+e^-\to\tilde{\gamma}\tilde{\gamma}$ cannot be too large. Since this process occurs via exchange of a scalar-electron, the absence of this process puts a limit on the scalar-electron: $M_{\tilde{e}} \gtrsim$ 100 GeV. By eq. (33), this implies that $M_{\tilde{q}} \gtrsim$ 100 GeV. Thus, in the case where the Higgsino is the LSP, supersymmetry could be the explanation for monojets if the gluino were very light, $M_{\tilde{g}} \lesssim$ 5 GeV.

7. CONCLUSIONS

We have analyzed the production of scalar-quarks and gluinos at the CERN $p\bar{p}$ Collider. Our analysis has included UA1 cuts and triggers from the 1984 running conditions.[16] Based on the newly reported UA1 data obtained during the 1984 run, we conclude that if there is any excess of monojet events after backgrounds are subtracted, it is unlikely to come from the production of scalar-quarks and gluinos. Under the assumption that the photino is lighter than gluinos and scalar-quarks and lives long enough to escape collider detectors, we rule out the following scalar-quark and gluino masses: $M_{\tilde{q}} \lesssim$ 60–70 GeV and $M_{\tilde{g}} \lesssim$ 50–60 GeV. We showed that a powerful new trigger in the 1984 run has virtually eliminated the so-called "window" for a light ($M_{\tilde{g}} \approx 5$ GeV) gluino. Even if one allows for a few monojets above background, one cannot explain these events as being due to supersymmetry. If such an explanation were viable (i.e. monojets due to supersymmetric particles with a mass of 60 GeV), then one would necessarily predict at least twice as many dijets, which is inconsistent with the observed data. Actually, two dijets with very large E_T^{miss} were observed in the 1984 data. These are not inconsistent with even heavier scalar-quarks or gluinos (in the mass region 70–90 GeV, where no monojets due to supersymmetry should be observed). However intriguing these two events are, there may be a long wait for adequate statistics to learn more about their origin.

If we change one of our basic assumptions, and assume that the Higgsino is the lightest supersymmetric particle, then our analysis changes somewhat. Because the photino would decay, the missing energy in events with scalar-quarks and/or gluinos would be softened, implying that fewer events would pass the UA1 cuts and triggers. New mass limits would then be obtained which would be less restrictive than those quoted above: We can only rule out $5 \lesssim M_{\tilde{g}} \lesssim$ 40 GeV and $M_{\tilde{q}} \lesssim$ 45–60 GeV. In particular, the light gluino window, $M_{\tilde{g}} \lesssim$ 5 GeV, has reappeared.

It is certainly disappointing that no evidence for sypersymmetry is indicated by the present CERN data! Does this mean that supersymmetry is becoming a less viable approach which could describe possible new physics beyond the Standard Model? I would say that the answer is certainly no! Note that although the mass limits we have found are quite large, they are still below m_W. If supersymmetry is to explain the origin of the electroweak scale, then it is natural to assume that this is the relevant scale for determining the masses of the various supersymmetric particles. But such a statement is at best qualitative and does not tell us whether to expect $M_{\tilde{q}} = 40$ GeV or 400 GeV. In fact, I believe that it is more reasonable to anticipate a mass somewhat above m_W rather than below m_W. The reason for this can be best seen by studying various low energy supergravity models. When it was thought that monojets were explained by gluinos or scalar-quarks with mass around 40 GeV, many groups proceeded to construct models which would give such masses.[38–39)] Two things were found. First, such models were very constraining; a slight increase in the experimental limit on scalar-electron masses was sufficient to rule out a number of such models. Second, many models of this type required a small gravitino mass (e.g. $m_{3/2} = 15$ GeV in ref. 39). Since one might expect $m_{3/2} \sim O(m_W)$, the message to be drawn from the models above is that scalar-quark masses around 40 GeV are too low from theoretical considerations. Thus, the limits we have found in this paper should come as no surprise to supersymmetry model builders. Supersymmetric masses of order 100–200 GeV are probably more likely.

Again, one should emphasize that these arguments are not precise. But they do suggest that it may be more likely to find evidence for supersymmetry at the Tevatron (rather than at the CERN Collider). Of course, we could be unlucky and need a machine like the Superconducting Super Collider to unlock the mysteries of the origin of electroweak symmetry breaking. On the other hand, we could be lucky and begin to see departures from the Standard Model at the CERN Collider as more data are accumulated. In all cases, it is important to perservere and to test the Standard Model as hard as possible at both present and future colliders. Evidence for new physics, and perhaps supersymmetry, may not be too far away.

ACKNOWLEDGEMENTS

I would like to acknowledge invaluable conversations with UA1 Collaboration members Richard Batley, Alan Honma, Witold Kozanecki, Aurore Savoy-Navarro and Felicitas Pauss. Also, many thanks to Guido Altarelli, Vernon Barger, John Ellis, Jack Gunion, Zoltan Kunszt, Frank Paige and Dave Soper for numerous useful conversations. I am expecially grateful to V. Barger for his invitation to speak at the "New Particles" Conference at Wisconsin, and to N. Paver, C. Verzegnassi and B.W. Lynn for their invitation to speak at the Trieste workshop on "Tests of Electroweak Theories." In addition, the hospitality of the SLAC Theory Group, where some of this work was carried out, is greatly appreciated. Finally, special thanks go to my co-workers Mike Barnett and Gordy Kane for a very enjoyable and productive collaboration.

REFERENCES

1. G. Arnison, et al., CERN-EP/85-108 (1985); L. Di Lella, invited talk given at the 1985 Lepton-Photon Conference, Kyoto, Japan, August 1985.

2. H.E. Haber and G.L. Kane, *Phys. Rep.* **117**, 75 (1985).

3. J. Ellis, in "Proceedings of the Yukon Advanced Study Institute on the Quark Structure of Matter", edited by N. Isgur, G. Karl and P.J. O'Donnell (World Scientific, Singapore, 1985) p. 256.

4. G. Arnison, et al., *Phys. Lett.* **139B**, 115 (1984).

5. J. Ellis and H. Kowalski, *Phys. Lett.* **142B**, 441 (1984); *Nucl. Phys.* **B246**, 189 (1984); *Phys. Lett.* **157B**, 437 (1985).

6. J. Ellis and H. Kowalski, *Nucl. Phys.* **B259**, 109 (1985).

7. V. Barger, K. Hagiwara and J. Woodside, *Phys. Rev. Lett.* **53**, 641 (1984); V. Barger, K. Hagiwara and W.-Y. Keung, *Phys. Lett.* **145B**, 147 (1984); V. Barger, K. Hagiwara, W.-Y. Keung and J. Woodside, *Phys. Rev.* **D31**, 528 (1985); **D32**, 806 (1985).

8. V. Barger, S. Jacobs, J. Woodside and K. Hagiwara, Wisconsin preprint MAD/PH/232 (1985).

9. E. Reya and D.P. Roy, *Phys. Rev. Lett.* **51**, 867 (1983) (E:**51**, 1307 (1983); *Phys. Rev. Lett.* **53**, 881 (1984); *Phys. Lett.* **141B**, 442 (1984); *Phys. Rev.* **D32**, 645 (1985); Dortmund preprint DO-TH 85/23 (1985).

10. R.M. Barnett, H.E. Haber and G.L. Kane, *Phys. Rev. Lett.* **54**, 1983 (1985).

11. R.M. Barnett, H.E. Haber and G.L. Kane, Berkeley preprint LBL-20102 (1985).

12. M.J. Herrero, L.E. Ibañez, C. Lopez and F.J. Yndurain, Phys. Lett. **132B**, 199 (1983); **145B**, 430 (1984).

13. A.R. Allan, E.W.N. Glover and A.D. Martin, *Phys. Lett.* **146B**, 247 (1984); A.R. Allan, E.W.N. Glover and S.L. Grayson, *Nucl. Phys.* **B259**, 77 (1985); F. Delduc, H. Navelet, R. Peschanski, and C.A. Savoy, *Phys. Lett.* **155B**, 173 (1985); X.N. Maintas and S.D.P. Vlassopulos, *Phys. Rev.* **D32**, 604 (1985).

14. N.D. Tracas and S.D.P. Vlassopulos, *Phys. Lett.* **149B**, 253 (1984);
15. G. Altarelli, B. Mele and S. Petrarca, Rome preprint (1985).
16. J. Rohlf, invited talk at the 1985 Division of Particles and Fields Conference, Eugene, Oregon, August 1985; C. Rubbia, invited talk at the 1985 Lepton-Photon Conference, Kyoto, Japan, August 1985.
17. P. Fayet, *Phys. Lett.* **69B**, 489 (1977); G. Farrar and P. Fayet, *Phys. Lett.* **76B**, 575 (1978).
18. J. Ellis, J.S. Hagelin, D.V. Nanopoulos, K. Olive and M. Srednicki, *Nucl. Phys.* **B238**, 453 (1984).
19. G.L. Kane and J.P. Leveille, *Phys. Lett.* **112B**, 227 (1982).
20. P.R. Harrison and C.H. Llewellyn Smith, *Nucl. Phys.* **B213**, 223 (1983) (E: **B223**, 542 (1983)).
21. I. Antoniadis, L. Baulieu and F. Delduc, *Z. Phys.* **C23**, 119 (1984).
22. S. Dawson, E. Eichten and C. Quigg, *Phys. Rev.* **D31**, 1581 (1985).
23. J. Ellis and S. Rudaz, *Phys. Lett.* **128B**, 248 (1983).
24. F. Herzog and Z. Kunszt, *Phys. Lett.* **157B**, 430 (1985).
25. B.A. Campbell, J. Ellis and S. Rudaz, *Nucl. Phys.* **B198**, 1 (1982); I. Antoniadis, C. Kounnas and R. Lacaze, *Nucl. Phys.* **B211**, 216 (1983); C. Kounnas and D.A. Ross, *Nucl. Phys.* **B214**, 317 (1983); S.K. Jones and C.H. Llewellyn Smith, *Nucl. Phys.* **B217**, 145 (1983); M.J. Herrero, C. Lopez and F.J. Yndurain, *Nucl. Phys.* **B244**, 207 (1984).
26. M. Mohammadi, CERN-EP/85-52 (1985), presented at the Conference on Collider Physics at Ultra High Energies, Aspen, Colorado, January 1985.
27. A. De Rujula and R. Petronzio, CERN-TH 4070/84 (1984).
28. G. Arnison, et al., *Phys. Lett.* **132B**, 214 (1983); J. Sass, in *Antiproton-Proton Physics and the W. Discovery*, Proc. of the International Colloquium of the CNRS, Third Moriond Workshop, March 1983, ed. by J. Tran Thanh Van (Editions Frontières, France, 1983), p. 295.
29. G. Arnison, et al., CERN-EP/82-122 (1982).
30. A. Honma, private communication.
31. E. Eichten, I. Hinchliffe, K. Lane, and C. Quigg, *Rev. Mod. Phys.* **56**, 579 (1984).
32. A.M. Cooper-Sarkar et al., CERN/EP 85-97 (1985).

33. H.P. Nilles, *Phys. Rep.* **110C**, 1 (1984); P. Nath, R. Arnowitt and A.H. Chamseddine, *Applied N=1 Supergravity*, ICTP Series in Theoretical Physics, Vol. I, (World Scientific, Singapore, 1985).

34. L. Alvarez-Gaumé, J. Polchinski and M. Wise, *Nucl. Phys.* **B221**, 495 (1983); J. Ellis, J. Hagelin, D.V. Nanopoulos and K. Tamvakis, *Phys. Lett.* **125B**, 275 (1983); M. Claudson, L. Hall and I. Hinchliffe, *Nucl. Phys.* **B228**, 501 (1983); S. Jones and G.G. Ross, *Phys. Lett.* **135B**, 69 (1984).

35. L. Ibañez and C. Lopez, *Phys. Lett.* **126B**, 54 (1983); *Nucl. Phys.* **B228**, 501 (1983).

36. C. Kounnas, A.B. Lahanas, D.V. Nanopoulos and M. Quiros, *Phys. Lett.* **132B**, 95 (1983); *Nucl. Phys.* **B236**, 438 (1984).

37. G. Arnison, et al., *Phys. Lett* **147B**, 493 (1984).

38. J. Ellis and M. Sher, *Phys. Lett.* **148B**, 309 (1984); L.J. Hall and J. Polchinski, *Phys. Lett.* **152B**, 335 (1985); M. Gluck, E. Reya and D.P. Roy, *Phys. Lett.* **155B**, 284 (1985); S. Nandi, *Phys. Rev. Lett.* **54**, 2493 (1985); K. Enqvist, D.V. Nanopoulos and A.B. Lahanas, *Phys. Lett.* **155B**, 83 (1985); J. Ellis, J.S. Hagelin and D.V. Nanopoulos, *Phys. Lett.* **159B**, 26 (1985).

39. L. Ibañez, C. Lopez and C. Muñoz, *Nucl. Phys.* **B256**, 252 (1985).

40. H.E. Haber, G.L. Kane and M. Quiros, *Phys. Lett.* **160B**, 297 (1985); Univ. of Michigan preprint UM TH 85-8 (1985).

41. H. Komatsu and J. Kubo, *Phys. Lett.* **157B**, 90 (1985).

42. S. Dawson, Berkeley preprint LBL-19460 (1985).

43. W. Bartel et al., *Phys. Lett.* **139B**, 327 (1984).

GLUINO PRODUCTION

Z. Kunszt

Institute for Theoretical Physics
University of Bern
Sidlerstrasse 5, CH-3012 Bern, Switzerland

ABSTRACT

Gluino pair production at hadron colliders is discussed with emphasis on hard gluon bremsstrahlung effects. We point out that due to kinematical and dynamical reasons the gg-> $\tilde{g}\tilde{g}g$ subprocess gives important contribution to the missing p_T signatures at the CERN $p\bar{p}$ collider for gluino mass values $m(\tilde{g}) < 35$GeV. At higher mass values the gluon bremsstrahlung effect is less important giving corrections of about 20%-30%.

The CERN UA1 collaboration[1] has demonstrated that the experimental study of missing p_T ($\not{p}_T$) events provides us a powerful method in the effort to find evidence for existence of supersymmetric partners of the known particles.

Phenomenological supersymmetric models have been suggested[2-5] as a possible way out of the hierarchy problem. They are designed to explain the relative smallness of the Fermi scale with respect to the Planck scale and they obtain mass values for the super partners of O(1 TeV). Provided that so called R-parity is exactly conserved, the lightest supersymmetric particle is absolutely stable. It is obtained by diagonalizing the mass matrix of the neutral gauginos and higgsinos. We shall assume, however, that lightest supersymmetric particle is the photino. In some supersymmetric models at the tree level gaugino masses are generated by non-singlet goldstino, therefore the gluino and photino get mass values only though radiative corrections and their masses are related via:

$$m_{\tilde{g}}/m_{\tilde{\gamma}} = \frac{3}{8}\frac{\alpha_s}{\alpha} \approx 6 \tag{1}$$

We shall assume that $m_{\tilde{\gamma}} = 0$. Photinos are produced in squark $\tilde{q}$ and gluino $\tilde{g}$ decays and would remain undetected in collider experiments leading to events with large missing p_T.

Present collider data give lower limit on the squark and gluino masses $m_{\tilde{g}}$, $m_{\tilde{q}} > 40$ GeV with a possible light gluino "window" given by the mass range $15\text{GeV} > m_{\tilde{g}} > 3\text{GeV}$[6-7]. These mass domains have been obtained by assuming that the data with $\not{p}_T > 4\sigma$ can be interpreted in terms of standard model background [8] and that the production rates for gluinos and quarks can reliably be obtained by the leading order QCD estimate based on $(2\to 2)$ subprocesses.

In a recent paper with F. Herzog [9], however, we have pointed out that with the very specific experimental cuts applied to the analysis of the missing p_T events the $O(\alpha_s^3)$ contribution given by $(2\to 3)$ subprocesses

$$gg\to \tilde{g}\tilde{g}g \qquad (2a)$$
$$gq\to \tilde{g}\tilde{g}q \qquad (2b)$$
$$q\bar{q}\to \tilde{g}\tilde{g}g \qquad (2c)$$

are enhanced significantly. The effect is partly dynamical [10]. The argument goes as follows.

The cross-sections of $(2\to 2)$ subprocesses $(gg\to gg)$, $(gg\to\tilde{g}\tilde{g})$ and $(qq\to q\bar{q})$ have the simple forms

$$\frac{d\sigma^{gg\to gg}}{dt} = \left(\frac{\pi\alpha_s^2}{s^2}\right)\left(\frac{9}{4}\right)\left(3+\frac{t^2+u^2}{s^2}+\frac{u^2+s^2}{t^2}+\frac{s^2+t^2}{u^2}\right) \qquad (3a)$$

$$\frac{d\sigma^{gg\to gg}}{dt} = \left(\frac{\pi\alpha_s^2}{s^2}\right)\left(\frac{9}{4}\right)\left(1-\frac{ut}{s^2}\right) \qquad (3b)$$

$$\frac{d\sigma^{gg\to q\bar{q}}}{dt} = \left(\frac{\pi\alpha_s^2}{s^2}\right)\left(\frac{1}{6}\right)\left(1-\frac{9\,ut}{4\,s^2}\right) \qquad (3c)$$

At $t=u=-s/2$ we obtain the ratios

$$d\sigma^{gg\to gg} \;:\; d\sigma^{gg\to gg} \;:\; d\sigma^{gg\to q\bar{q}} = 9:1:7/162 \qquad (4)$$

Note that there is strong positive interference for the $gg\to gg$ subprocess while the two other subprocesses have strong negative interference. The cross-section of subprocess $gg\to\tilde{g}\tilde{g}$ is suppressed by an order of magnitude relative to the cross-section of gg- gg subprocess. Furthermore the colour factor of the subprocess $gg \to q\bar{q}$ is smaller than the colour factor of the other two subprocesses with an order of magnitude.

The (2→3) subprocess ($gg \to \tilde{g}\tilde{g}g$) in the pole approximation schematically can be given as

$$d\sigma^{gg \to f\bar{f}g} = d\sigma^{gg \to gg} \otimes P_{g \to f\bar{f}} + P_{g \to gg} \otimes d\sigma^{gg \to f\bar{f}} + \ldots \quad (5)$$

where f denotes quark or gluino and the symbol $\otimes$ indicates that the cross-section of (2→2) subprocess has to be folded with the corresponding Altarelli - Parisi function $P_{g \to f\bar{f}}$ or $P_{g \to gg}$, respectively. In the phase space region where pole dominance as given by the first term of eq.(5) is reasonable approximation we expect large positive interference hence large corrections, while in the phase space region where pole approximation given by the second term of eq.(5) is relevant we expect small corrections. We note that the combined colour factors of the two terms of eq.(5) have the same order of magnitude even for the $gg \to q\bar{q}$ subprocess since the Altarelli - Parisi functions are proportional to the colour factors $P_{g \to gg} \sim C_A = 3$ and $P_{g \to q\bar{q}} \sim T_R = 1/2$.

It is emphasised that the arguments are not valid at high quark and/or gluino mass values.

In the calculation of the missing transverse momentum spectrum of gluino production the main point is this: the experimental cuts applied by the UA1 collaboration favour the kinematical region where the gluino pair production with a hard gluon is resonably well approximated by pole dominance of the $gg \to gg$ subprocess. The transverse momenta of light gluinos produced by the (2->2) subprocess are nearly back-to-back. Since the gluinos are relatively fast the momenta of their decay products ($g \to q\bar{q}\tilde{\gamma}$) are also likely back-to-back. Since the total missing transverse energy is determined by the vector sum of the transverse momenta of the photinos

$$E_T^{miss} = \left| \vec{p}_T^{\,\tilde{\gamma}_1} + \vec{p}_T^{\,\tilde{\gamma}_2} \right|$$

these envents have the tendency to get lower E_T^{miss} values, consequently they fail to pass the large E_T^{miss} cut. However, in the case of (2->3) subprocesses, it is kinematically allowed that the gluinos have momenta in the same hemisphere. In this case the photinos are also often emitted in the same hemisphere therefore it is much easier for the event to pass the large E_T^{miss} cut. Since these configurations are unlikely for the leading order (2->2) processes, the unknown loop corrections can not modify this argument significantly.

This qualitative picture has been justified by calculating explicitely the coss-sections of (2-> 3) subprocesses (2a-2c) with massive squarks and gluinos. In Fig.1 the missing p_T spectrum of gluino pair production is given at various gluino mass values for $p\bar{p}$ collision at $\sqrt{s}$=540GeV energy (with infinitely large squark mass value). We can see that the large p_T^{miss} tail of the (2-> 3) subprocesses (solid lines) is flatter than the p_T^{miss} tail of the leading order contribution. At gluino mass $m_{\tilde{g}}$=15GeV the (2-> 3) contribution is dominant. With increasing the gluino mass the effect becomes less significant.

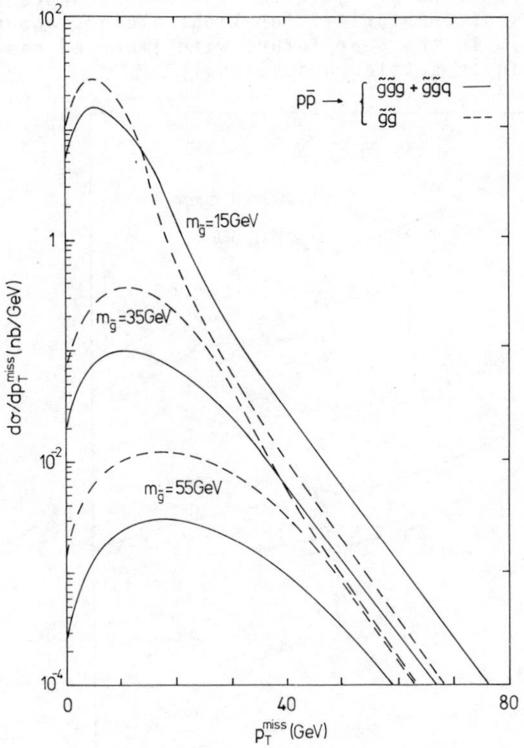

Fig.1: Missing p_T distributions at $\sqrt{s}$=540GeV for gluino pair production. The distributions are shown for differnt gluino masses. The solid lines show distributions for $p\bar{p}\to(\tilde{g}\tilde{g}g + \tilde{g}\tilde{g}q)$, $\tilde{g}\to q\bar{q}\tilde{\gamma}$, whereas the dashed lines show the contributions from $p\bar{p}\to\tilde{g}\tilde{g}$, $\tilde{g}\to q\bar{q}\tilde{\gamma}$.

In Fig.2 we can see the total cross-section values as a function of the gluino mass at $\sqrt{s}$=540GeV. In case of (2→3) subprocesses we have required p_T^{gluon} > 10GeV and that the emitted gluon satisfies Sterman-Weinberg type resolution cuts with respect to the incoming partons (ε=0.2, δ=30°). In the decay of final gluinos we have required that the emitted gluons be harder than $\varepsilon m(\tilde{g})/2$ and have opening angle with the quarks larger than δ. Applying the cuts $p_T > 4\sigma$, $p_T^{jet} > 25$GeV and $p_T^{miss} > \max(4\sigma, 35)$GeV, respectively, the contributions from the one-jet (1j),- two jet(2j)- and three jet(3j)- configurations to the total cross-section (tot) are shown for $p\bar{p} \to \{(\tilde{g}\tilde{g}g + \tilde{g}\tilde{g}q),\ \tilde{g} - q\bar{q}\tilde{\gamma}\} + 2\{\tilde{g}_1\tilde{g}_2,\ \tilde{g}_1 \to q\bar{q}\tilde{\gamma},\ \tilde{g}_2 \to q\bar{q}g\tilde{\gamma}\}$ as solid (—) lines and for $p\bar{p} \to \{\tilde{g}\tilde{g},\ \tilde{g} - q\bar{q}\tilde{\gamma}\}$ as dashed lines (---). With this enhanced production rate "the window of opprtunity" for light gluinos appears to be very unlikely. In the near future with improved new data this question can be settled unambigously.

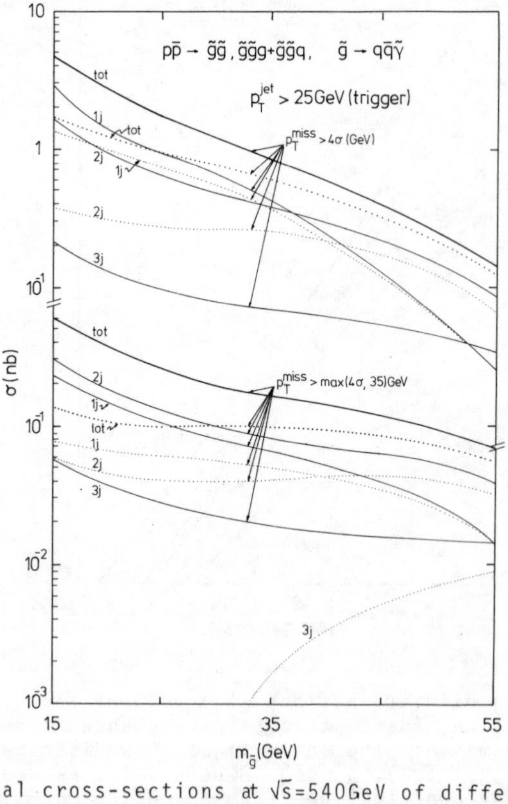

Fig.2: Total cross-sections at $\sqrt{s}$=540GeV of different jet topologies as a function of gluino mass with standard UA1 type cuts.

ACKNOWLEDGEMENT
I thank F. Herzog for an enjoyable collaboration. I also whish to thank V. Barger and F. Halzen for their hospitality.

REFERENCES

1) C. Rubbia, talk given at the Workshop on Proton-Antiproton Collider Physics, Saint-Vincent, Ed. M.Greco, World Scientific(1985). See also H. Hänni, UA2 collaboration, ibid.

2) J. Ellis and H. Kowalski, Phys.Lett.142B(1984)441, Nucl. Phys.B246(1984)189.

3) E. Reya and D. P. Roy, Phys.Rev.Lett.52(1983)867, (E:51(1983)1307), Phys.Rev.Lett.53(1984)881.

4) V. Barger, K. Hagiwara and J. Woodside, Phys.Rev.Lett.53 (1984)661, V. Barger and W.Y. Keung, Phys.Lett.145B(1984) 147, A.R. Allen, E.W.N. Glover and A.D. Martin, Phys. Lett.146B(1984)247.

5) R.M. Barnett, H.E. Haber and G.L. Kane, Phys.Rev.Lett. 54(1985)1983.

6) M.J. Herrero, L.E.Ibanez, C. Lopez and F.J. Yndurain, Phys.Lett.132B(1983)199.

7) A. DeRujula and R. Petronzio, CERN-TH 4070/84(1984).

8) See e.g. S.D. Ellis, R. Kleiss and W.J. Stirling, CERN-TH-4144-85.

9) F. Herzog and Z. Kunszt, Phys.Lett.157B(1985)430.

10) Z. Kunszt and E. Pietarinen, Nucl.Phys. B164(1980)45.

ASP: A SEARCH FOR SINGLE PHOTON EVENTS AT PEP*

R. HOLLEBEEK

Stanford Linear Accelerator Center
Stanford University, Stanford, California 94305

Abstract

The ASP search for events with a single photon and no other observed particles is reviewed. New results on the number of neutrino generations and limits on selectron, photino, squark and gluino masses are presented.

Introduction

During the past year, two experiments have been carried out at PEP to search for events with a photon and large missing transverse momentum . These event searches are particularly sensitive to contributions from supersymmetric photino production but can also be used to search for additional neutrino generations, supersymmetric weak charged currents or the production of any other neutral particle whose interactions in matter are of the order of the weak interaction. The MAC collaboration began modifications of their apparatus during the summer of 1983 to be able to detect photon events with no other observed charged or neutral particles. At the same time, a new experiment, ASP, was approved for installation at the PEP ring to do a high sensitivity search for such events. At the time of the New Particles Conference, the status of the ASP detector and its potential were discussed. In this paper are included those discussions and recent limits derived from the data analyzed during the summer of 1985.

Single Photon Sources

In the standard model of weak and electromagnetic interactions, events with a single photon and no other observed particles will be produced by radiative corrections to the production of neutrino pairs:

$$\sigma(e^+e^- \to \gamma\nu\bar{\nu}) \sim \alpha G_F^2 s \left(1 + \frac{N_\nu}{4}\right) \quad .$$

Since this cross section is of order αG_F^2 , one might think that it is too small to measure, but in fact, the total cross section for reasonable assumptions about the photon acceptance is a few times 10^{-2} pb at PEP and is therefore detectable.

* Work supported by the Department of Energy, contract DE – AC03 – 76SF00515.

This process receives contributions from the weak neutral currents (Fig. 1(a)) proportional to the number of neutrino generations and from the production of electron neutrinos through the charged weak currents (Fig. 1(b)). Because of the sensitivity of the cross section to the presence of all generations of light neutrinos, this process has been suggested[1] as a means of counting the number of different types of neutrinos and hence placing limits on the number of lepton generations. Unlike many other methods of counting generations, this technique has the advantage of being insensitive to the masses of the associated charged leptons which are observed to increase rapidly with generation number in our present examples.

In addition to the standard model sources of single photon events, many models of new physics contain particles which can be stable and which interact in matter with a cross section which is of the order of magnitude of the weak cross section. If the cross section for production of events containing only these particles is known, then the single photon rate can be calculated from the radiative corrections to that cross section. For example, if σ_0 is the cross section for the production of such a state, then the single photon rate is given by

$$\frac{d^2\sigma}{dx_\gamma d\cos\theta_\gamma} = \frac{2\alpha}{\pi}\frac{1}{x_\gamma}\frac{1}{\sin^2\theta_\gamma}\sigma_0(s')$$

where $s' = s(1 - x_\gamma)$ is the reduced center of mass energy squared.

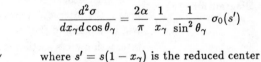

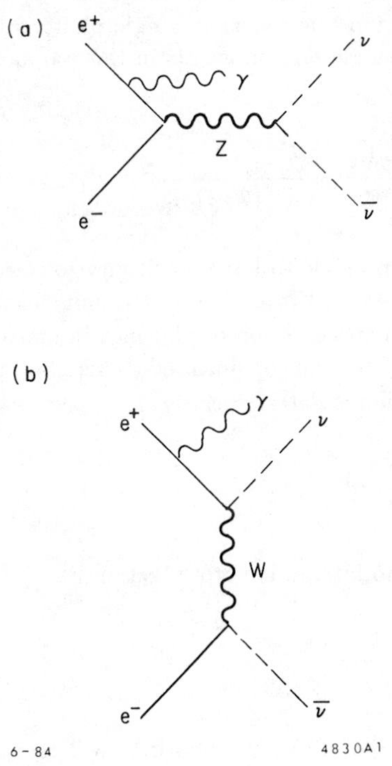

Fig. 1. Single photon contributions from the weak (a) neutral and (b) charged currents.

Supersymmetry is an example of a model of new physics in which there are many possibilities for events which contain only weakly interacting neutral particles. Final states which might behave this way are photino pairs, zino pairs, or sneutrino pairs. Because of a conserved quantum number (R-parity), the lightest supersymmetric particle would be stable, and cosmological arguments indicate that it is probably neutral.[2,3] Possible candidates for the lightest neutral sparticle are the photino ($\tilde{\gamma}$), the neutral shiggs ($\tilde{H}$) and the gravitino ($\tilde{G}$). Since these are all fermions, they may be protected by chiral symmetry from attaining a large mass due to supersymmetry breaking. Of course, the sneutrino ($\tilde{\nu}$), although it is a boson may also be the lightest neutral supersymmetric particle.

As pointed out by Fayet and others,[4,5] a search for single photon states is a particularly sensitive way of testing supersymmetry if the photino is light. The radiative correction to photino pair production is

$$\sigma(e^+e^- \to \gamma\tilde{\gamma}\tilde{\gamma}) \sim \alpha^3 \frac{s}{m_{\tilde{e}}^4}$$

where $m_{\tilde{e}}$ is the mass of the spin zero partner of the electron (selectron). Since the coupling constant of the supersymmetric particles is still α, the calculation proceeds as in QED except for spin factors and masses (see Fig. 2). If we compare the supersymmetric source of single photons to the cross section for single photon events in the standard model using

$$\alpha G_F^2 \sim \frac{\alpha g_{WS}^4}{m_W^4} \sim \frac{\alpha^3}{\sin^4(\theta_{WS}) m_W^4} \quad ,$$

we can see that it is possible to have sensitivity to selectron masses of order m_W. If the selectron is sufficiently heavy, the interaction cross section of photinos in matter will be small and they will not be detected. This photino interaction cross section is determined by[4]

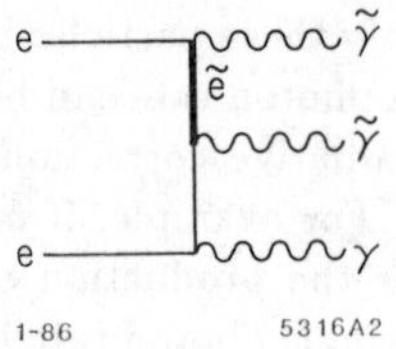

Fig. 2. Feynman diagram for $e^+e^- \to \gamma\tilde{\gamma}\tilde{\gamma}$.

$$\sigma(e\tilde{\gamma} \to e\tilde{\gamma}) = \frac{8\pi}{3}\, \alpha^2\, \frac{s}{m_{\tilde{e}}^4}$$

so that the ratio of the photino and neutrino interaction cross section is

$$\frac{\sigma_{\tilde{\gamma}}}{\sigma_\nu} \sim 50 \left(\frac{40 \text{ GeV}}{m_{\tilde{e}}^4} \right)$$

and thus if the selectron mass is a few tens of GeV, the photinos will escape detection.

Because the detected final states for radiative neutrino and photino pair production are the same and because the Feynman diagrams of Fig. 1(b) and Fig. 2 are similar, these two processes have very similar differential cross sections. Except for the addition of small corrections for photons radiated from the W

or selectron, the cross sections in the local limit ($s \ll m^2$) for photons with $x = 2E/\sqrt{s}$ are

$$\frac{d^2\sigma}{dx\,dy} = K\,\frac{1}{x}\,\frac{1}{1-y^2}\,s(1-x)\left[\left(1-\frac{x}{2}\right)^2 + \frac{x}{4}\,y^2\right]$$

where

$$K_{\gamma\nu\bar{\nu}} = \frac{G_F^2\alpha}{6\pi^2}\left[N_\nu\left(g_V^2+g_A^2\right)+2\left(g_V+g_A+1\right)\right] \quad ,$$

$$K_{\gamma\tilde{\gamma}\tilde{\gamma}} = \frac{4\alpha^3}{3m_{\tilde{e}}^4}$$

and $y = \cos\theta_\gamma$. A limit on the single photon cross section probes the sum of the neutrino pair and the supersymmetric sources, so one might think of the generation counting process as a "background" to the search for new physics. In this case, for $m_{\tilde{e}} > \sqrt{2}\,m_W$ the weak cross section for $\gamma\nu\bar{\nu}$ dominates and the only way to separate the two is a careful study of the $\sqrt{s}$ dependence since the weak cross section has a resonance and the supersymmetric cross section does not. The possibility of new physics is also a background for the generation counting experiment. Because of this, we will not be able to interpret the results of SLC and LEP studies of the Z width without using the lower energy (PEP) data.

If the sneutrino is the lightest supersymmetric particle,[6] single photon events will be produced by the standard weak neutral current and the supersymmetric charged currents (see Fig. 3). In the neutral current diagram, the size of the cross section is determined by the number of light sneutrinos. Due to spin factors, each sneutrino generation adds to the cross section an amount equivalent to half that of a neutrino generation. For the charged currents, the cross section is determined by the masses of the charged weak fermions $\tilde{W}^\pm$ which may be different and by the mixing angles[7] between them and the higgsinos. It is still possible, however, to place a model independent limit on the cross section for such processes since the photon spectrum is insensitive to these parameters. The interpretation of the cross section in terms of masses will depend on the mixing angle assumptions.

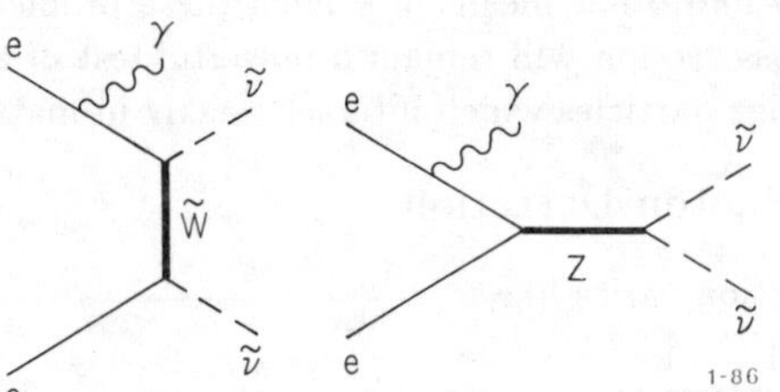

Fig. 3. Feynman diagrams for $e^+e^- \to \tilde{\nu}\tilde{\nu}\gamma$.

Supersymmetric theories postulate a new symmetry of nature between bosons and fermions and hence predict many new particles. Within this theory, every particle has a supersymmetric partner with opposite spin statistics and since no pair of particles in our current particle table is known to be a particle-sparticle pair, there are as many new particles as old ones! Clearly the theoretical community must be highly motivated to propose such an idea . In order to understand this, we should look carefully at the standard model of weak and electromagnetic interactions. Despite the obvious successes of the Weinberg-Salam-Glashow model, there is one outstanding problem related to the Higgs mass. The Higgs boson is crucial within the theory for generating the masses of the weak bosons, yet many particle searches have failed to locate such a particle leading to speculations that its mass may be large. Theoretically, the mass is very large since higher order corrections cause it to be ultraviolet divergent. There are two possible solutions to the problem: give the Higgs internal structure so that large corrections to its mass are cut off by a form factor, or introduce new interactions which cancel the higher order terms. Supersymmetry is an example of the latter approach because for every boson loop contribution to the Higgs mass, there is an opposite sign contribution from a partner fermion loop. This cancellation would be complete if the masses of the fermions and bosons were degenerate, but in general the correction to the Higgs mass has the form[8]

$$\delta_{M_H^2} \sim \frac{g^2}{16\pi^2}\left(m_B^2 - m_F^2\right) \quad .$$

Note that if the superpartners are too heavy relative to normal matter, supersymmetry can no longer be looked on as a solution to the Higgs mass problem.

While supersymmetry is of great interest at the moment because of the way in which it solves divergence problems in the standard model and within quantum gravity, we may in the future find other means of solving these problems. In this case, the single photon cross section will remain a powerful test of any model of new physics which contains particles which interact weakly in matter.

Single Photon Detection

Because the radiative cross section varies like

$$\frac{1}{x_\gamma}\,\frac{1}{\sin^2\theta_\gamma} \quad ,$$

it is important to detect photons at as small an energy and angle relative to the beam energy and angle as possible. In the energy dependence, the integrated rate will vary as $\ln x_{min}$ so that the difference between a 2 and a .25 GeV threshold at PEP will be a factor of two in the observable rate. Similar factors can be obtained from the $\sin\theta$ dependence. This is illustrated in Fig. 4 which shows Monte Carlo data folded with efficiency for events within the ASP detector as a function of the energy and angle of the detected photon. Note the rapid increase in the cross section for low energy and angles.

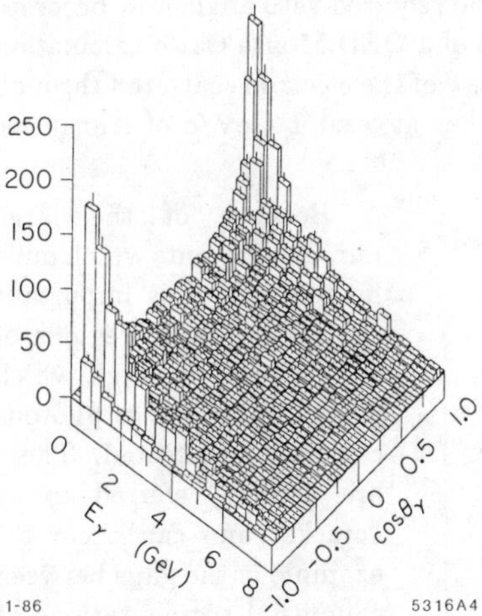

Fig. 4. Monte Carlo data in the ASP acceptance $\sigma(ee \to \gamma\nu\bar{\nu})$.

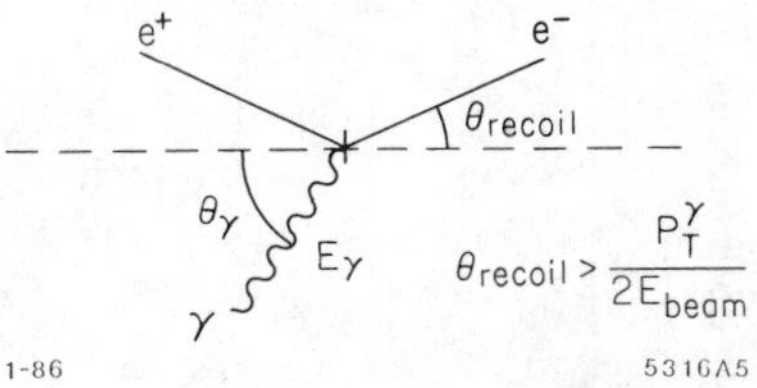

Fig. 5. Kinematics used to eliminate QED backgrounds.

In addition to having as large an acceptance as possible, it is also necessary to show that the detected events are accompanied only by neutrinos or other weakly interacting particles. This is particularly important because of the presence of radiative corrections to QED production of electron, photon, muon, and tau pairs. The cross section for $ee \to \gamma ee$ is the largest background and requires a rejection of $\sim 10^{-4}$ to reach a sensitivity to neutrino pair production. Fortunately this level of rejection can be achieved by using the kinematics of the three-body final state. As shown in Fig. 5, the transverse momentum of the detected photon relative to the beam line must be balanced by the electron pair. Thus for a detected photon with transverse momentum p_t^γ, at least one of the other particles will be at an angle larger than

$$\sin\theta_{recoil}^{min} = \frac{E_t^\gamma}{2E_{beam} - E_\gamma} \quad .$$

As an example, for a photon with 1 GeV/c of transverse momentum, detection of additional particles must extend to a veto angle less than 36 mrad at PEP. In practice, the veto angle for QED processes can be almost a factor of two larger because for soft photons the QED matrix elements are dominated by the case where the photon tends to be balanced by only one of the other particles. This results in a decrease of the cross section between the above recoil angle and

$$\theta_{recoil}^{min} \sim \frac{E_t^\gamma}{E_{beam}} \quad .$$

For a 1 GeV/c transverse momentum cut , the required veto angle now becomes roughly 69 mrad. Figure 6 shows the results of a QED Monte Carlo calculation of the $ee\gamma$ cross section as a function of the $\cos\theta$ of the electron scattered through the maximum θ when the detected photon has at least 1 GeV/c of transverse momentum.

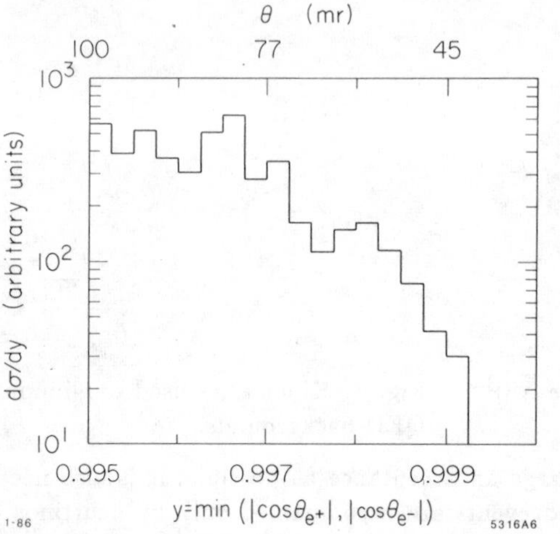

Fig. 6. QED Monte Carlo calculation of $\sigma(ee\gamma)$ versus $\cos\theta_e$.

Because of the large number of events which must be rejected, it is important that there be no regions of the detector through which charged particles or photons can pass undetected. These are usually referred to as "cracks" and can occur for example in the gaps between azimuthal segmentations of a calorimeter or in the transition region between calorimeters at large and small values of theta. Special care must be taken to either cover all such areas with detection equipment or to have a design which has no cracks. Covering the cracks can sometimes be difficult since the background from $\gamma\gamma\gamma$ events requires that the detector be thick enough so the non-conversion probability of photons is small.

ASP – A Detector Optimized for Single Photons

The ASP detector was designed specifically for the single photon search and was optimized to give both a large acceptance for the photons and a high degree of certainty that any additional particles in the event at small angles would be detected. As discussed before, good photon acceptance requires small angle and low energy detection. The ASP detector is constructed from a lead-glass array which can detect photons down to 20° and can veto other particles down to $\sim 10°$. Lead-glass is well suited to low energy photon detection both because of its good intrinsic resolution, and also because a phototube based system has much less electronic noise than a proportional wire system which allows a clean trigger at low energies. The full apparatus is shown in Fig. 7. The central region is the lead-glass array, and the forward detectors are arranged so that scintillators and calorimeters cover the region from 30° to 100 mrad, and four planes of drift chambers and a calorimeter cover from 100 mrad to 21 mrad.

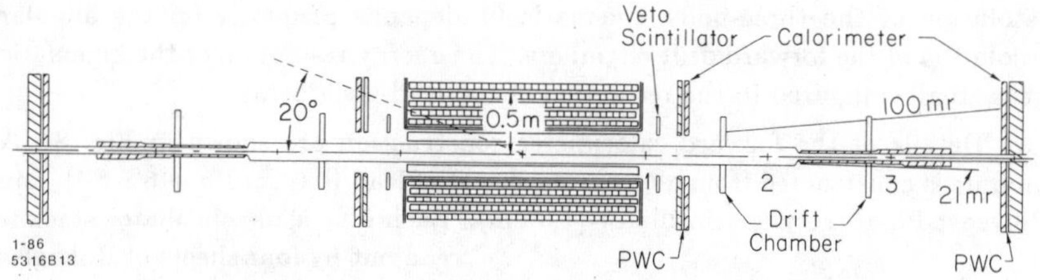

Fig. 7. View along the beam axis of the ASP detector. The apparatus is approximately 8 meters long and 1 meter wide.

ASP – Forward Detectors

The forward region contains a tungsten mask which sits in a special indentation in the vacuum chamber. This indentation allows the mask to cover the region between 12 and 20 mrad and thus shield the central array from synchrotron backgrounds. The indentation is also used to minimize the amount of material in front of particles below about 27 mrad. This window can be used to verify that the QED production of $ee\gamma$ events behaves as expected near the kinematic limit for a 0.75 GeV photon transverse momentum. The materials in the vacuum chamber are summarized in Table 1.

Table 1. ASP Vacuum Chamber Materials

Angle (mrad)	Material
> 100	120 mil AL
50 – 100	100 mil AL
45 – 50	Al-stainless weld
30 – 45	stainless flange (3.5 X_0)
27 – 30	60 mil stainless
21 – 27	60 mil stainless @ 30°

Drift chambers are placed in the region in front of the low angle calorimeters in order to measure the exit angles of charged particles in three-body QED final states. By measuring the angles and energies of forward electron pairs in $ee\gamma$ events, the properties of the photon can be determined by a constrained kinematic fit. The results of such a fit can be used to determine the efficiency for photon reconstruction as well as the resolution for all of the photon parameters. Because angles are in general measured much more precisely than energies, the resolution of the three-body kinematic fit depends primarily on the angular resolution of the forward drift chambers. The energy resolution of the kinematic fit is small compared to the resolution of the lead-glass array.

Details of the forward calorimeter construction are shown in Fig. 8. A module is constructed from alternating sheets of lead (0.6 cm Pb + 6% Sb), and Polycast PS-10 acrylic scintillator (1.3 cm). Each $6X_0$ lead-scintillator stack is read out by four sheets of Rohaglas GS1919 wavelength shifter viewed by an Amperex S2212A phototube. The modules are constructed in left and right halves to allow easy assembly around the beam pipe, but care has been taken that there are no gaps in the coverage of the modules. As shown in Fig. 8, this is accomplished by having a 4 cm overlap of the joint in the front and back halves of a module. Two modules are used at 1.5 meters and three are used at 4 meters. The larger number of radiation lengths at small angles is needed to assure that there is no background from QED production of $\gamma\gamma\gamma$ events with non-conversion of the two forward photons. Between modules of each calorimeter are proportional wire chambers used to determine the position of showers in the calorimeter.

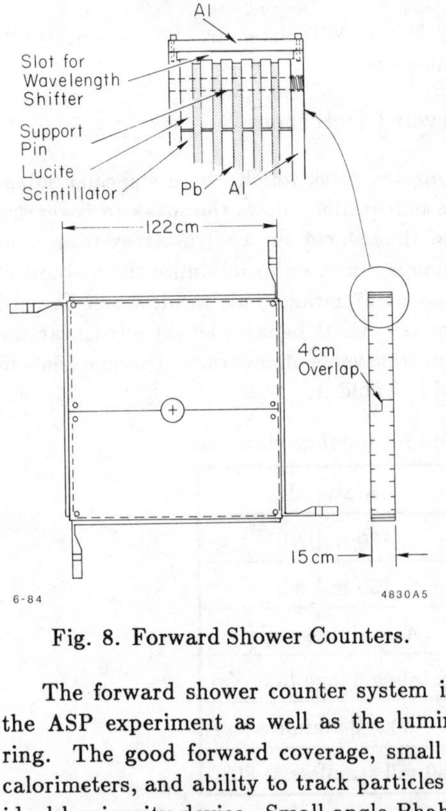

Fig. 8. Forward Shower Counters.

The forward shower counter system is a veto and calibration system for the ASP experiment as well as the luminosity monitor for the PEP storage ring. The good forward coverage, small amount of material in front of the calorimeters, and ability to track particles in the small angle region make it an ideal luminosity device. Small angle Bhabha events can also be used to verify the veto performance of the device. Figure 9(a) shows the mean energy of such events as a function of the theta angle with respect to the beam. Coverage extends to angles of about 20 mrad with good uniformity. At approximately 0.12 rad one can see the effect of the transition from showers which are contained in the modules nearest the central calorimeter to those which are contained in the calorimeters at 4 meters. The behavior at .04 rad is due to the presence of a vacuum flange. Figure 9(b) shows the response of the system as a function of the azimuthal angle and illustrates that despite the fact that the modules are constructed in two halves, there is no gap in the coverage. The energy resolution of the forward system is $25\%/\sqrt{E}$ when averaged over the region used as a luminosity region (50 mrad $< \theta <$ 90 mrad). The angular difference between the two electrons in the forward Bhabha events can be used to determine the angular resolution of the system. (See Fig. 10). The total integrated luminosity for the search is determined to be 68.7 pb^{-1}. A comparison of the measured and calculated rates for $\sigma(ee \rightarrow ee)$ is shown in Fig. 11.

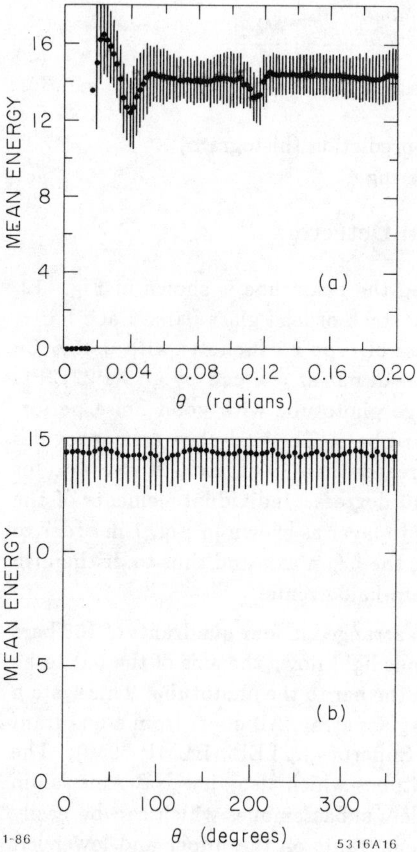

Fig. 9. Mean energy of Bhabha events in the forward calorimeters as a function of (a) θ and (b) ϕ.

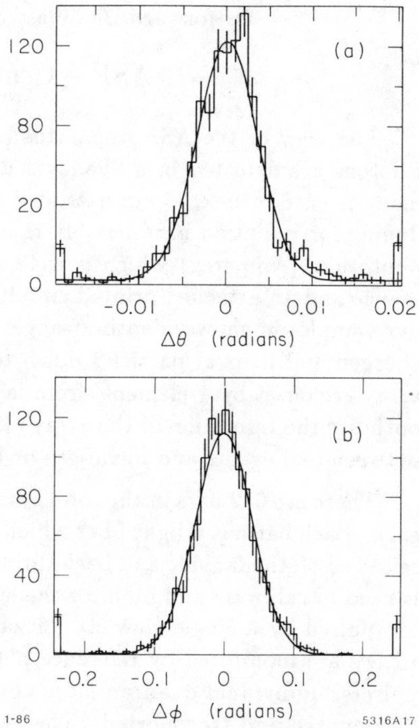

Fig. 10. The $\Delta\theta$ and $\Delta\phi$ for Bhabha events in the forward calorimeters. $\sigma_\theta = 2.5$ mrad $\sigma_\phi = 28$ mrad.

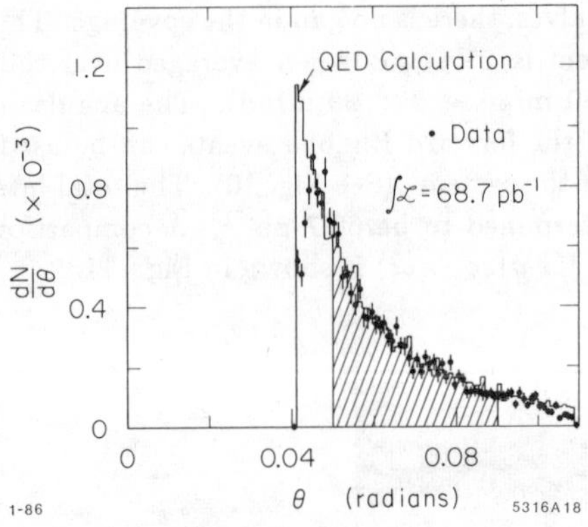

Fig. 11. Data and QED prediction (histogram) for forward Bhabha scattering.

ASP – Central Detectors

The view of the ASP apparatus along the beam line is shown in Fig. 12. Photons are detected in a five layer deep stack of lead glass bars. Each bar is made from 6 × 6 × 75 cm extruded glass of type F2 (Schott) with 0.35% Ce doping for radiation hardness. Bars are read out at one end by an XP2212PC phototube (Amperex) which is a 12 stage phototube with good noise performance and an attached printed circuit card base. The lead-glass array provides five samples in shower depth at angles greater than 30 degrees. It can veto for charged and neutral particles down to 10 degrees. Individual elements of the array are offset by $\frac{1}{2}$ elements from layer to layer as shown in Fig. 7 in order to optimize the resolution of the array along the beam axis and thus to distinguish between real events and beam-gas or beam-halo events.

There are 632 bars in the total system arranged in four quadrants of 158 bars each. Each bar has a light fiber which sends light down the axis of the bar to be reflected off the far side and back through the bar to the phototube. This system is used to calibrate and monitor the lead-glass array. All fibers from a quadrant are pulsed by a single Hewlett Packard Superbright LED (HLMP-3950). The LED's are monitored by reference phototubes which also view NaI-Americium pulsers. Individual quadrants are complete subassemblies which can be easily dismounted and transported. The two quadrants on the upper and lower left are mounted together on rails as are the two quadrants on the right. The entire central apparatus can be split apart with a hydrolic drive system to allow easy access to detector elements around the beam pipe and to protect the lead-glass from excess radiation exposure during injection into the storage ring. Each layer of lead-glass is followed by proportional wire chambers constructed from

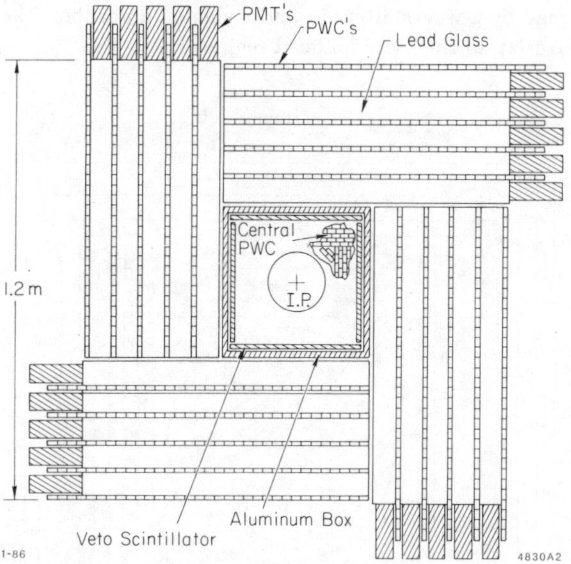

Fig. 12. View along the beam axis of the ASP central detector.

aluminum extrusions. The extrusions are eight-cell closed structures with 1.23 × 2.36 × 200 cm channels with 0.18 cm walls. The wires are 48μ gold-plated tungsten, and four extrusions are used to form a PWC plane. The PWC planes provide photon pattern recognition in the xy plane.

Inside the photon calorimeter are two systems, each of which is designed to adequately reject events with accompanying charged particles and to distinguish between electrons and photons. The innermost system (central tracker, Fig. 13) is made from .9 in × .4 in × 88 in aluminum tubes which are thinned by etching to a wall thickness of .012 in. The wire used is Stablohm 800 and the tubes are read at each end so that the coordinate along the wire can be determined by charge division. The extrusions are glued together to form two L shaped modules which are mounted on a Hexcell backplate and then assembled around the beam pipe. The tubes are arranged so that radial lines from the beam axis do not pass through tube walls, and extra tubes are added at the corners to ensure that charged particles pass through at least five layers. The use of the tube design is intended to ensure that the chamber does not have correlated inefficiencies which would result for example from wires which draw current and the resulting bad field configurations which can occur in open geometries.

Surrounding the central tracker is the second veto system: a 2 cm thick scintillator. Each of the four sides is made from two sheets of 33.5 × 225 × 1 cm Kiowa scintillators. The two sheets could be read separately but at the moment are read by a waveshifter bar and a single phototube. The edges of the scintillators overlap so there are no dead regions.

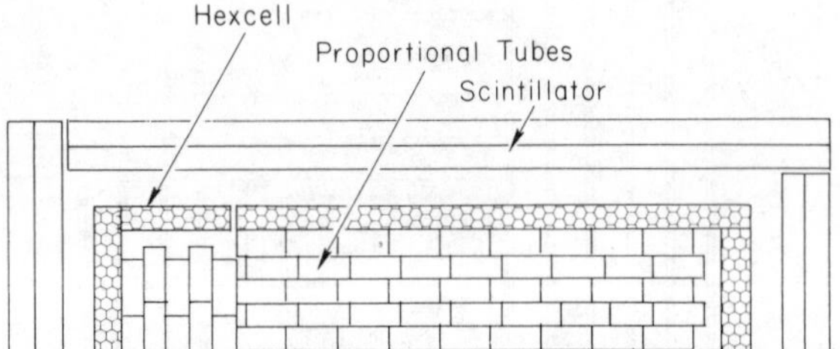

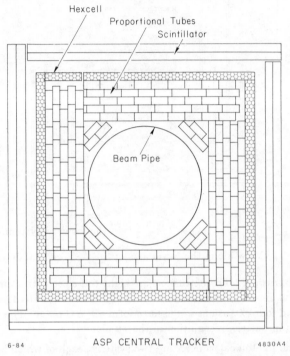

Fig. 13. Central tracker and veto scintillators.

ASP Trigger

Triggers for the detector consist of two types: monitor triggers and single photon triggers. The single photon triggers are based mainly on analog sums[9] of the pulse heights found in the total lead-glass array, individual quadrants, combinations of layers, or groups of eight bars in a layer. The signals from individual glass bars are sent through passive transformer splitters. One of the signals after the splitter goes to a SHAM-BADC system for the primary readout of the calorimeter. The other signal goes both to the trigger system and a second independent ADC system which is used to verify that missing signals in the central calorimeter are not the result of electronics failures. (These transformers are also used to break the ground loops between the detector and the summing circuits and produce a lower achievable threshold.) The highest threshold single photon trigger is formed from the analog sum of all 632 bars of lead glass. The threshold for this trigger is however only 1.6 GeV which for photons concentrated around 30° translates to a p_t threshold of 0.8 GeV. The distribution of trigger energies from this trigger on a typical run is shown in Fig. 14(a). The lowest trigger threshold is obtained by requiring fewer than three central veto scintillators, $E_{tot} > 0.4$ GeV with at least 0.15 GeV in layers 2 through 5, and energy in the forward system either less than 1 GeV or more than 7 GeV (see Fig. 14(b)). The threshold for this trigger is about 700 MeV or transverse momenta of 350 MeV at 30 degrees.

Monitor triggers consist of randoms, cosmics, forward luminosity (Bhabha) triggers and a special trigger for $ee\gamma$ and $\gamma\gamma\gamma$ events. The random triggers are used to determine the level of occupancy in each detector system. These occupancies are in turn used to determine the efficiency of each veto requirement used in the single photon analysis. Typical occupancies are 1% for $E > 40$ MeV in the lead glass and 5% for $E > 100$ MeV in the forward system. Cosmic ray triggers are formed by the coincidence of two central veto scintillators in a narrow gate 15 nsec prior to beam crossing. This yields a sample of minimum ionizing tracks roughly in-time with the beam crossing which can be used to monitor the calibration of the lead-glass and also the response of the central system to minimum ionizing tracks. The forward Bhabha triggers are used to determine the luminosity for the search, and finally the $ee\gamma$ triggers are used to provide a sample of single photon events which have all of the same characteristics as signal events except that they have two forward tracks. An example of such an event is shown in Fig. 15.

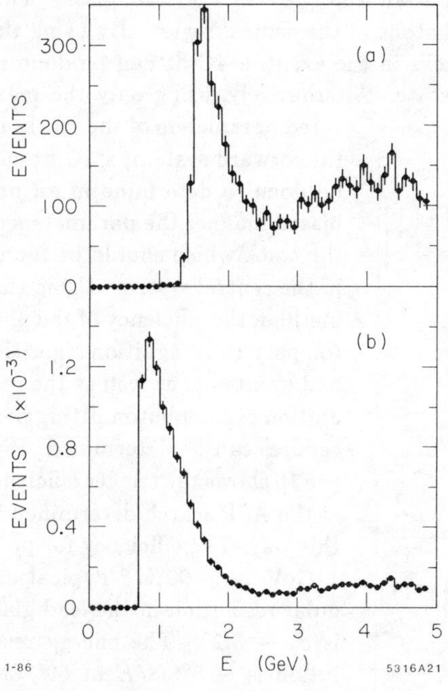

Fig. 14. Number of events versus energy for (a) total energy trigger (b) lowest threshold trigger – ASP.

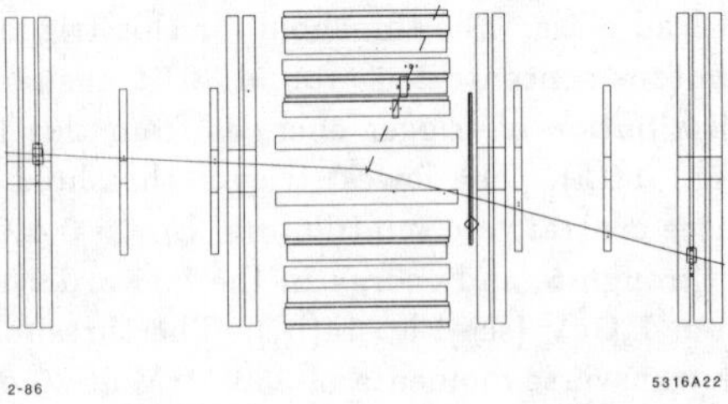

Fig. 15. Typical $ee\gamma$ event used to provide a source of constrained single photons. The vertical scale has been increased by a factor of three.

The $ee\gamma$ Sample

The $ee\gamma$ trigger provides a sample of over 130,000 triggers with two forward tracks and an energy deposit greater than 200 MeV in the lead glass. This sample contains both electrons and photons in the central region. By using the measured parameters of all of the tracks in the event, a 4C fit can be done to the hypothesis of a three-body final state. Alternatively, using only the measured parameters of the tracks in the forward system, a 1C fit can be done to determine in an unbiased manner the parameters of the track which should be found in the central system. Using this method, the efficiency of the photon pattern recognition algorithm and event cuts as well as the resolution of the photon fitting procedures can be determined. Figure 16 shows the trigger efficiency of the ASP search determined in this way. The efficiency for $p_t^\gamma >$ 1 GeV is $> 99\%$. Typical angular resolution in the lead glass is $\sigma_\theta \sim 3.2°$. The energy resolution is $\sim 8\%/\sqrt{E}$ at 60° and $\sim 15\%/\sqrt{E}$ at 20 – 25° without correction for energy leakage into the forward calorimeters. The same procedure can be applied to determine the efficiency of all photon pattern recognition cuts. This analysis efficiency is shown in Fig. 17 as a function of the photon energy.

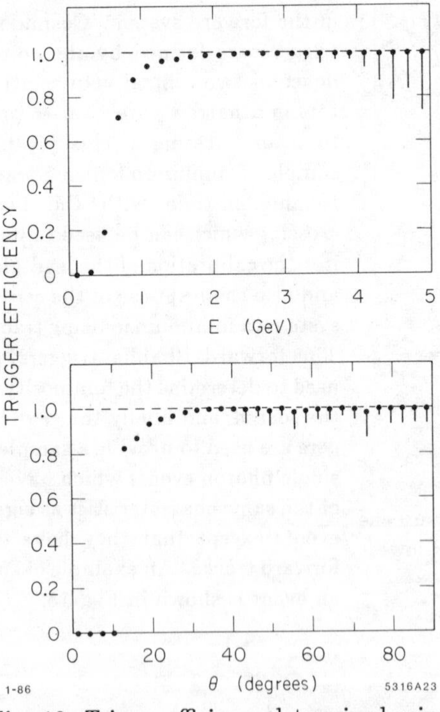

Fig. 16. Trigger efficiency determined using constrained $ee\gamma$ events as a function of (a) Energy and (b) θ. The error bars represent 95% CL limits.

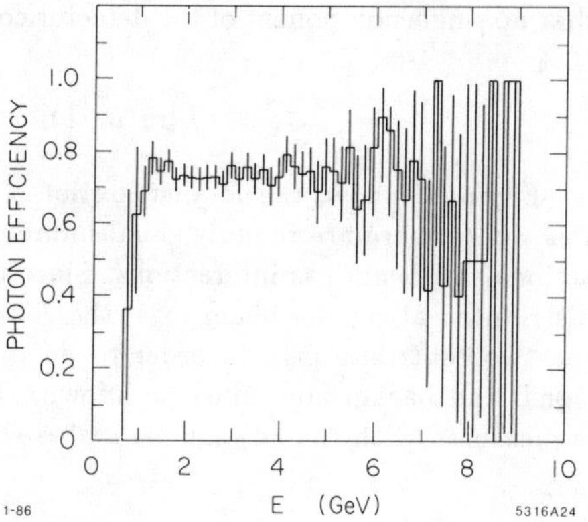

Fig. 17. Efficiency of all cuts applied to the photon candidate versus energy.

Event Selection

Photon candidates are required to have a cluster of lead-glass bars whose pattern (i.e. which bars are above threshold) is consistent with photon patterns determined from the $ee\gamma$ sample. The time of the lead-glass total energy sum signal (relative to the beam crossing time) as well as the time of each layer of the lead-glass is used to form a time for the event. This time is required to be within $\pm 3\sigma_T$ of the known beam crossing time. The resolution is 2.4 ns at 1 GeV and slightly better at higher energies. Candidate showers are fit to a straight line in the XY plane and the XZ or YZ planes. The projected distance of closest approach to the beam axis in XY is required to be less than 20cm to eliminate cosmic rays, and a value is extracted for R_0, the signed projected distance of closest approach to the beam collision point along the beam axis in the XZ or YZ plane. Photon showers are distinguished from other energy deposits by the loose requirements that the width in each layer be consistent with a photon shower and that the ratio of energy deposited in the front half of the shower to that in the back half be less than 0.5. The average efficiency of all of these cuts is measured to be 75% with little variation in E_γ and θ_γ. The majority of the inefficiency occurs due to reconstruction inefficiency in the pattern recognition software at azimuths where showers span two lead-glass quadrants.

In addition to having a valid photon candidate, an event must have no other charged or neutral particles visible in the detector. The ability to veto against events which do have additional particles depends crucially on the electronic noise levels and occupancies of the components of the detector. Random triggers are used to determine the efficiency for the veto cuts, and $ee\gamma$ events and $\gamma\gamma$ events are used to study occupancies which are correlated to the presence of a photon such as backsplash from the central calorimeter into the central veto scintillators and tracker and leakage into the forward shower modules. The efficiency for all veto cuts is determined to be 60% with the cuts arranged so that no single component of the detector contributes an inefficiency greater than 10%.

The R_0 Distribution

Figure 18 shows the R_0 distribution of events at an early stage of the analysis where there are roughly equal numbers of events coming from QED interactions and beam-gas interactions. Since the latter are to first order uniformly distributed along the beam axis, the R_0 distribution can be used to separate the two contributions. In order to do this, the shape of the distribution for signal and background must be known. The R_0 distribution for signal events is measured with the $ee\gamma$ sample and is shown in Fig. 19(a). The resolution is

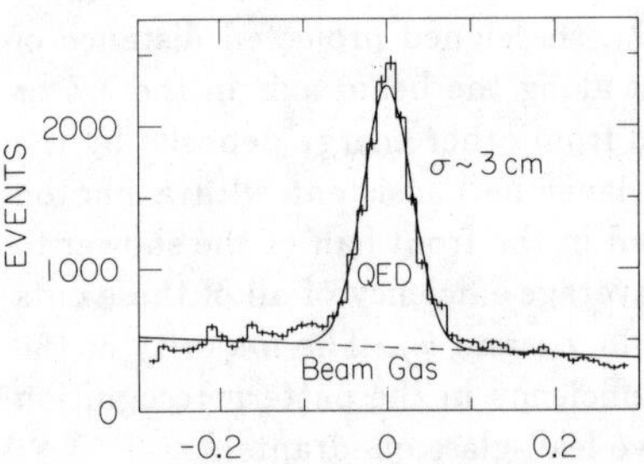

Fig. 18. The R_0 distribution for a mixture of QED and beam-gas interactions.

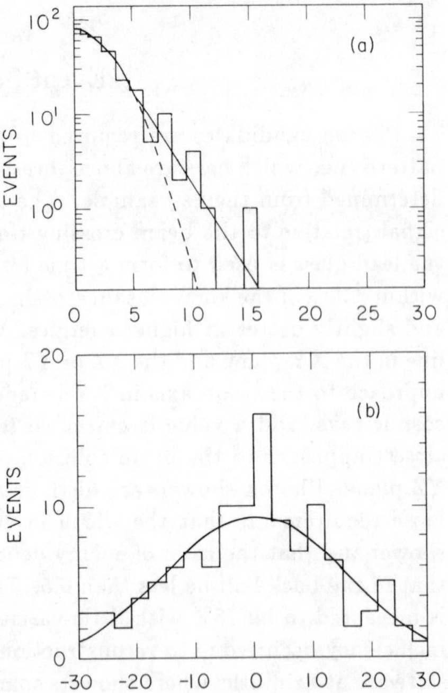

Fig. 19. (a) The R_0 distribution of photons from $ee\gamma$ events. The line is a Gaussian with $\sigma = 3$ cm. The non-Gaussian contributions are estimated by an exponential tail starting at $R_0 = 6$ cm. (b) The distribution of background events with $p_t^\gamma > 0.6$ GeV/c. The line is a Gaussian with $\sigma = 12$ cm.

$\sigma = 3$ cm with a small non-Gaussian tail which is approximated by an exponential. The resolution is found to be independent of the transverse momentum of the photon. Electrons have slightly better resolution ($\sigma \sim 2.9$ cm) than photons due to their earlier shower development. The background from beam-gas interactions is observed to be flat in R_0 before the application of several photon pattern cuts that are biased to accept showers from $R_0 < 30$ cm. The shape of the final background is measured by relaxing cuts other than these pattern cuts. (See Fig. 19(b).)

Final Event Sample

Three events with single photon energies consistent with the beam energy were observed in the data sample. These are interpreted as $ee \to \gamma\gamma$ events in which one photon escapes due to non-conversion. A study of observed $\gamma\gamma$ events predicts 1.5 single photons from this source. The requirement $E_\gamma < 12$ GeV determined from an analysis of ee and $\gamma\gamma$ final states eliminates this background with negligible loss of signal acceptance. The final event sample is shown in Fig. 20 for those events with transverse momentum greater than 0.5 GeV/c and polar angle greater than 20 degrees. For the present analysis, we determine a limit using only those events with $p_t^\gamma > 1$ GeV/c. Several methods of using tighter cuts on the identification of candidate energy deposits as photon showers are being studied and should eventually allow the use of the lower energy data to improve the sensitivity of the search.

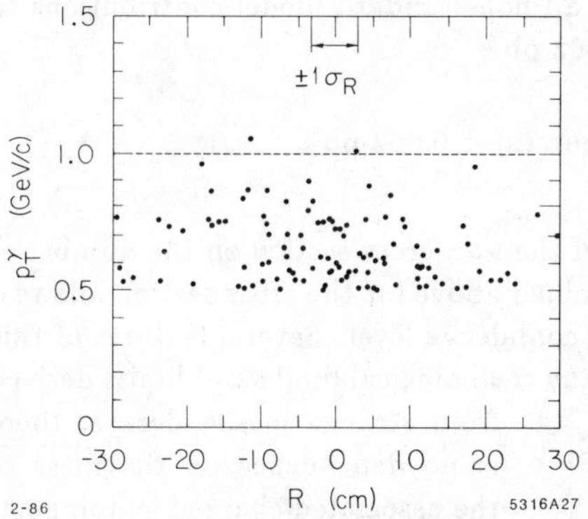

Fig. 20. Final sample of single photon candidates with $p_t^\gamma > 0.5$ GeV/c and $\theta_\gamma > 20°$.

To obtain the best estimate of the possible number of signal (S) and background (B) events, a maximum-likelihood fit is done to the measured distributions of signal and background events in R_0. For a given true number of signal and background events, the confidence level of this experiment is computed by Monte Carlo as the fraction of equivalent experiments which would estimate a value larger than S. The 90% and 95% confidence level upper limits for the observed distributions are 2.9 and 3.9 events respectively.

ASP Limits

The upper limit of 2.9 events together with the measured luminosity and photon event efficiencies implies that the 90% CL upper limit for the sum of all contributions to the single photon cross section is

$$\sigma(ee \to \gamma + \text{weakly interacting particles}) < 0.094 \text{ pb}$$

for a photon acceptance defined by $E_\gamma < 12$ GeV, $p_t^\gamma > 1.0$ GeV/c and $\theta_\gamma > 20°$. (Note that this limit is actually more stringent than the MAC limit of 57 fb because the photon acceptance of the ASP apparatus is much larger.) Since the detection efficiency for photons is nearly constant over the signal region in E_γ and θ_γ, the extension of the limit to processes in which the photon energy or angular distribution differs from $\gamma\nu\bar{\nu}$ or $\gamma\tilde{\gamma}\tilde{\gamma}$ is straightforward. The cross section for radiative pair production of three neutrino generations within the acceptance is 0.032 pb, so the sum of all non-standard model contributions to the cross section must be less than 0.062 pb

$$\sigma(ee \to \gamma + \text{new sources}) < 0.062 \text{ pb} \quad .$$

Figure 21 shows the dependence of the $\nu\bar{\nu}\gamma$ cross section on the number of neutrino generations. The limit determined above for the cross section allows a maximum of 14 generations with 90% confidence level. Several features of this limit should be noted. First, unlike many cosmological limits and limits derived from strange meson decays, there is no dependence on the mass of the associated charged lepton partner of the neutrino. Second, the mass of the neutrinos could be of order a few GeV without affecting the limit. The validity of the limit requires no assumption other than the standard model coupling of the Z_0 to a single generation. By contrast, the method of Deshpande et al.[10] extracts N_ν from the ratio of W to Z production in $\bar{p}p$ interactions and requires cancellation of QCD k factors, the branching ratio for both $Z \to \nu\bar{\nu}$ and $W \to e\nu$, and the mass of the top quark.[11]

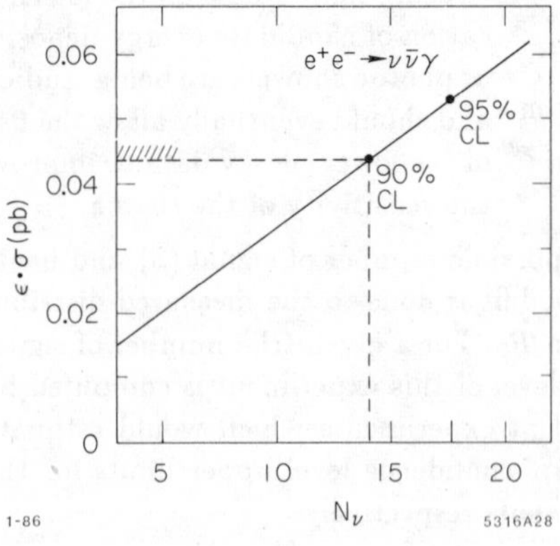

Fig. 21. Corrected $\nu\bar{\nu}\gamma$ cross section versus the number of neutrino generations.

As can be seen from Fig. 22, all contributions to the γ rate observed at PEP must sum to something slightly less than twice the $\gamma\nu\bar{\nu}$ cross section. Supersymmetry is the most topical but not necessarily the only source for such additions. In the case of $\gamma\tilde{\gamma}\tilde{\gamma}$ production, the magnitude of the cross section is determined by the mass of the exchanged virtual selectron (see Fig. 2) and the mass of the final state photinos. The excluded region in these parameters is shown in Fig. 23 both for the case where the left and right handed coupling selectrons ($\tilde{e}_L, \tilde{e}_R$) are degenerate in mass and the case where only one contributes to the observed rate. Note that the supersymmetric decay of the Z into pairs of selectrons is excluded for photinos masses up to 6 GeV.

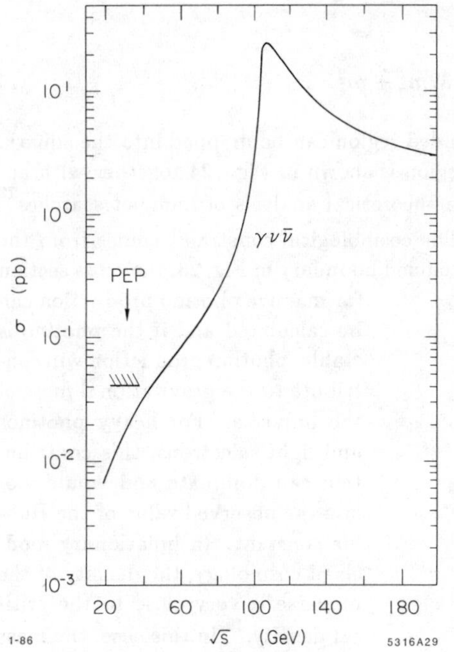

Fig. 22. $\gamma\nu\bar{\nu}$ cross section versus $\sqrt{s}$ within the ASP acceptance and the ASP limit from PEP.

If the lightest supersymmetric particle is the sneutrino (see Fig. 3), then the ASP limit constrains the mass of the wino. From the model of Ref. 6, the 90% CL limit on the mass of the wino is $m_{\tilde{W}} > 48$ GeV/c^2 with the assumptions of one massless sneutrino, no mixing between the wino and higgsinos, and only one light wino ($m_{\tilde{\nu}} = 0$, $O^+ = 1$, and $m_1 \ll m_2$). Other limits can be easily determined by scaling the cross section to the above assumptions. For the case of light gravitinos, limits on $M_{\tilde{G}}$ can be found using (see Ref. 4)

$$\sigma(\gamma\tilde{\gamma}\tilde{G}) \simeq 2.8 \times 10^{-3} \text{ pb} \left(\frac{10^{-5} \text{ eV}}{m_{\tilde{G}}}\right)^2 \frac{s}{(40 \text{ GeV})^2} \quad .$$

Finally, if one assumes that at some scale the gauge couplings for strong and electromagnetic interactions become equal and that at that scale the squarks and sleptons are related by $m_{\tilde{q}} = m_{\tilde{e}}$, then the relations between these masses can be calculated at any scale. In particular, at present energies we would have[12]

$$m_{\tilde{g}} \simeq \frac{3}{8}\,\frac{\alpha_s}{\alpha_{em}}\; m_{\tilde{\gamma}} \simeq 6.3 m_{\tilde{\gamma}}$$

and

$$m_{\tilde{q}}^2 \simeq 32 m_{\tilde{\gamma}}^2 + m_{\tilde{e}}^2 \quad .$$

Using these relations, the ASP excluded region can be mapped into the squark, gluino mass plane. The excluded region is shown in Fig. 24 together with approximate limits determined from a theoretical analysis of monojet searches[13] and cosmological constraints.[14,15] The cosmological constraint comes from the fact that for every point along the excluded boundary in Fig. 23, the cross section for massive photino production can be calculated and if the photino is stable, photino production will contribute to the gravitational mass of the universe. For heavy photinos and light selectrons, this contribution can dominate and would violate the observed value of the Hubble constant. In inflationary models of cosmology, the density of the universe is very close to the critical density.[16] In this case, the mass of the photino and selectron can be related by the requirement that the photinos supply the missing dark matter. The allowed values then lie along a line[17] in Fig. 23.

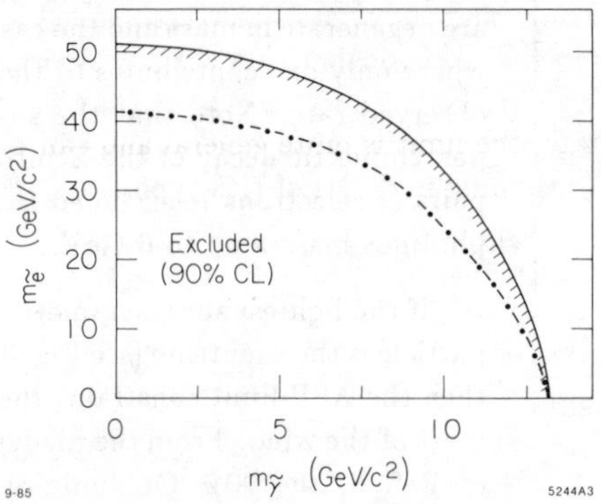

Fig. 23. Region of $m_{\tilde{e}}$ and $m_{\tilde{\gamma}}$ excluded by the present ASP data for $m_{\tilde{e}_L} = m_{\tilde{e}_R}$ (solid curve) and $m_{\tilde{e}_{L,R}} \gg m_{\tilde{e}_{R,L}}$ (dot dashed curve).

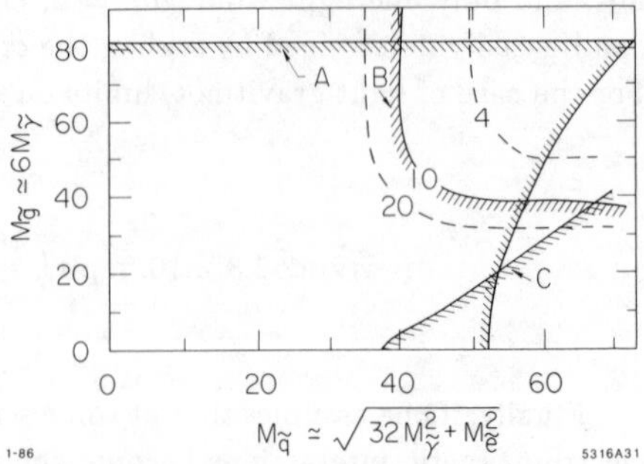

Fig. 24. The excluded region in $m_{\tilde{g}}$ and $m_{\tilde{q}}$ from (A) ASP, (B) monojets (see Ref. 15) and (C) cosmology.

Summary

The technique of detecting a single photon whose transverse momentum can be shown to be unbalanced by any detected particles can provide significant constraints both on our standard model and on important extensions of this model which predict the presence of new particles. The experimental identification of such final states requires that particular care be given both to the acceptance over which the photon can be detected as well as the ability of the detector to observe other particles over a large acceptance with small occupancy. Previous data from the MAC detector and new data from the ASP detector have been presented. The ASP data have reached a sensitivity which excludes any new contributions greater than twice the contribution of $\gamma\nu\bar{\nu}$ events and therefore provide new limits on the number of neutrino generations, the masses of selectrons, photinos, and winos and important restrictions on the masses of squarks and gluinos. Since the method requires only that the photon be accompanied by neutral particles whose interactions are small enough that they do not interact in tens of radiation lengths of material, and since the photon spectrum is rather insensitive to the details of the final state, the limit is quite general and can be easily extended to any future models containing particles of this type.

Acknowledgements

I would like to thank J. Ellis, J. Hagelin, P. Fayet, and M. Sher for many useful theoretical discussions. The SLAC technical staff for their support of the ASP experiment, and all the members of the ASP collaboration for their work on both the apparatus and the data analysis.

REFERENCES

1. E. Ma and J. Okada, Phys. Rev. Lett. 41, 287 (1978); K. Gaemers, R. Gastmans and F. Renard, Phys. Rev. D19, 1605 (1979).
2. S. Wolfram, Phys. Lett. 82B, 65 (1979).
3. See also H. Haber and G. Kane, Phys. Rept. 117, 75 (1985).
4. P. Fayet, Phys. Lett. 117B, 460 (1982).
5. J. Ellis and J. Hagelin, Phys. Lett. 122B, 303 (1983).
6. J.S. Hagelin, G.L. Kane and S. Raby, Nucl. Phys. B241, 638 (1984).
7. J. Ellis, J. Frere, J. Hagelin, G. Kane and S. Petcov, Phys. Lett. 132B, 436 (1983).
8. J. Ellis, Nuffield Workshop, 1982, p. 91.
9. R. Wilson, SLAC-PUB-3838, November 1985.
10. N. Deshpande, G. Eilam, V. Barger and F. Halzen, Phys. Rev. Lett. 54, 1757 (1985).
11. See for example L. di Lella, *1985 International Symposium on Lepton and Photon Interactions at High Energies*, Kyoto, Japan.
12. M. Sher, private communication and J. Polchinski, Phys. Rev. D26, 3674 (1982).
13. J. Ellis, *1985 International Symposium on Lepton and Photon Interactions at High Energies*, Kyoto, Japan, CERN-TH-4277/85.
14. H. Goldberg, Phys. Rev. Lett. 50, 1419 (1983).
15. J. Ellis, J. Hagelin, D. Nanopoulos, K. Olive and M. Srednicki, Nucl. Phys. B241, 381 (1984).
16. A. Guth, Phys. Rev. D23, 347 (1981).
17. J. Silk and M. Srednicki, Phys. Rev. Lett. 53, 624 (1984).

SUPERGRAVITY CONSTRAINTS ON MONOJETS

S. Nandi
Center for Particle Theory, The University of Texas at Austin
Austin, Texas 78712

ABSTRACT

In the standard model, supplemented by N = 1 minimal supergravity, all the supersymmetric particle masses can be expressed in terms of a few unknown parameters. The resulting mass relations, and the laboratory and the cosmological bounds on these superpartner masses are used to put constraints on the supersymmetric origin of the CERN monojets. The latest MAC data at PEP excludes the scalar quarks, of masses up to 45 GeV, as the origin of these monojets. The cosmological bounds, for a stable photino, excludes the mass range necessary for the light gluino-heavy squark production interpretation. These difficulties can be avoided by going beyond the minimal supergravity theory. Irrespective of the monojets, the importance of the stable $\tilde{\gamma}$ as the source of the cosmological dark matter is emphasized.

1. INTRODUCTION

The observed monojet events[1] at the CERN proton-antiproton collider have been interpreted as the production of supersymmetric particles.[2-10] Three such interpretations have been proposed. They are: A) the production of a scalar quark-scalar antiquark ($\tilde{q}\tilde{\bar{q}}$) pair with the scalar quark mass, $m_{\tilde{q}}$ between 25 and 45 GeV, and the mass of the gluino, $m_{\tilde{g}}$ greater than $m_{\tilde{q}}$;[2] B) the production of a pair of gluinos with $m_{\tilde{g}}$ between 30 and 50 GeV, and $m_{\tilde{q}} >> m_{\tilde{g}}$;[3] C) the production of a light gluino and a heavy scalar quark with $m_{\tilde{g}} \lesssim 10$ GeV, and $m_{\tilde{q}} \sim 100$ GeV.[4]

In the currently popular SU(3) × SU(2) × U(1), grand unified, N = 1 supergravity theory (which we shall call Supergravity Standard Model, SSM), the masses of all the supersymmetric particles can be expressed in terms of a few unknown parameters. In particular, one obtains interesting mass relations among the scalar quarks, the scalar leptons, the photino and the gluinos.[11-13] There are also various bounds on these superparticles masses, obtained from laboratory experiments and cosmology. In this talk, we shall discuss whether the above scalar quark and gluino mass ranges (from monojets) are compatible with the constraints from the supergravity theory and the various laboratory and the cosmological bounds. In particular, we shall see that the monojet interpretation C (light gluino and heavy scalar quark) does not satisfy the cosmological energy density bound. For the other two interpretations, the allowed monojets regions are further narrowed down. We mention that the cosmological considerations of these works[14,15] were motivated by the monojets, but not limited to it. We shall see that a stable photino plays a very important role in cosmology. Its contribution to the present energy density of the universe is, at least, comparable to that of the baryons, and could be much more. Thus the photinos could be an important source of the dark matter of the universe. This talk is organized as follows. In Section 2, we discuss the mass spectrum of the superpartners in the minimal supergravity theory. In Section 3, we discuss the constraints from the various laboratory experiments, and how they narrow down the allowed monojet regions. Section 4 is devoted to cosmological considerations, and the constraints on superparticle masses derived from these considerations. In Section 5, we discuss briefly what happens if we go beyond minimal N = 1 supergravity theory, and then we conclude by summarizing the main points of this talk. The results of the sections 3 and 5 is contained in Ref. 14, whereas those of section 4 is contained in Refs. 14 and 15.

2. SUSY MASS SPECTRUM

We work within the framework of $SU(3) \times SU(2) \times U(1)$, N = 1 supergravity theory.[16] The three interactions are assumed to be grand unified around (or somewhat lower) the Planck's scale. In the last few years, a consistent picture for the supersymmetry breaking, as well as for the breaking of weak interaction symmetry has emerged.[16] This is the so-called hidden sector theory, with the supergravity as the messenger of supersymmetry breaking. The idea is the following. All the particles are divided into three sectors. 1) The Matter sector: This contains the quarks, the leptons, the Higgs, the gauge fields and their superpartners. 2) The Hidden sector: This contains typically some chiral superfields, and 3) the Supergravity sector: This contains the graviton and the gravitino. The matter sector and the hidden sector do not interact directly, but only through the supergravity sector. The superpotential of the hidden sector is chosen in such a way that the supersymmetry in the hidden sector is broken spontaneously at a scale, m_I. This generates the gravitino mass, $m_{3/2} \sim m_I^2/M_P$ where M_P is the Planck mass. Due to the indirect interaction between the matter and the hidden sector through the supergravity sector, soft SUSY breaking terms are generated in the effective potential of the matter sector at low energy. The weak interaction symmetry, $SU(2) \times U(1)$ is also broken in the matter sector at a scale $m_{3/2}$ via these soft SUSY breaking terms. Taking m_I to be about 10^{11} GeV, one obtains the correct scale of the weak symmetry breaking, $m_{3/2} \sim m_W \sim 100$ GeV.

The scalar fermions and the gaugino masses in the above theories depends on two unknown functions of the chiral superfields. These are $f_{\alpha\beta}(\phi_i)$ and $G(\phi_i,\phi_i^*)$, where the chiral superfields ϕ_i can be from the matter or the hidden sectors. The function $f_{\alpha\beta}(\phi_i)$ multiplies the kinetic terms for the gauge fields, $G(\phi_i,\phi_i^*)$ multiplies the mass terms for the scalar fields, whereas the gaugino mass terms are proportional to $\partial f^*_{\alpha\beta}(\phi_i)/\partial\phi^*_j$. In the case of $f_{\alpha\beta} = \delta_{\alpha\beta}$, the gaugino mass terms vanishes at the tree level. We shall make the following assumptions for the functions $f_{\alpha\beta}(\phi_i)$ and $G(\phi_i,\phi^*_i)$.

1) $f_{\alpha\beta}(\phi_i)$ depends only the hidden sector chiral superfields (z_i), i.e., $f_{\alpha\beta}(\phi_i) = f(z_i)\delta_{\alpha\beta}$. 2) The Kähler potential, $G(\phi_i,\phi^*_j) \sim \phi_i\phi^*_j$. The N = 1 supergravity models satisfying the above two assumptions will be called the minimal models. In these minimal models, assuming $g_1 = g_2 = g_3$ at the grand unification point M, we obtain $\tilde{m}_1 = \tilde{m}_2 = \tilde{m}_3$ at M (1,2,3 refers to the gauge groups U(1), SU(2) and SU(3) respectively $\tilde{m}_3 \equiv m_{\tilde{g}}$). Using this, the photino and the gluino masses, at the grand unification point, are obtained to be

$$m_{\tilde{\gamma}} = \frac{8}{3}\frac{\alpha}{\alpha_0}\tilde{m}_0$$
$$m_{\tilde{g}} = \frac{\alpha_3}{\alpha_0}\tilde{m}_0 \tag{1}$$

where $m_0(\alpha_0)$ is the common gaugino mass (coupling constant) at M, and $\alpha(\alpha_3)$ is the electromagnetic (strong) coupling constant. From (1), we obtain a very important minimal supergravity mass relation

$$\frac{m_{\tilde{\gamma}}}{m_{\tilde{g}}} = \frac{8}{3}\frac{\alpha}{\alpha_3} \simeq \frac{1}{6} \quad . \tag{2}$$

Note that the gauginos satisfy the renormalization group equation

$$\frac{\partial}{\partial t}\left(\frac{\tilde{m}_a}{\alpha_a}\right) = 0 \quad , \qquad a = 1,2,3 \tag{3}$$

so that the first equality in Eq. (2) is valid at all energies. The approximate value 1/6 is at $\mu \sim m_W$. The scalar fermions obtain their masses from two sources:

i) The F-terms (the soft SUSY breaking terms)

$$\sim \frac{\partial^2 V(\phi_i,\phi^*_i)}{\partial\phi_i\partial\phi^*_i}\phi^*_i\phi_i \sim \mu^2_i\,\phi^*_i\phi_i \quad . \tag{4}$$

According to assumption 2), for the minimal model, at the unification point M, μ_i is the same for all the scalar fermions and equal to the gravitino mass, $m_{3/2}$.

ii) The D-term contribution

$$\sim \sum_{a=1}^{3} \frac{g_a^{\,2}}{2} \left(\sum_{i,j} \phi_i^* T_{ij}^a \phi_j \right)^2$$

$$\Rightarrow r m_Z^2 \; (-I_3 + Q \sin^2\theta_W) \tag{5}$$

where T^a (a = 1,2,3) are the generators of the gauge groups, m_Z is the mass of the Z-boson, and $r \equiv (v_2^2 - v_1^2)/(v_2^2 + v_1^2)$, v_1 and v_2 being the vacuum expectation values of the up and down type Higgs bosons. Using (4) and (5), the scalar fermions masses, m_i are given by

$$m_i^2 = \mu_i^2 \pm r \; (-I_3 + Q \sin^2\theta_W) \tag{6}$$

where +(-) signs are for the scalar partner of the left (right) handed fermions respectively. The mass parameters μ_i^2 and $\tilde{m}_a$ satisfy the appropriate renormalization group equations.[17] We solve them by using the boundary conditions that all μ_i have a common value $m_{3/2}$, and all $\tilde{m}_a$ have a common value m_0 at the unification scale, $\mu = M_{GUT}$. Then, using Eq. (3), we replace m_0 in terms of the gluino mass, $m_{\tilde{g}}$ at ordinary energies ($\mu \simeq m_W$). Thus, all the scalar fermion masses at ordinary energies can be expressed in terms of three parameters which we choose to be $m_{3/2}$, $m_{\tilde{g}}$ and r. Then, using $\sin^2\theta_W = 0.22$ in Eq. (6), the scalar fermion masses at ordinary energies ($\mu \sim m_W$) are given by

$$\begin{aligned}
m_{\tilde{e}_R}^2 &= m_{3/2}^2 + J_1 m_{\tilde{g}}^2 + 0.22\; r m_Z^2 \;, \\
m_{\tilde{e}_L}^2 &= m_{3/2}^2 + \left(J_2 + \tfrac{1}{4} J_1 \right) m_{\tilde{g}}^2 + 0.28\; r m_Z^2 \;, \\
m_{\tilde{\nu}}^2 &= m_{3/2}^2 + \left(J_2 + \tfrac{1}{4} J_1 \right) m_{\tilde{g}}^2 - 0.50\; r m_Z^2 \;, \\
m_{\tilde{d}_R}^2 &= m_{3/2}^2 + \left(J_3 + \tfrac{1}{9} J_1 \right) m_{\tilde{g}}^2 + 0.07\; r m_Z^2 \;, \\
m_{\tilde{u}_R}^2 &= m_{3/2}^2 + \left(J_3 + \tfrac{4}{9} J_1 \right) m_{\tilde{g}}^2 - 0.15\; r m_Z^2 \;, \\
m_{\tilde{d}_L}^2 &= m_{3/2}^2 + \left(J_3 + J_2 + \tfrac{1}{36} J_1 \right) m_{\tilde{g}}^2 + 0.43\; r m_Z^2 \;, \\
m_{\tilde{u}_L}^2 &= m_{3/2}^2 + \left(J_3 + J_2 + \tfrac{1}{36} J_1 \right) m_{\tilde{g}}^2 - 0.35\; r m_Z^2 \;.
\end{aligned} \tag{7}$$

The quantities J_a depends on the values of the coupling constants at

$\mu = m_W$ and M_{GUT}, and also on the number of light generations. For 3 generation, the values of J_1, J_2 and J_3 are 0.017, 0.056, and 0.787 respectively, where for 4 generation, the values are 0.048, 0.143, and 1.70 respectively.[13] In Eq. (7), the masses are in increasing order for r = 0 (which is the case in some SUSY models). In deriving Eqs. (7), we have ignored the contributions of the Yukawa couplings. So Eq. (7) is valid only for the first three generations, except the scalar top quark. Since we do not know the value of r, we do not know which of the scalar fermions is the lightest. We shall assume that scalar quarks produced at CERN collider (if the monojets are due to the scalar quarks) have masses equal to the root mean square of the scalar quark masses given in Eq. (7). The mean square of the scalar quark masses, $m_{\tilde{q}}$, the charged scalar lepton masses, $m_{\tilde{\ell}}$, are given by

$$m_{\tilde{q}}^2 = m_{3/2}^2 + \left(J_3 + \frac{1}{2} J_2 + \frac{11}{72} J_1\right) m_{\tilde{g}}^2 \quad ,$$
$$m_{\tilde{\ell}}^2 = m_{3/2}^2 + \left(\frac{1}{2} J_2 + \frac{5}{8} J_1\right) m_{\tilde{g}}^2 + 0.25\ r m_z^2 \quad . \tag{8}$$

From Eqs. (8), and also using $m_{\tilde{\nu}}^2 \geq 0$ in Eq. (7), we obtain,

$$m_{\tilde{q}}^2 \geq \frac{2}{3} m_{\tilde{\ell}}^2 + \beta\ m_{\tilde{g}}^2 \quad , \tag{9}$$

where $\beta \equiv J_3 - J_2/6 - 25J_1/72 \simeq 0.77$ (1.66) for three (four) light generations. Eqs. (2) and (9) are very important minimal supergravity mass constraints.

3. LABORATORY CONSTRAINTS ON MONOJETS

In this section, we shall discuss how the current bounds on the SUSY particle masses from other laboratory experiments can be used to restrict the monojet regions for the $\tilde{q}\tilde{\bar{q}}$ production.[14] In particular, we shall use the laboratory bounds on $m_{\tilde{\gamma}}$ and $m_{\tilde{e}}$, together with the two minimal supergravity mass constraints, Eqs. (2) and (9).

The lower bound on $m_{\tilde{e}}$ from the process $e^+e^- \to \tilde{e}^+\tilde{e}^-$ at PETRA is 22 GeV.[18] The process $e^+e^- \to \gamma\tilde{\gamma}\tilde{\gamma}$ is a very interesting SUSY reaction,[19] the phase space suppression is minimal. This annihilation is mediated by the scalar electrons $\tilde{e}_L$ and $\tilde{e}_R$. The cross-section[19] for this process is proportional to $(1/m_{\tilde{e}_L}^4 + 1/m_{\tilde{e}_R}^4)$ and also is a function of the photino mass. The MAC group at PEP ($\sqrt{s}$ = 29 GeV) did not observe a photon signal for this process, and gives an upper limit for the cross-section to be less than 57 fb.[20] Using this limit, we obtain

$$2^{1/4}\left(m_{\tilde{e}_L}^{-4} + m_{\tilde{e}_R}^{-4}\right)^{-1/4} \geq \tilde{m} \tag{10}$$

where $\tilde{m}$ on the right hand side is a function of the photino mass. Writing $m_{\tilde{e}_L} = m_{\tilde{e}_R}(1 + \varepsilon)$, Eq. (7) gives

$$m_{\tilde{e}}^2 \geq \tilde{m}^2\left(1 + \frac{15}{8}\,\varepsilon^2\right) \tag{11}$$

where $\varepsilon \ll 1$. Using Eq. (8) and the expressions for $m_{\tilde{e}_L}$ and $m_{\tilde{e}_R}$ in Eq. (7), we get

$$\varepsilon \simeq \frac{0.04\, m_{\tilde{g}}^2 + 0.06\, r m_Z^2}{2(m_{\tilde{q}}^2 - 0.78\, m_{\tilde{g}}^2 + 0.25\, r m_Z^2)}\,. \tag{12}$$

For a given scalar quark mass, one can find the allowed region in the r and $m_{\tilde{g}}$ plane using $m_{\tilde{e}} \geq 22$ GeV, and $m_{\tilde{\nu}}^2 \geq 0$ in Eq. (7) and (8). We find that for the case A, $\varepsilon \leq 0.1$; for the case B, ε lies between 0.03 - 0.15; whereas for the case C, $\varepsilon \simeq 0.03$. Thus, from Eq. (11), we find that the lower bound on $m_{\tilde{e}}$ is essentially given by the MAC bound, $\tilde{m}$; and the $\tilde{e}_L$ and $\tilde{e}_R$ masses differ by, at most, 10 to 15%. The MAC result gives bounds on $m_{\tilde{e}}$ which are bigger than 22 GeV for the values of the photino mass up to 8.5 GeV.

The lower bound obtained for the scalar quark mass, from Eq. (9), for the various gluino masses is shown in Fig. 1, for the case of three light generations.[14] For a given gluino mass, the photino mass was calculated from Eq. (2), and then the lower bound on $m_{\tilde{e}}$ was obtained from the MAC curve via Eq. (11). The MAC result gives the bound on $m_{\tilde{q}}$ all the way up to $m_{\tilde{g}} \leq 50$ GeV. For $m_{\tilde{g}} > 50$ GeV, the bound on $m_{\tilde{e}}$ ($m_{\tilde{e}} \leq 22$ GeV) from the reaction $e^+e^- \to \tilde{e}^+\tilde{e}^-$ gives the better bound on $m_{\tilde{q}}$. Fig. 1 shows that up to $m_{\tilde{q}} = 45$ GeV, $\tilde{q}\bar{\tilde{q}}$ production as the source of the observed monojets, is excluded. The shaded region is the allowed monojet region for the $\tilde{q}\bar{\tilde{q}}$ production for $m_{\tilde{q}}$ up to 50 GeV. We give results up to 50 GeV in case the future data with better statistics would allow such a $m_{\tilde{q}}$ range.

In the above analysis, we assumed the photino to be a pure gaugino in which case the Eq. (2), relating the photino and the gluino masses, is exact. The photino mass can be slightly bigger than that given by Eq. (2) for two reasons: a) In the general case, the photino may contain some Higgsino component, (coming from the mixing through mass matrix) in which case the right hand side of the Eq. (2) is modified by the addition of a small term proportional to the Higgsino mixing contribution. In most of the popular supergravity models, the effect of this mixing is to increase the photino mass[21] by about 10%. b) Also, the factor 1/6 in Eq. (2) can be made

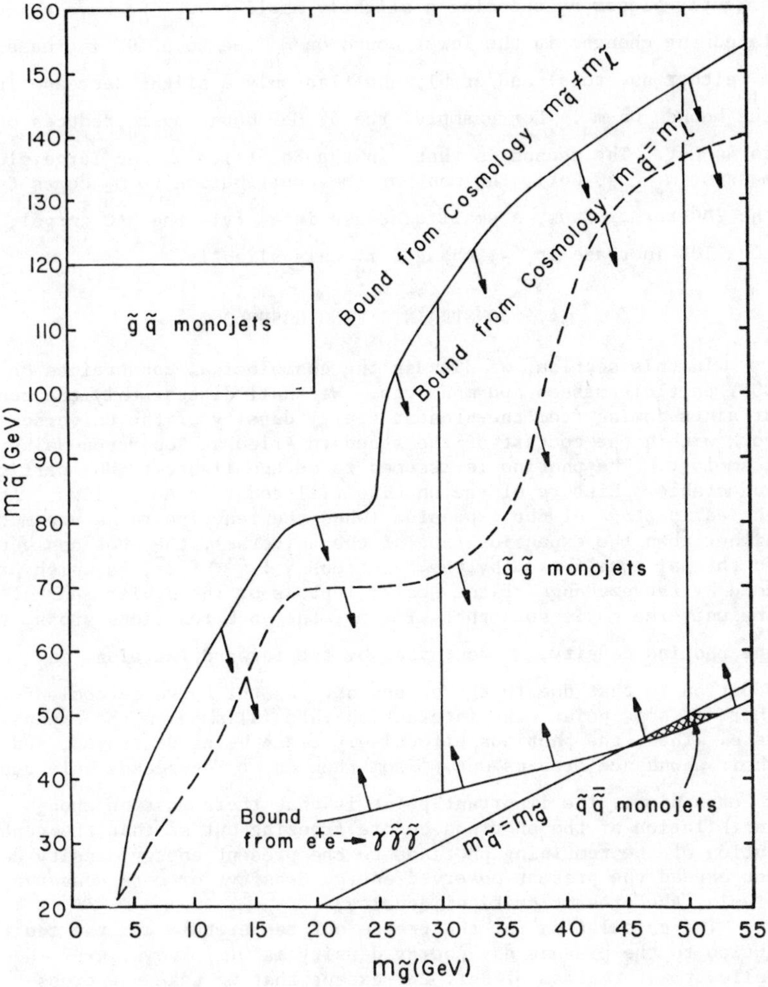

Fig. 1. Bounds from laboratory and cosmology in the $m_{\tilde{q}}$ - $m_{\tilde{g}}$ plane for minimal N = 1 supergravity theory with three light generations. Only the regions in the directions of the arrows are allowed. The two vertical lines, at $m_{\tilde{g}}$ = 30 and 50 GeV, are the allowed $m_{\tilde{g}}$ range from the monojet rates for $\tilde{g}\tilde{g}$ production. The rectangle at the left is the allowed region (Ref. 7) for $\tilde{g}\tilde{q}$ monojets.

slightly bigger by choosing a slightly smaller α_3. We have calculated the changes in the lower bound on $m_{\tilde{q}}$ due to a 10% increase in $m_{\tilde{\gamma}}$ either due to a) and/or b), and find only a slight decrease in the bounds on $m_{\tilde{q}}$. For example, the 45 GeV bound on $m_{\tilde{q}}$ reduces only to 44 GeV. The reason is that, in the Eq. (9), for the large gluino masses ($m_{\tilde{g}} \geq 40$ GeV), the most of the contribution to $m_{\tilde{q}}$ comes from the 2nd term. Thus, a small decrease in $m_{\tilde{\ell}}$ (via the MAC curve), due to a 10% increase in $m_{\tilde{\gamma}}$, changes $m_{\tilde{q}}$ only slightly.

4. CONSTRAINTS FROM COSMOLOGY

In this section, we discuss the cosmological constraints on the SUSY particle masses and monojets. We shall discuss only the constraints coming from the present energy density of the universe. We work within the context of the standard Friedman-Robertson-Walker cosmology. The photino is assumed to be the lightest SUSY particle and stable. History of the photino till today is as follows. At the early stage of the expansion (when the reaction rates are much higher than the expansion rate of the universe), the photinos were in thermal equilibrium by the reactions $\tilde{\gamma} + \tilde{\gamma} \rightleftharpoons \ell^+\ell^-$, $q\bar{q}$ which proceed by the exchange of the scalar leptons or the scalar quarks. As the universe cools such that $kT < m_{\tilde{\gamma}}$, the back reactions stops; then the photino density, $\eta_{\tilde{\gamma}}$ decreases by the forward reaction, in addition to that due to the expansion. As the universe cools further, at some point, the interaction rate falls below the expansion rate. Then, the photinos effectively cease being destroyed, and their abundance freezes out. From then on, $\eta_{\tilde{\gamma}}$ decreases only due to expansion. The important point is that there must be enough annihilation of the photinos before freezing out so that the contribution of the remaining photinos to the present energy density does not exceed the present observed energy density, or more conservatively, the present critical density.

The calculation for the freeze out temperature and the contribution to the present day energy density is well known.[22,23] We follow the formalism of Ref. 23, except that we take the cross-section appropriate to the Majorana $\tilde{\gamma}$, and also the appropriate number of degrees of freedom for the fermions, f.[24,25] The annihilation rate for the process $\tilde{\gamma} + \tilde{\gamma} \to f\bar{f}$ (assuming the photino to be pure gaugino) is given by[26]

$$\sigma v_{rel.} = \sum_f \frac{Q_f^4 e^4 p_{\tilde{\gamma}}}{2\pi m_{\tilde{f}}^4 m_{\tilde{\gamma}}} \left[\frac{1}{3}\left(m_{\tilde{\gamma}}^2 - m_f^2\right) v_{rel.}^2 + m_f^2\right] \theta(m_{\tilde{\gamma}} - m_f) \quad , \qquad (14)$$

where $p_{\tilde{\gamma}} = (m_{\tilde{\gamma}}^2 - m_f^2)^{1/2}$.

Here, Q_f is the charge of the fermion f, and θ is the step function. Notice that the annihilation cross-section is $\propto m_{\tilde{\gamma}}^2/m_{\tilde{f}}^4$ so that the photino mass must be big enough and scalar fermion masses must be small enough to have sufficient annihilation before decoupling. The freeze out temperature (T_F) is obtained by equating the interaction time ($\tau_{\tilde{\gamma}}$) with the life-time of the universe.[27]

$$\tau_{\tilde{\gamma}} \equiv \frac{\langle \eta_{\tilde{\gamma}} \rangle}{\langle \eta_{\tilde{\gamma}} \eta_{\tilde{\gamma}} \sigma v_{rel.} \rangle} = \frac{10^{20}}{T_F^2} \text{ sec.} \tag{15}$$

where $\langle \ \rangle$ represents a thermal average, and

$$\langle \eta_{\tilde{\gamma}} \rangle = \frac{4\pi}{h^3} \int \frac{p^2 dp}{e^{E/kT} + 1} \ . \tag{16}$$

Finally, the contribution of the remaining photinos to the present day energy density is

$$\rho_{\tilde{\gamma}} = m_{\tilde{\gamma}} \eta_F \left(\frac{T}{T_F}\right)^3 \left(\frac{4}{11} \chi\right) \tag{17}$$

where η_F is the number density of the photinos at the freeze out temperature T_F, and T is the present photon temperature which we take to be 2.8°K. ($4\chi/11$) is the appropriate reheating factor which takes into account the rise in photon temperature relative to the photino temperature due to fermion-antifermion annihilation. For example, $\chi = 1$ for $m_e < kT_F < m_\mu$; $\chi = 43/57$ for $m_\mu < kT_F < m_\pi$, $\chi = 43/69$ for $m_\pi < kT_F < T_c$ (where T_c is the quark-gluon plasma temperature), $\chi = 43/247$ for $T_c < kT_F < m_c$ (and $m_s < T_c$); and so on. For the quark masses, we used $m_u = m_d = 10$ MeV, $m_s = 150$ MeV, $m_c = 1.5$ GeV and $m_b = 5$ GeV. The quark-gluon plasma temperature was taken to be 200°K.

The photino energy density today, $\rho_{\tilde{\gamma}}$ can now be calculated using (14) - (17), once we specify the masses of the scalar fermions. Taking the present value of the Hubble constant, H_0 to be 100 km/sec/Mpc, we get $\rho_c \equiv 3H_0^2/8\pi G = 2 \times 10^{-29}$ gm/cm^3. Then, demanding $\rho_{\tilde{\gamma}} \leq \rho_c$ gives a lower bound on the photino mass for a given set of scalar fermion masses. For example, if we take all the scalar masses equal and equal to 22 (100) GeV, then we obtain $m_{\tilde{\gamma}} \gtrsim 0.5$ (6) GeV. Conversely, for a given photino mass, one gets an upper bound for the scalar fermion mass.

Our results for the upper bounds on the scalar quark masses for the various gluino masses are given by the solid curve in Fig. 1.[14] In obtaining these bounds, we took all the scalar quark and the charged lepton masses to be equal to their root mean square values. Furthermore, we used the two important minimal supergravity mass relations given by Eq. (2) and (8). Eq. (2) gives the gluino mass in terms of the photino mass while Eq. (8) relates the scalar quarks and the charged scalar lepton masses as

$$m_{\tilde{q}}^2 = m_{\tilde{\ell}}^2 + \gamma m_{\tilde{g}}^2 - 0.25\ r m_Z^2 \tag{18}$$

where $\gamma \equiv (J_3 - 17J_1/36) \simeq 0.78$, and $|r| \leq 1$. We used $r = -1$ in Eq. (18) (other values of r makes the bound lower), together with the experimental constraint, $m_{\tilde{\ell}} \geq 22$ GeV. In fig. 1, only the region below the curve is allowed. We see that this excludes the $m_{\tilde{g}}$ and $m_{\tilde{q}}$ mass range necessary for the light gluino-heavy scalar quark interpretation of the monojets (shown by the rectangle). The lower value of ρ_c or the higher value of T_c lower the bounds. For completeness, we also give the bounds for the case of $m_{\tilde{q}} = m_{\tilde{\ell}}$, the dashed curve (the value of T_c used for this curve is 300 MeV).

The minimal supergravity models with a stable photino has interesting cosmological implications.[14,15] Using the experimental bound, $m_{\tilde{\ell}} \geq 22$ GeV ($\ell = e, \mu, \tau$), and the Eq. (18) with $r = -1$, we obtain, $\rho_{\tilde{\gamma}} \geq 0.002\ \rho_c$ for $H_0 = 100$ km/sec/Mpc, and $\rho_{\tilde{\gamma}} \geq 0.01\ \rho_c$ for $H_0 = 50$ km/sec/Mpc, for the photino mass up to 10 GeV (or gluino mass up to 60 GeV, Eq. (2)). Thus, the stable photinos contribute almost as much to the energy density of the present universe as the baryons. They contribute much more than the baryons for some region of the $m_{\tilde{g}} - m_{\tilde{q}}$ plane, and thus could be an important source of the dark matter. Their possible contribution to the observed cosmic ray antiproton flux has recently been explored.[28] For 60 GeV $< m_{\tilde{g}} <$ 100 GeV, $\rho_{\tilde{\gamma}} \geq 0.0005\ \rho_c$ ($\rho_{\tilde{\gamma}} \geq 0.0025\ \rho_c$) for $H_0 = 100$ (50) km/sec/Mpc.

5. CONCLUDING REMARKS

In all of the above analysis, we have assumed that, at the unification scale, the F-term contributions to the scalar quarks and the scalar lepton masses are equal, and equal to the gravitino mass. This need not be true if we go beyond minimal supergravity models. For example, a non-trivial Kähler potential, $G(\phi_i, \phi_i^*)$ may split them. In that case, the first term, $m_{3/2}^2$ in Eq. (7) will be different for each scalar fermions. As a result, Eqs. (8), (9) and (18) need not be true, and thus a scalar quark of mass less than 45 GeV may be allowed to produce the $\tilde{q}\bar{\tilde{q}}$ monojets. Also, since the scalar quark and the scalar lepton masses are not constrained by the Eq. (18), we can have light scalar leptons ($m_{\tilde{\ell}} \geq 22$ GeV), but heavy scalar quarks. Thus, the cosmological bound for a stable photino could also be satisfied for sufficiently light charged scalar lepton masses. In Fig. 2, we give[14] the upper bound on the charged scalar

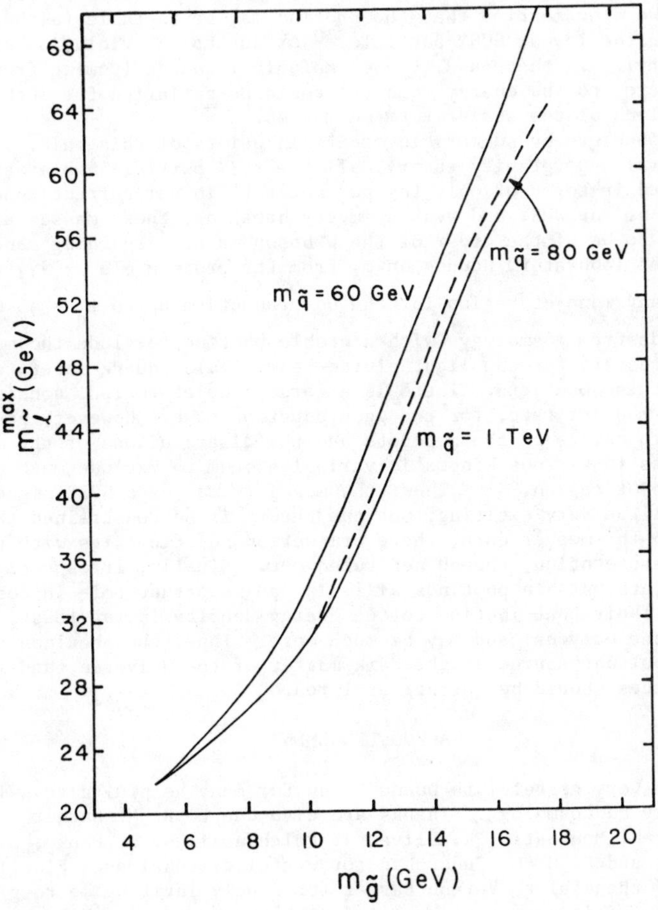

Fig. 2. Maximum values of the scalar-lepton masses that satisfy the cosmological bound, for the nonminimal N = 1 supergravity theory, in the $\tilde{g}\tilde{q}$ monojet region. For any value of $m_{\tilde{q}}$ between 60 and 1000 GeV, the corresponding bound lies between these two curves.

lepton masses as a function of the gluino masses, for $m_{\tilde{q}} = 60$, 80 and 1000 GeV, corresponding to $\rho_c = 2 \times 10^{-29}$ gm/cm^3, and $T_c = 200$ MeV. For example, with $m_{\tilde{g}} = 10$ GeV, and $m_{\tilde{q}} = 80$ GeV, $m_{\tilde{\gamma}} \leq 31$ GeV will satisfy the cosmological bound. Another possibility to satisfy the monojet constraints within the minimal supergravity models is to modify the supergravity relation between the photino and the gluino mass, Eq. (2). Attempts along this direction does not seem promising.[29] We also mention that the photino may be unstable (either by not being the lowest SUSY particle,[30] or due to the violation of the R-symmetry). In that case, the cosmological bounds (coming from the contribution to the energy density) could be satisfied for much lower values of the scalar fermion masses.

We conclude by summarizing the main points of this talk. In the minimal supergravity theory, all the SUSY particle masses are determined in terms of only few parameters. In our current understanding of the SUSY and weak symmetry breaking, these masses are expected to be of the order of the W-boson mass. Regarding monojets, the latest laboratory bounds on $m_{\tilde{e}}$ from the process $e^+e^- \to \gamma\tilde{\gamma}\tilde{\gamma}$ exclude any monojet region for the $\tilde{q}\bar{\tilde{q}}$ production up to $m_{\tilde{q}} = 45$ GeV. The bounds from cosmology (with a stable photino) exclude the mass range necessary for the light gluino-heavy scalar quark interpretation of these monojets. There is a large monojet region, consistent with the monojet rate, for the $\tilde{g}\tilde{g}$ production case. However, pencil like sharp nature of the monojets and the distributions of the monojet events in various kinematic variables seem to exclude much of this monojet region.[8-10] Thus, the monojets may be a SUSY signature which will be very exciting, but the theory is so constrained that even with the meager data, there are serious difficulties with the SUSY interpretation, though not ruled out. Finally, irrespective of the monojets, stable photinos will play an important role in cosmology. Their contribution to the energy density is, at least, as much as the baryons, and may be much more. Thus, the photinos could be an important source of the dark matter of the universe, and its consequences should be further explored.

ACKNOWLEDGEMENT

I am very grateful to Duane Dicus for many helpful discussions, especially on cosmology. Thanks are also due to R. M. Barnett, M. Kreuz, R. N. Mohapatra, K. Olive, J. Polchinski, W. W. Repko, M. Srednicki and E. C. G. Sudarshan for useful discussions. Finally, I am very thankful to Vernon Barger for kindly inviting me to give this talk. This paper was prepared with the support of U.S. Department of Energy Grant No. DE-FG05-85ER40200; however, any opinions, findings, conclusions, or recommendations expressed herein are those of the author and do not necessarily reflect the views of DOE.

REFERENCES

1. G. Arnison et al., Phys. Lett. 139B, 115 (1984).
2. J. Ellis and H. Kowalski, Phys. Lett. 142B, 441 (1984).
3. E. Reya and D. P. Roy, Phys. Rev. Lett. 51, 867 (1983) (E: 51, 1307 (1983); ibid. 53, 881 (1984); Phys. Lett. 141B, 442 (1984); Dortmund preprint 84/9 (1984); J. Ellis and H. Kowalski, Nucl. Phys. B246, 189 (1984).
4. V. Barger, K. Hagiwara, W.-Y. Keung, and J. Woodside, Phys. Rev. Lett. 53, 641 (1984); M. J. Herrero, L. E. Ibanez, C. Lopez, and F. J. Yndurain, Phys. Lett. 132B, 199 (1983); 145B, 430 (1984).
5. A. R. Allan, E. W. N. Glover and A. D. Martin, Phys. Lett. 146B, 247 (1984); A. R. Allan, E. W. N. Glover and S. L. Grayson, Durham preprint DTP/84/28 (1984).
6. V. Barger, K. Hagiwara and W.-Y. Keung, Phys. Lett. 145B, 147 (1984); ibid. Wisconsin preprint MAD/PH/197 (1984).
7. R. M. Barnett, H. E. Haber and G. L. Kane, Phys. Rev. Lett. 54, 1983 (1985).
8. A. DeRujula and R. Petronzio, CERN preprint, CERN-TH.4070/84, Dec. 1984.
9. H. E. Haber, Talk at this Conference.
10. K. Hagiwara, Talk at this Conference.
11. L. Ibanez and C. Lopez, Nucl. Phys. B233, 511 (1984); C. Kounnas, A. B. Lahanas, D. V. Nanopoulos and M. Quiros, Nucl. Phys. B236, 438 (1984).
12. J. Ellis and M. Sher, Phys. Lett. 148B, 309 (1984).
13. L. J. Hall and J. Polchinski, Phys. Lett. 152B, 335 (1985).
14. S. Nandi, Phys. Rev. Lett. 54, 2493 (1985).
15. J. Ellis, J. S. Hagelin and D. V. Nanopoulos, CERN-TH.4157/85, MIU-THP-85/014 preprint, 1985.
16. For example, see the reviews by P. Nath, R. Arnowitt and A. H. Chamseddine, Applied N = 1 Supergravity, the ICTP series in Theoretical Physical - Vol. 1 (World Scientific, Singapore, 1984); H. P. Nilles, Phys. Rep. 110, 1 (1984), and the references cited therein.
17. K. Inoue, A. Kakuto, H. Komatsu and S. Takeshita, Prog. Theo. Phys. 68, 927 (1982).
18. S. Wagner, talk at Santa Fe DPF meeting (1984).
19. P. Fayet, Phys. Lett. 117B, 460 (1982); J. Ellis and J. S. Hagelin, Phys. Lett. 122B, 303 (1983); K. Grassie and P. N. Pandita, Dortmund University preprint, DO-TH83/23 (1983); T. Kobayashi and M. Kuroda, Phys. Lett. 139B, 208 (1984); J. D. Ware and M. E. Machacek, Phys. Lett. 142B, 300 (1984).
20. E. Fernandez et al., Phys. Rev. Lett. 54, 1118 (1985).
21. K. Enqvist, D. V. Nanopoulos and A. B. Lahanas, CERN preprint, CERN-TH.4095/85 (1985).
22. B. W. Lee and S. Weinberg, Phys. Rev. Lett. 39, 165 (1977); D. A. Dicus, E. W. Kolb and V. L. Teplitz, Phys. Rev. Lett. 39, 168 (1977).
23. D. A. Dicus, E. W. Kolb and V. L. Teplitz, Astrophysical Jour. 221, 327 (1978).
24. H. Goldberg, Phys. Rev. Lett. 50, 1419 (1983).

25. J. Ellis, J. S. Hagelin, D. V. Nanopoulos, K. Olive and M. Srednicki, Nucl. Phys. B238, 453 (1984).
26. Our annihilation rate agrees with Ref. 25.
27. This gives a very good approximation to actually integrating the Boltzmann rate equation.
28. J. Silk and M. Srednicki, Phys. Rev. Lett. 53, 624 (1984); J. S. Hagelin and G. L. Kane, MIU preprint, MIU-THP-85/012; F. W. Stecker, S. Rudaz and T. F. Walsh, Univ. of Minnesota preprint, UMN-TH-520/85.
29. J. Ellis, K. Enqvist, D. V. Nanopoulos and K. Tamvakis, CERN preprint, CERN-TH.4108/85.
30. H. E. Haber, G. L. Kane and M. Quiros, Univ. of Michigan preprint, UM TH 85-12, 1985.

STANDARD MODEL SOURCES OF MISSING p_T

Ken-ichi Hikasa

Physics Department, University of Wisconsin, Madison, WI 53706

ABSTRACT

Cross section for high-p_T Z production with jet(s) is calculated. Uncertainties in the calculation are studied in detail. Other Standard Model sources of missing-p_T events are discussed.

A year ago we heard of the observation[1] of events with a large missing transverse momentum associated with a jet (so-called monojet events) at the CERN $p\bar{p}$ collider. What excited us so much was the claim that those events cannot be explained within the Standard Model.

There is a good source of missing momentum in the Standard Model—neutrinos. For example, if a Z boson is produced at high p_T with a recoiling jet and if the Z decays into a neutrino-antineutrino pair, the event looks like a monojet event. So, in a sense, what was unexpected about monojets is not the event configuration itself, but the magnitude of the cross section—or the number of events.

In the 1983 run, the UA1 experiment had an integrated luminosity of 0.113 pb^{-1} and they observed five events. The observed cross section was thus about 50 pb. Well, this is quite a small number. Why is this too large? To understand the situation, let me remind you of the general description of hard processes in $p\bar{p}$ interactions. The cross section in $p\bar{p}$ collisions can be decomposed:

$$\sigma \sim \hat{\sigma} \cdot \mathcal{L} \,, \tag{1}$$

or, more precisely,

$$\frac{d\sigma}{d\ell n \hat{s}} = \hat{\sigma}(\hat{s}) \cdot \tau \frac{d\mathcal{L}}{d\tau} \,. \tag{1'}$$

Here $\hat{\sigma}$ is the c.m. energy of the final system in interest (in the above example $Z+$jet c.m. energy), $\hat{\sigma}(\hat{s})$ is the cross section at the quark-lepton level ($q\bar{q} \to Zg$ or $qg \to Zq$ cross section), $\tau = \hat{s}/s$, and $\mathcal{L}$ is the relevant parton-parton luminosity function.

The parton luminosity for $u\bar{u}$ is shown in Fig. 1. It ranges between 0.1 and 1 at energies under consideration ($\sqrt{\hat{s}} \sim 100$ GeV). Other luminosities ($u\bar{d}, d\bar{d}, \ldots$) are less than that for $u\bar{u}$ in this region. Thus the luminosity function can be at most one for monojet production.

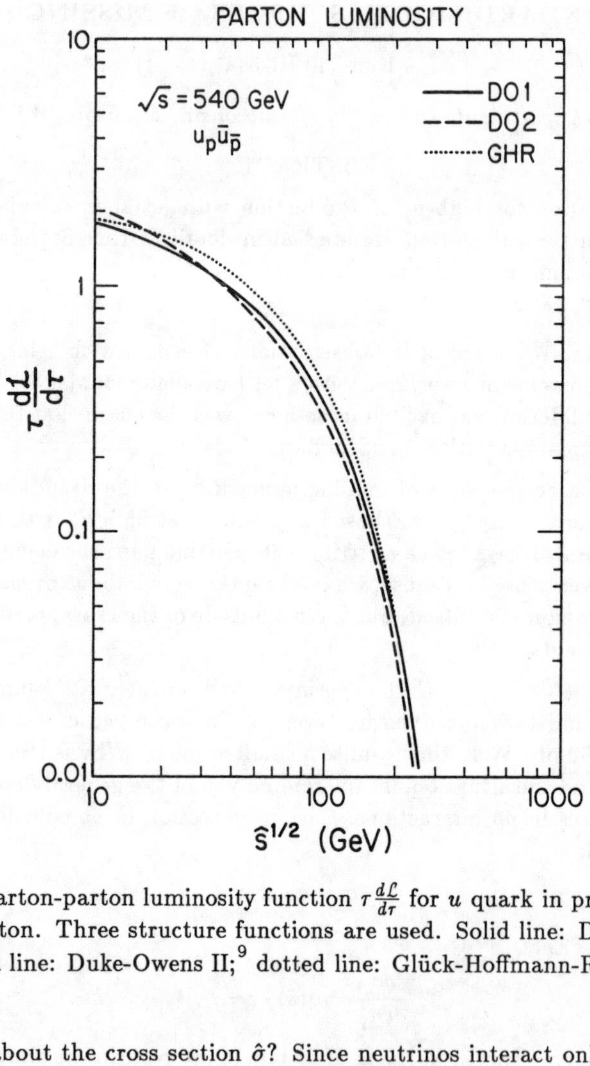

Fig. 1. Parton-parton luminosity function $\tau\frac{d\mathcal{L}}{d\tau}$ for u quark in proton and $\bar{u}$ in antiproton. Three structure functions are used. Solid line: Duke-Owens I;[9] dashed line: Duke-Owens II;[9] dotted line: Glück-Hoffmann-Reya.[10]

How about the cross section $\hat{\sigma}$? Since neutrinos interact only via weak bosons, the process should be electroweak. We have

$$\hat{\sigma} \sim \frac{\alpha^2}{\hat{s}} \lesssim 10 \text{ pb}\,, \tag{2}$$

given the energy scale $\hat{s} \sim (100 \text{ GeV})^2$. Thus we obtain

$$\sigma \lesssim 10 \text{ pb}\,. \tag{3}$$

This must be multiplied by branching ratio(s) and detection efficiencies, leading to a much smaller cross section than observed. This conclusion is quite independent of the detail of the production mechanism.

A number of scenarios have been proposed to explain the monojets. From the above point of view, they should contain some enhancement factor to get the right cross section. The luminosity function cannot be changed much. The cross section $\hat{\sigma}$ can be pushed up by employing a larger coupling factor than α^2 in Eq. (2):

Coupling	$\not{p}_T$ source	scenario
α_s^2	$\tilde{\gamma}$	SUSY[2]
1	ν	composite[3]
α	h^0	Z decay[4]

In addition, a smaller mass scale than $\hat{s}$ may be involved.

We have made a semi-quantative argument that the Standard Model gives too small a cross section for missing p_T. But the disagreement is within an order of magnitude. So in this talk I reexamine[5] the Standard Model predictions of missing-p_T sources, putting emphasis on the uncertainties in the calculation. I will mainly concentrate on the process $p\bar{p} \to Z +$jet(s) $+X$ as a source of monojets but discuss other processes briefly at the end of the talk.

We want to calculate the p_T distribution for the process

$$p\bar{p} \to Z + \text{jet(s)} + \text{anything}\,. \tag{4}$$

The most important elementary process for this reaction is

$$q\bar{q} \to Zg\,. \tag{5}$$

The crossed process

$$qg \to Zq \tag{6}$$

contributes less than 25% of the cross section.

The uncertainties in the evaluation can be divided into two categories: theoretical and semi-experimental. I discuss these in turn.

1. Theoretical uncertainties

These are purely theoretical problems in the calculation.

1a. *Mass scale* Q^2. We have a mass scale which enters in the running coupling constant $\alpha_s(Q^2)$ and parton densities $f(x, Q^2)$. In the present stage we have no means to fix this mass scale, as we do not have a complete next-order calculation. We take $Q^2 = p_T^2$ (of Z) and $\hat{s}$ ($q\bar{q}$ or qg c.m. energy) to represent the uncertainty of this quantity. The results of these two choices differ typically by 60%. Other choices such as $Q^2 = M_Z^2$ would give a cross section between the two, while $Q^2 = \hat{s}\hat{t}\hat{u}/(\hat{s}^2 + \hat{t}^2 + \hat{u}^2)$ gives a higher value than others.

1b. *Higher-order corrections.* As I mentioned before, there exists no complete calculation of the next-order ($O(\alpha_s^2)$) correction to the processes (5), (6). (There is a calculation for the valence part only.[7]) To estimate the corrections, let us take a look at the closely related process $q\bar{q} \to Z$. Here we know how large the correction is[8]—it is the famous K factor. The largeness of the correction comes from the following facts.

(1) We use the structure functions experimentally obtained from deep inelastic scattering where the momentum transfer is space-like.

(2) In the above process (or the Drell-Yan process), the momentum transfer is timelike. The logarithmic factor coming from mass singularity has an imaginary part π which remains after the singularity is absorbed in the renormalization of structure functions. This results in a large factor of π^2.

We use the same factor for the estimate of the corrections for (5), (6) because the momentum transfer is also time-like in these processes. This gives an increase of the cross section by 30–40%. It is not as large as 100% at lower energies because α_s gets smaller at high Q^2.

2. Semi-experimental uncertainties

By this I mean the inputs which should be measured by other experiments.

2a. *QCD coupling constant or* Λ. This is the basic parameter of the strong interactions and we have no prediction of this quantity so far. The experimental determination of Λ suffers from various difficulties, and what we can say quite surely is that Λ lies somewhere between 50 and 500 MeV. We treat this uncertainty together with the next one.

2b. *Structure functions.* The quark/gluon densities in a proton are in principle calculable from quantum chromodynamics, but we have no means to do so at present. They are determined from deep inelastic scattering experiments and inevitably have a certain error. We use several parametrizations of (Q^2-dependent) parton distributions including Duke-Owens I (Λ = 200 MeV),[9] Duke-Owens II (Λ = 400 MeV),[9] Glück-Hoffmann-Reya(Λ = 400

MeV),[10] Eichten-Hinchliffe-Lane-Quigg (Set 1, $\Lambda = 200$ MeV).[11] There is a 25% difference among the results.

I would like to note that all of the above parametrizations are derived using the best fits to the experiments. Although they surely reflect the uncertainties in structure functions, the actual uncertainty may be larger. It would be helpful to clarify this point if we could have several sets corresponding to, say, 1σ upper (lower) limit structure functions.

These uncertainties are summarized in Table I. Adding up these errors gives an overall uncertainty of a factor 3. The band in Fig. 2 shows the range of the results for the p_T distribution of Z and W. The upper boundary corresponds to the set ($Q^2 = p_T^2$, with K factor, Glück-Hoffmann-Reya), and the lower ($Q^2 = \hat{s}$, without K factor, Duke-Owens I). The integrated p_T distribution is shown in Fig. 3. The cross section for $p\bar{p} \to Z + \text{jet}$, $Z \to \nu\bar{\nu}$ can be as large as 10 pb for $p_T > 35$ GeV.

TABLE I. Uncertainties in $\dfrac{d\sigma}{dp_T}(p\bar{p} \to Z + \text{jet} + X)$

Mass scale	60%
Higher-order corrections	30–40%
α_s and structure function	25%
Overall	2.6–2.8

There is an important factor when we compare the prediction and experimental data. Since the p_T spectrum is a steeply falling function of p_T, the resolution for the missing p_T can have a large effect. For example, a Gaussian $\not{p}_T$ resolution of 10 GeV would cause an apparent shift of the distribution by ~ 5 GeV, equivalent to an increase of the cross section by 50–60%.

Finally, the cross section for $p\bar{p} \to Z + \text{jet(s)}$ can in principle be measured using other decay modes of Z. The hadronic decay modes (BR $\sim 70\%$) typically leads to three-jet final states. Although the branching ratio is much larger than the neutrino mode ($\sim 20\%$), the separation of the signal from the QCD three-jet background may not be easy. Leptonic decay modes of Z, especially e^+e^- and $\mu^+\mu^-$ final states, could give a clean signature. The

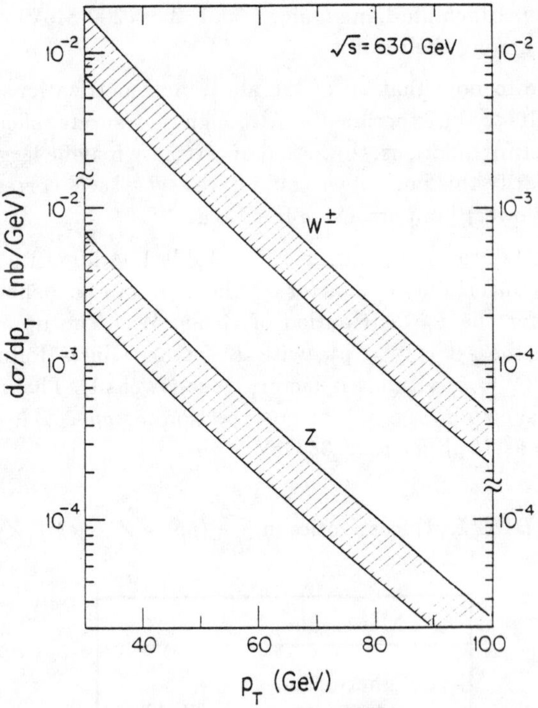

Fig. 2. Transverse momentum distribution for $W^{\pm}$ and Z in $p\bar{p}$ collisions at $\sqrt{s} = 630$ GeV. The bands represent the uncertainty in the calculation.

statistics in the future, this process should give a good calibration of the Z^0 p_T distribution.

Now I briefly discuss other sources of monojets.

(1) $W \to \tau\nu$, $\tau \to$ hadrons $+\nu$.[12] This process gives a monojet signature especially when the τ decays into several hadrons, since the second neutrino is likely to be soft. The cross section is ~ 20 pb for $\not{p}_T > 35$ GeV. Note that this process can be separated using the charged multiplicity cut.

(2) $p\bar{p} \to W+$ jet $+X$, $W \to (\ell)\nu$.[6] This looks like a monojet when the charged lepton escapes detection, mainly buried within the jet. The cross section is calculated to be 5–10 pb for $\not{p}_T > 40$ GeV. However, it would not be very difficult to separate this contribution because (i) the muon in jets can be measured, giving a cross section for all leptons when multiplied by 3; (ii) a jet with a hard electron in it would not look like a typical hadronic jet.

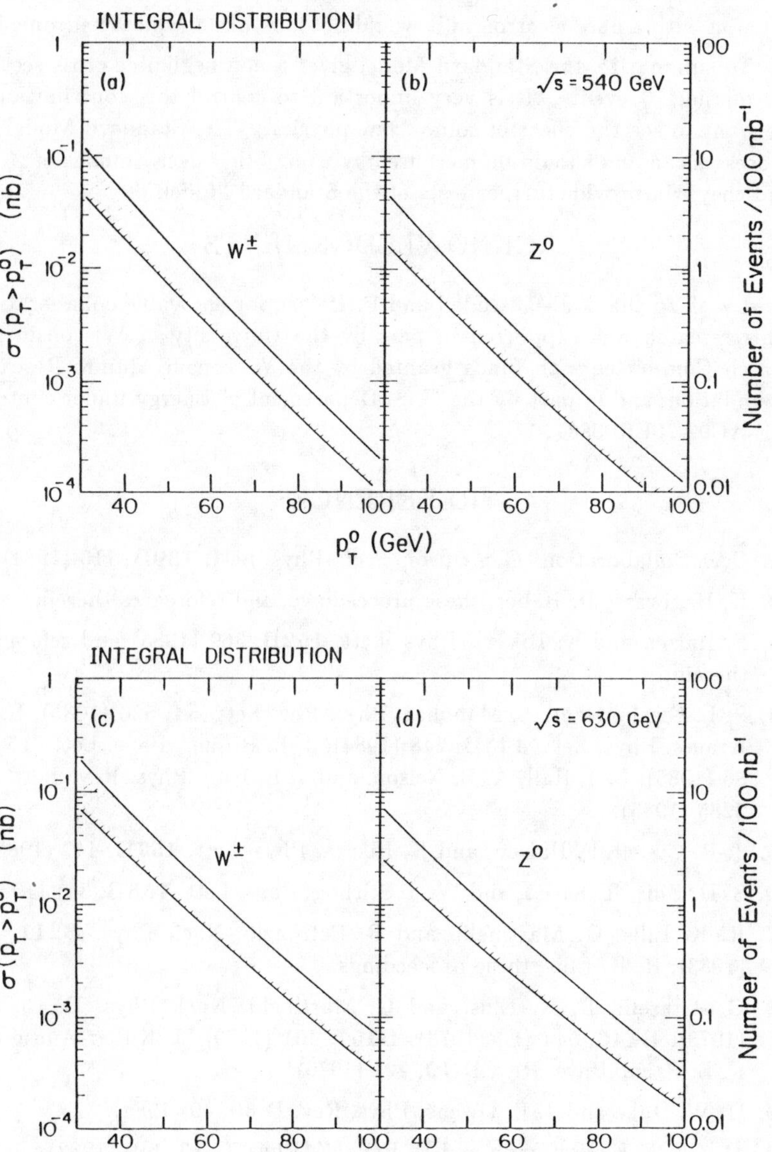

Fig. 3. Cumulated p_T distribution for $W^{\pm}$ and Z at $\sqrt{s} = 540$ and 630 GeV.

(ii) a jet with a hard electron in it would not look like a typical hadronic jet.

To summarize, the Standard Model gives a non-negligible cross section for missing-p_T events. It is very important to control this contribution if we want to see the effect of some "new physics". The Standard Model $\not{p}_T$ processes can and should be measured by using other decay modes of Z, W and they will provide further tests of the Standard Model.

ACKNOWLEDGEMENTS

I wish to thank J.-R. Cudell and F. Halzen for enjoyable collaboration. This research was supported in part by the University of Wisconsin Research Committee with funds granted by the Wisconsin Alumni Research Foundation, and in part by the U. S. Department of Energy under contract DE-AC02-76ER00881.

REFERENCES

1. UA1 Collaboration: G. Arnison *et al.*, Phys. Lett. **139B**, 115 (1984).
2. K. Hagiwara, H. Haber, these proceedings, and references therein.
3. F. Halzen and K. Hikasa, Phys. Lett. **152B**, 369 (1985) and references therein.
4. S. L. Glashow and A. Manohar, Phys. Rev. Lett. **54**, 526 (1985); L. M. Krauss, Phys. Lett. **143B**, 248 (1984); J. L. Rosner, Phys. Lett. **154B**, 86 (1985); L. J. Hall, A. E. Nelson, and J. E. Kim, Phys. Rev. Lett. **54**, 2285 (1985).
5. J.-R. Cudell, F. Halzen, and K. Hikasa, Phys. Lett. **157B**, 447 (1985).
6. S. D. Ellis, R. Kleiss, and W. J. Stirling, Phys. Lett. **158B**, 341 (1985).
7. R. K. Ellis, G. Martinelli, and R. Petronzio, Nucl. Phys. **B211**, 106 (1983); R. K. Ellis, these proceedings.
8. G. Altarelli, R. K. Ellis, and G. Martinelli, Nucl. Phys. **B143**, 521 (1978), **B146**, 544 (E) (1978), **B157**, 461 (1979); J. Kubar-André and F. E. Paige, Phys. Rev. D **19**, 221 (1979).
9. D. W. Duke and J. F. Owens, Phys. Rev. D **30**, 49 (1984).
10. M. Glück, E. Hoffmann and E. Reya, Z. Phys. C **13**, 119 (1982).
11. E. Eichten, I. Hinchliffe, K. Lane, and C. Quigg, Rev. Mod. Phys. **56**, 579 (1984).
12. P. Aurenche and R. Kinnunen, Z. Phys. C **28**, 261 (1985); E.W.N. Glover and A. D. Martin, Z. Phys. C **29**, 399 (1985).

DIMUON EVENTS AT THE CERN $p\bar{p}$ - COLLIDER

Karsten Eggert
III. Physikalisches Institut, Aachen, Germany

ABSTRACT

The observation of 433 muon pairs at the CERN $p\bar{p}$ - Collider is reported. The isolated unlike-sign pairs show a clear Υ signal above the Drell-Yan continuum. The non-isolated dimuons from heavy flavour decays indicate a large $b\bar{b}$ production. The like to unlike-sign ratio of 0.48 ± 0.08 in the heavy flavour sample cannot be interpreted by second generation decays alone, but suggests a sizeable $B^0 - \bar{B}^0$ - mixing.

INTRODUCTION

Shortly after the discovery of Z^0 decays into muons at the CERN $p\bar{p}$ - collider [1] interest also concentrated on lower mass muon pairs. Muons offer the experimental advantage that they can still be identified at low momenta, even in the presence of hadronic activity. Hence the search for dimuons has been extended down to muon pair masses of 6 GeV/c^2, where the background, mainly due to pion and kaon decays, is still acceptable (~ 10%).

Lepton pairs in $p\bar{p}$ - collisions originate from several competitive sources which are often difficult to disentangle.

In the Drell-Yan mechanism a muon pair is produced via a virtual photon:

$$p\bar{p} \to \gamma^* + X$$
$$\hookrightarrow \mu^+ \mu^-$$

The process is characterised by two oppositly charged muons which are isolated, i.e. not accompanied by hadronic activity. The underlying event should be similar to that in a minimum bias event. Unlike sign muon pairs can also arise from J/ψ and Υ - decays.

However, the majority of the dimuon events comes from semileptonic decays of heavy quarks, like CHARM and BOTTOM. Because of the large centre-of-mass energy $\sqrt{s}$=630 GeV at the $p\bar{p}$ - collider these heavy quarks are copiously produced in pairs by various QCD processes, e.g. gluon - gluon fusion.

$$p\bar{p} \rightarrow QQ + X \qquad Q = c, b, t$$
$$Q \rightarrow q\,\ell\,\nu$$

Furthermore heavy quark pairs can also arise from the weak decays of the Intermediate Vector Bosons in which case the muons are produced together with large p_t jets. Once sufficient statistics are accumulated, the $W \rightarrow t\bar{b}$ decay with two same sign high p_t muons and obvious jets represent one of the most convincing signatures for the TOP quark.

Fig.1 schematically indicates the heavy quark decay cascades leading to multilepton final states (from Ref.2). First generation decays of pair produced b- and c-quarks lead to unlike-sign muon pairs, where the muons are in general accompanied by hadronic activity. The production cross-sections of b- and c-quarks are comparable at transverse momenta above the b-quark mass. However, when a threshold of p_t >3 GeV/c is applied to the muons, the dominant contribution to the dimuon sample comes from $b\bar{b}$ final states rather than $c\bar{c}$. This is mainly a consequence of the harder fragmentation function ($\langle z \rangle \sim 0.8$) of the b-quark. Muons can also come from second generation decays leading to same sign dimuons. Second generation muons have a softer p_t spectrum and are therefore suppressed once a p_t cut is applied to the muon.

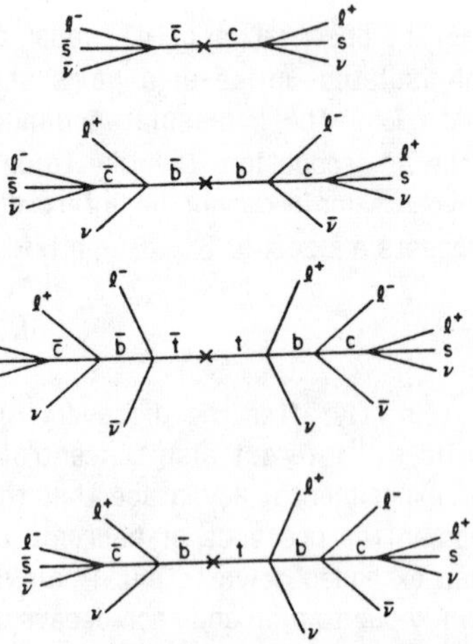

Fig. 1 : Fish bone diagrams, indicating the heavy quark decays, which lead to multilepton final states.

There is an exciting additional source of like-sign muons from first generation decays through $B^0 - \bar{B}^0$ oscillations [3], analogous to $K^0 - \bar{K}^0$ oscillations. In the standard model, the degree of the $B^0 - \bar{B}^0$ mixing is determined by the Kobayashi - Maskawa matrix [4] which largely favours mixing in the B^0_s channel compared to the B^0_d channel. Data from e^+e^- machines give stringent limits only on the B^0_d channel, but no limits on the B^0_s channel, where mixing may be almost complete.[5] The large $b\bar{b}$ cross-section at the $p\bar{p}$ collider which dominates the dimuon sample makes it feasible to measure the degree of $B^0 - \bar{B}^0$ oscillations from the ratio of like-sign to unlike-sign events. Since mixing is expected to occur predominantly in the B^0_s channel, it effects only the small fraction of the b quarks which hadronize into a B^0_s. Thus the effect on the ratio of like-sign to unlike-sign muon pairs is about 0.2 for full mixing.

Whereas on an event-by-event basis it is difficult to distinguish the different sources for dimuons, there are several criteria by which the dimuon sources can be statistically identified : the mass of the muon pair, the hadronic activity near the muon, the charges of the two muons, and the mass of the μ-μ-jet-jet system. Muons from heavy flavour decays are in general accompanied by hadronic activity, in the Drell Yan process they are isolated. However, due to the bias of demanding a large p_t muon, this muon often takes the largest momentum in the semileptonic decay of a low mass QQ pair and appears therefore isolated, just resembling the Drell-Yan process.

DETECTOR, TRIGGER AND EVENT SELECTION

The UA1 detector has been described in detail elsewhere [6]. In brief, a muon from the $p\bar{p}$ - interaction traverses the Central Detector, where the momentum is determined by the deflection in the central magnetic dipole field of 0.7 T. After penetrating the electromagnetic and hadronic calorimeters and the additional iron absorber (a total of more than 9 interaction lengths) the muon is finally detected in the muon chambers, which form the outer shell of the UA1 detector and cover 70% of the pseudorapidity range $|\eta| < 2.0$. A fast hardware muon trigger, based on the hit information in the drift tubes of the muon chambers, selects muon candidates by requiring a track pointing to the interaction region within a cone of aperture ± 150 mrad. Data were recorded using an inclusive muon trigger with a pseudorapidity acceptance of $|\eta| < 1.5$ and a dimuon trigger within $|\eta| < 2.0$.

The data presented in this paper were collected during the three collider periods from 1983 to 1985. The total integrated luminosity amounts to 692 nb^{-1}, 108 nb^{-1} at $\sqrt{s}$=546 GeV (1983) and 584 nb^{-1} at $\sqrt{s}$=630 GeV (1984/85). During data taking the most interesting events with a large p_t muon or with two muons identified by the muon hardware trigger were selected on a special tape, the so called express-line tape. Whereas the analysis of the 1983/84 data is completed, only the express-line data from 1985 were analysed. Because of the high dimuon trigger efficiency, most of the dimuon data should already be contained in this express-line sample. Dimuon events were selected with the following cuts:

$p_t(\mu) > 3$ GeV/c for each muon

$M(\mu\mu) > 6$ GeV/c^2

A seperate analysis without the above mass cut is in progress. After some technical cuts to reduce the background and a scan of all events on a graphic display 433 dimuon events (Z^0 - candidates were removed) remain in the final sample.

The background to this dimuon sample has been studied extensively. The probability that a hadron simulates a muon either by penetrating the absorber without interaction or by leakage of the induced hadron shower was found to be small ($<10^{-4}$). Severe background arises from pions (kaons) which decay in flight in the central detector with a probability of $0.02/p_t$ $(0.11/p_t)$. We selected an inclusive sample with a $p_t > 3$ GeV/c muon and an additional high p_t particle which could have decayed into a muon. The decay in the central detector was then simulated assuming 50% pions and 25% kaons and the same cuts as for the dimuon sample were applied. Thus the overall background to the dimuon sample was estimated to be 50 events (preliminary for 1985). For isolated muons the background is about 5 times smaller.

CLASSIFICATION OF THE EVENTS

With the aim to disentangle the various dimuon sources, we classify the events according to the pair charge (like or unlike sign) and the degree of muon isolation, which is an important tool to distinguish between the Drell-Yan process and heavy flavour decays. After the energy deposited by the muon in the calorimeters was subtracted the isolation is defined by the transverse energy around the muon, i.e. the sum ΣE_t of the transerve energy of all calorimeter cells within a cone $\Delta R < 0.7$ about the muon in pseudorapidity η - azimuth ϕ - space (ϕ about the beam axis).

$$\Sigma E_t \text{ in } \Delta R = (\Delta\eta^2 + \Delta\phi^2)^{1/2} < 0.7$$

This isolation quantity is plotted in fig.2 for like - and unlike - sign pairs. Both muons are considered as isolated, if

$$S = \sum_{0.7} E_t(\mu_1)^2 + \sum_{0.7} E_t(\mu_2)^2 < 9 \text{ GeV}^2$$

From examining muons from W and Z decays, which should be isolated, we deduced a $\geqq 80$ % efficiency of the above cut for events with two isolated muons. For unlike-sign events the clearly visible enhancement at $S < 9$ GeV^2 arises from the Drell-Yan process and Υ - decays. The wide distribution of events outside the above circle indicates the large contribution from heavy flavour decays. The only contribution to like-sign events is heavy flavour decays. The distribution in S is rather flat and similar to the unlike-sign events, except for an obvious depletion for isolated events. Applying the shape of the S-distribution of like-sign events to the unlike sign event indicates a small, but not negligible, contribution from heavy flavour decays to isolated muon pairs. However, the two distributions do not have to be identical, since like-sign events can also come from the second generation decays which may show a hadronic activity different from that of the first generation decays.

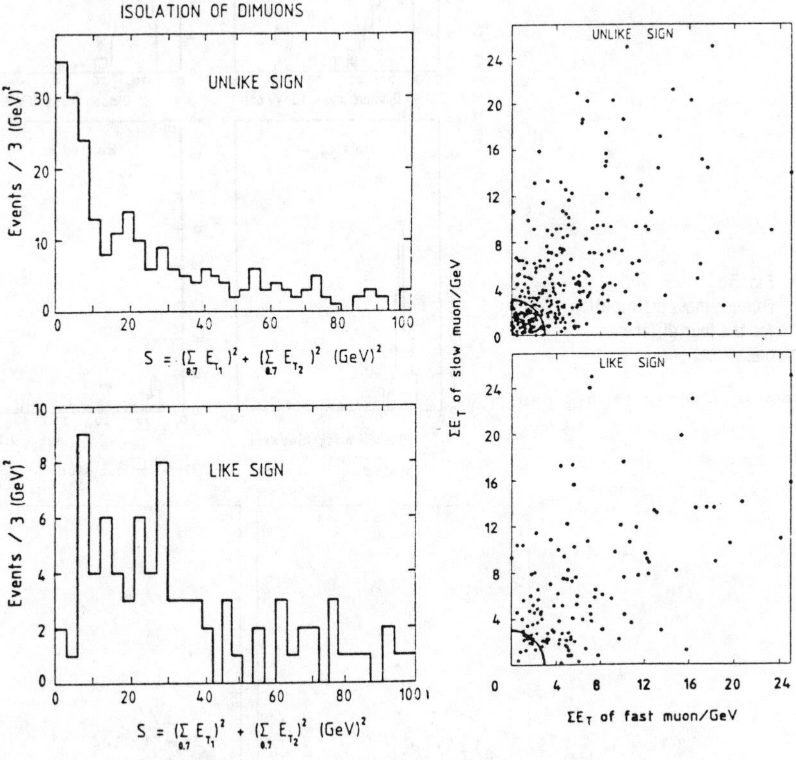

Fig. 2 : Scatter plot of the muon isolation defined by ΣE_t in a cone of $\Delta R < 0.7$ about the muon, for like- and unlike-sign muon pairs. Distribution of the quantity S, which is the radius squared in the scatter plot.

We classify the data in like- and unlike-sign and isolated and not isolated events. Fig. 3 shows the dimuon mass distribution and Fig. 4 the scatter plot of the transverse momenta of the two muons for the above four classes. Heavy flavour decays dominate the non-isolated and like-sign events. We observe a remarkable large fraction of like-sign events, some of them even being isolated.

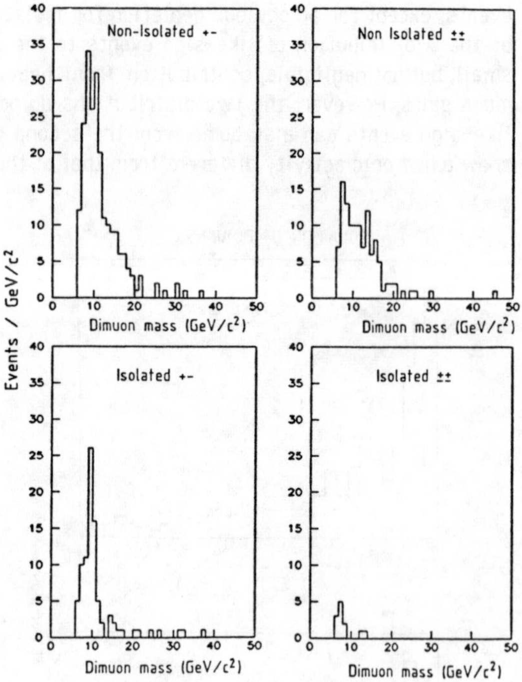

Fig. 3 :
Dimuon mass distribution for the four different classifications

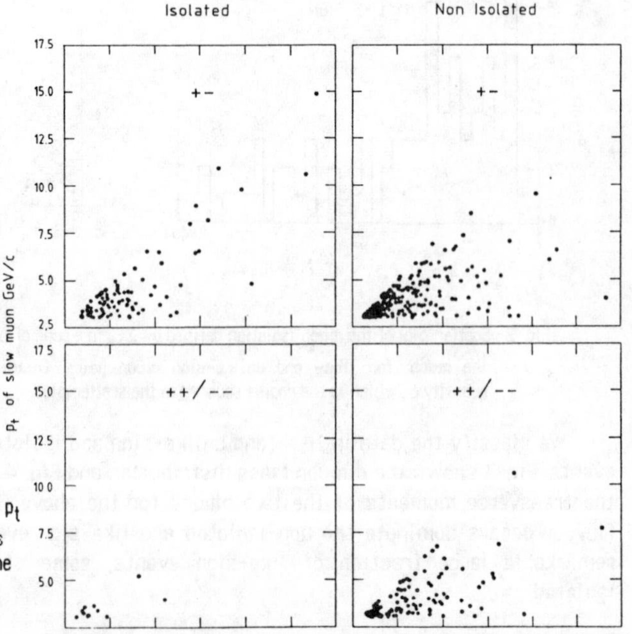

Fig. 4 :
Transverse momentum p_t of the fast muon versus the slow muon p_t for the four classifications

DRELL-YAN AND UPSILON PRODUCTION

We observe 89 isolated unlike-sign events, mainly due to the Drell-Yan mechanism and Υ - decays with some contamination from heavy flavours. The dimuon mass distributions in Fig. 5 shows a clear Υ - peak. Since the detector resolution is not sufficient to separate the individual Υ - states, a ratio of 1 : 0.3 : 0.15 was assumed for the Υ, Υ', and Υ'' states. Allowing for the detector resolution, the mass fit agrees well with the observed peak.

The contributions from Drell-Yan and heavy flavours are difficult to disentangle. The mass dependence for the two processes is similar and dominated at low masses by the acceptance due to the cuts. The shape of the mass distribution was determined using the ISAJET Monte Carlo program[7] and a simulation of the UA1 detector. The Drell-Yan fit in Fig. 5 was calculated under the assumption that at large masses, $m_{\mu\mu} > 11$ GeV/c^2, the only contribution is from Drell-Yan. The heavy flavour and Υ contributions are then fitted to the Drell-Yan subtracted mass distribution. There are large uncertainties in the cross-section estimate due to this method. If the isolation of muons from heavy flavours is better understood, the heavy flavour contribution can be estimated from Fig. 2. With the above method the production cross-sections for Drell-Yan and Υ are as follows :

$$\sigma_{DY}(m_{\mu\mu} > 11 \text{ GeV/c}^2) = 287 \pm 65 \text{ pb}$$

$$BR \cdot \sigma(p\bar{p} \rightarrow \Upsilon, \Upsilon', \Upsilon'') = 860 \pm 206 \text{ pb}$$

The systematic errors are of the same order as the quoted statistical ones. In Fig. 6 our measurements are compared with results from low energy data[8,9]. They are consistent with an extrapolation based on a scaling behaviour. A recent theoretical calculation [10] yields $\sigma_{DY}(m_{\mu\mu} > 11 \text{ GeV/c}^2) =$ 290 pb.

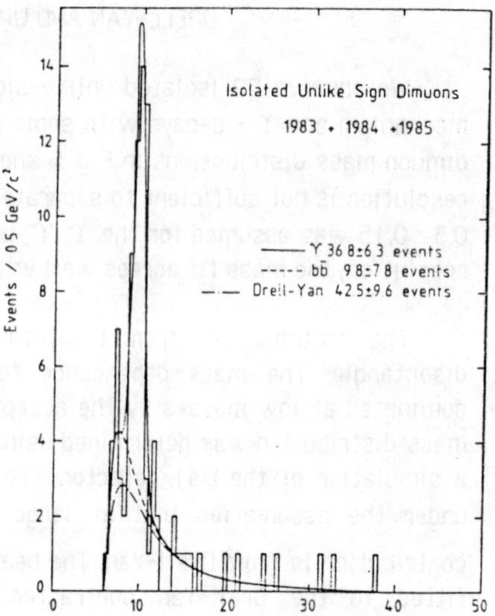

Fig. 5 : Dimuon mass distribution for isolated unlike-sign muons. The simultaneously fitted contributions of the Drell-Yan mechanism, heavy flavour decays, and Υ - decays are indicated.

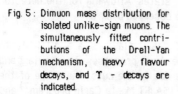

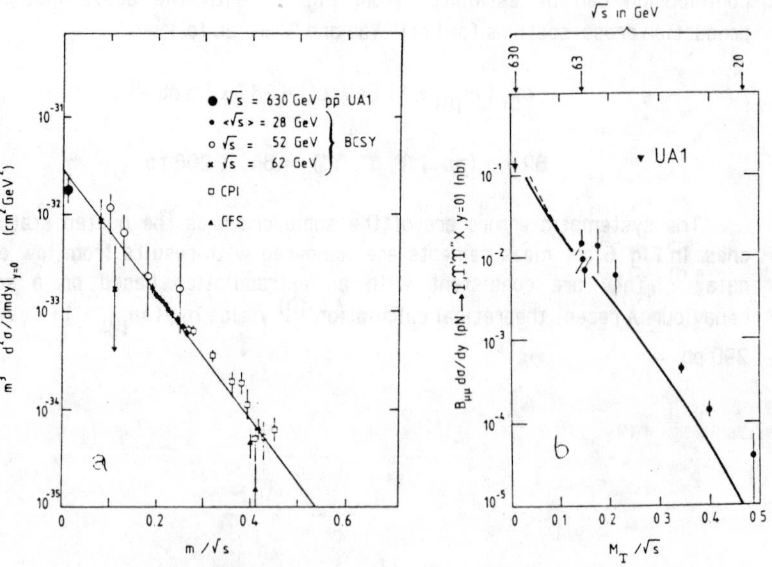

Fig. 6 : Comparison of UA1 measurements of a) Drell-Yan and b) Υ production cross-sections with extrapolations from low energy data.

HEAVY FLAVOUR PRODUCTION

The non-isolated events (216 unlike-sign and 116 like-sign) and the 12 isolated like-sign events are expected to come from heavy flavour decays. Monte Carlo studies with the EUROJET program [11] which includes second order QCD processes, have shown that the majority of the events come from $b\bar{b}$ - decays (82 %). Charm is suppressed because of its softer fragmentation.

An approximation to the mass of the produced heavy quark pair can be obtained by calculating the invariant mass of the muon-muon-jet-jet system. The nearest jet to each muon within $\Delta R < 1$ is used. If there is no identified jet, the energy vectors from all cells in $\Delta R < 0.7$ around the muon are summed and the resulting vector is used as the jet momentum. Fig. 7 shows this mass distribution together with the expectation of the COJET Monte Carlo[12], normalized to the data. Taking into account that the Monte Carlo distribution has not been smeared with the detector resolution the agreement is remarkable. The high mass events mainly come from a configuration with a forward and backward jet and may reflect a better trigger efficiency for muons in the forward direction.

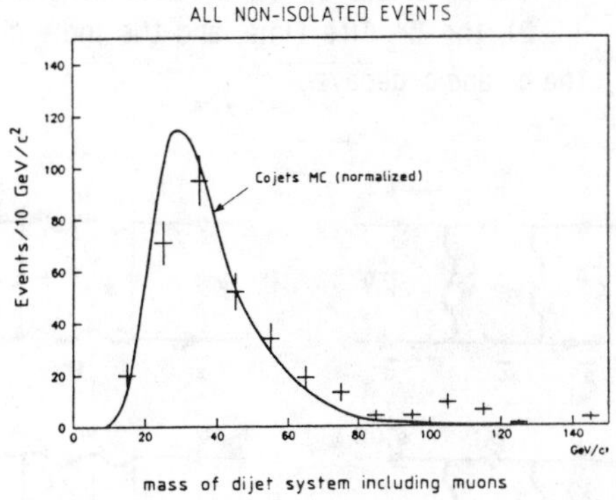

Fig. 7 : Distribution of the muon-muon-jet-jet mass. The curve is a Monte Carlo prediction from COJET.

Using the Monte Carlo and including the geometrical acceptance of the chambers, the track finding efficiency in the Central Detector and corrections for background subtraction we obtain the following $b\bar{b}$ cross-section :

$$\sigma(p\bar{p} \to b\bar{b}) = 1.2 \pm 0.07_{stat} \pm 0.2_{syst}\ \mu b$$

$$p_t(b) > 5\ GeV/c \quad \text{for both quarks}$$

$$|\eta(b)| < 2 \quad \text{for both quarks}$$

No allowance is made for uncertainties in the Monte Carlo calculation.

B^0-$\bar{B}^0$ - Oscillations

B^0-$\bar{B}^0$ - oscillations are expected in close analogy to the well-known phenomenon of K^0-K^0 - oscillations. The relevant box diagrams are shown in Fig. 8. As a consequence of the K-M-Matrix the parameters of which are now quite well constrained by experiments, the mixing in the B^0_d-$\bar{B}^0_d$ - system should be small, whereas it is probably large in the B^0_s-$\bar{B}^0_s$ - system. New experimental data from e^+e^- machines exclude a significant mixing in the B^0_d - system, but not in the B^0_s- system.

The ratio of like- to unlike-sign dimuons $R = \mu^{\pm}\mu^{\pm}/\mu^+\mu^-$ depends on the degree of B^0-$\bar{B}^0$ mixing, but also on details of the b-fragmentation and decay, which can yield like-sign dimuons via second generation decays. We have used several Monte Carlo calculations (ISAJET, EUROJET, M.C. of Barger and Phillips, M.C. of Halzen and Martin) to determine the ratio R. All calculations agree with a value R = 0.2-0.25 in the case of no mixing, and 0.45-0.5 for full mixing in the B^0_s channel. It has to be stressed that the above results depend on the choice of various parameters, like the strangeness content in the sea (we assumed s : d = 1 : 2), the B^0_s life time, and the individual semi-leptonic branching ratios in the b- and c-decays.

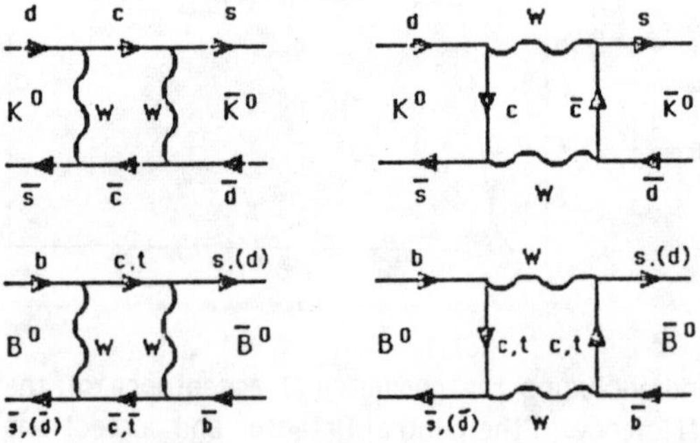

Fig. 8 : Box diagrams responsible for B^0-$\bar{B}^0$ and K^0-K^0 mixing.

Experimentally the ratio R can be obtained in two slightly different ways : either by considering only the non isolated events or with the total event sample with Drell-Yan and Υ contributions subtracted. The two results are in good agreement.

$$R = 0.48 \pm 0.08 \text{ for non isolated events}$$
$$R = 0.51 \pm 0.08 \text{ including the isolated events}$$

The above errors contain the systematics. We have assumed that the 50 background events are evenly distributed between ±± and +- signatures, and also included misassociations of muon and CD track.

We can study the dependence of the ratio R on different cuts, which could reduce second generation decays. Fig. 9 shows R as a function of the transverse momentum cut on the muons together with Monte Carlo calculations from EUROJET. The data are in good agreement with full mixing in the B^0_s channel and indicate that the effect also persists at larger muon p_t. In Fig. 10, R is plotted versus the μ-μ-jet-jet mass, which is a good approximation to the $b\bar{b}$ - mass. The line is an EUROJET calculation for the case of no mixing. The effect of mixing can be best seen at low $b\bar{b}$ masses, whereas at large masses second generation decays dominate.

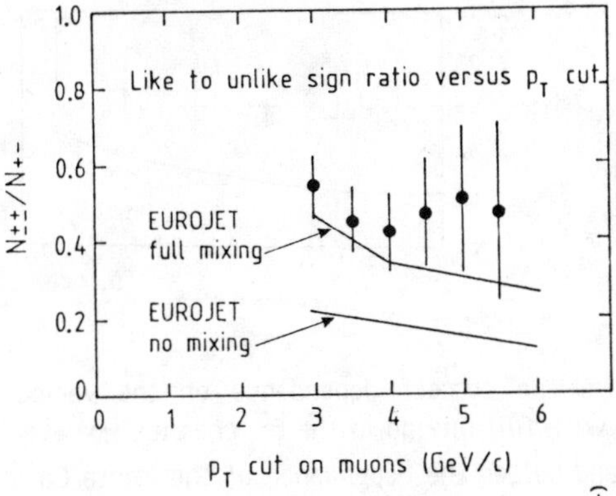

Fig. 9 : The ratio R as a function of the cut on the muon p_t

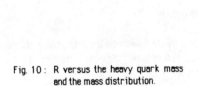

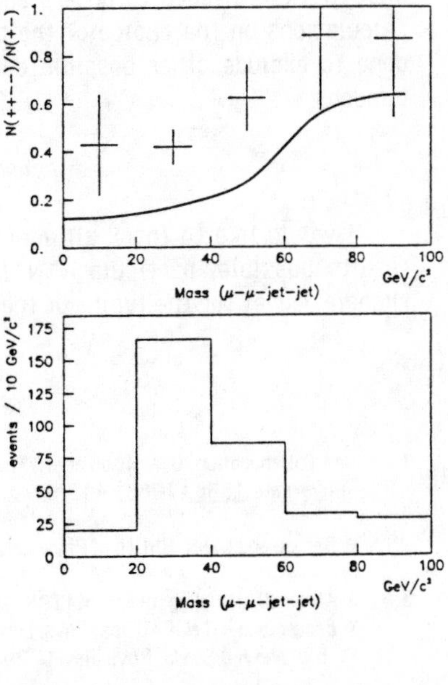

Fig. 10 : R versus the heavy quark mass and the mass distribution.

Fig. 11 shows the dependence of R on the muon isolation variable S (defined earlier) and on the isolation of the slow muon, which should rarely be isolated, if it comes from second generation decays. The curve represents the EUROJET calculation for no mixing.

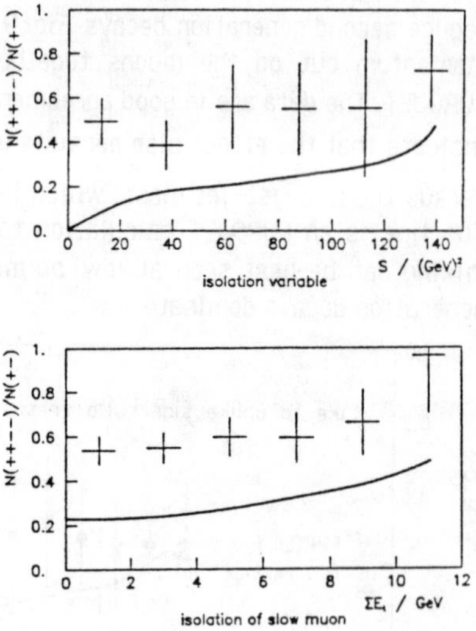

Fig. 11 : R versus the isolation variable S and the isolation of the slow muon (as defined earlier)

The observed ratio R shows the correct dependence on the various parameters, and is in agreement with full mixing in the $B^0{}_s$ channel. However, more work is needed to understand better the dependance of the Monte Carlo calculations on the choice of the parameters. Also further studies have to be done to exclude other possible decay models which could lead to same sign dimuons.

Acknowledgements

I would like to thank all my colleagues in UA1 who helped to make these results possible, particularly N. Ellis and H.G. Moser. Special thanks go to Michèle Jouhet for the typing of the manuscript.

REFERENCES

1. UA1 Collaboration, G. Arnison et al., Phys. Lett. 147B (1984) 241, Phys. Lett. 155B (1985) 443.

2. V. Barger and K.J.N. Phillips, Phys. Lett. 143B (1984) 259.

3. A. Ali and C. Jarlskog, preprint CERN-TH 3896 (1984).
V. Barger and R.J.N. Phillips, Phys. Lett. 143B (1984) 259.
I.I. Bigi and A.I. Sanda, Phys. Rev. D. 29 (1984) 1393.

4. M. Kobayashi and M. Maskawa, Prog. Theor. Phys. 49 (1973) 652.

5. P. Avery et al., Phys. Rev. Lett. 53 (1984) 1309.
T. Schaad et al., Phys. Lett. 160B (1985) 188.

6. M. Barranco et al., Nucl. Instrum. and Methods 176 (1980) 175
M. Calvetti et al., Nucl. Instrum. and Methods 176 (1980) 225
M.J. Corden et al., Rutherford preprint RL-83-116 (1983)
K. Eggert et al., Nucl. Instrum. and Methods 176 (1980) 217 and 233.

7. F.E. Paige and S.D. Protopescue, ISAJET program, BNL 31987 (1982).

8. C.W. Fabian, Acta Phys. Austriaca, Suppl. XIX (1978) 621.

9. R.J.N. Philipps, in Proc. of the XX International Conf. on High Energy Physics, Madison (1980).

10. G. Altarelli et al., preprint CERN-TH 4015 (1984)

11. B. van Eijk, these proceedings.
B. van Eijk, Proceedings of the 5th Topical Workshop on Proton-Antiproton Collider Physics, St Vincent (1985) 165, ed. by M. Greco.

12. R. Odorico, CERN-TH 3760, Nucl. Phys. B228 (1983) 381.

PRODUCTION PROPERTIES OF THE INTERMEDIATE VECTOR BOSONS AT THE SPS COLLIDER

UA1 Collaboration, CERN, Geneva, Switzerland

Aachen – Amsterdam (NIKHEF) – Annecy (LAPP) – Birmingham – CERN – Harvard – Helsinki – Kiel – London (Queen Mary College) – Padova – Paris (Coll. de France) – Riverside – Rome – Rutherford Appleton Lab. – Saclay (CEN) – Vienna – Wisconsin

Presented by E. Tscheslog, RWTH Aachen, Germany

Geneva, 13 July 1985

Abstract: The production properties of Intermediate Vector Bosons produced at the CERN Super Proton Synchrotron (SPS) Collider and observed in the electron and the muon decay channels are described. The longitudinal and transverse momentum distributions, and the properties of the hadronic jet activity produced in association with the weak bosons are in agreement with the expectations of the QCD improved Drell – Yan mechanism.

1. INTRODUCTION

Data taken at the CERN $p\bar{p}$ Collider during 1982/1983 have enabled the two experiments UA1 and UA2 [1] to prove the existence of the Intermediate Vector Bosons. UA1 has observed 66 W^{+-} decays and 9 Z^0 decays in these periods. With the the addition of the data from 1984 run UA1 has now observed a sample of 216 W^{+-} events (172 $W \rightarrow e\nu$, 44 $W \rightarrow \mu\nu$) and a sample of 31 Z^0 events (22 $Z^0 \rightarrow e^+e^-$, 9 $Z^0 \rightarrow \mu^{+-}$). These event yields have allowed the study of the production properties of the IVB, the W in particular, in detail [2]. The longitudinal momentum of the IVB's is presented in chapter 2 of the following paper and will be discussed with respect to the structure functions of the annihilating quarks. Chapter 3 shows the transverse momentum distribution of the IVB's and how it can be understood in the framework of QCD. QCD effects also give a good description of the jets in IVB events. These are discussed and shown to be in agreement with QCD expectations for initial state bremsstrahlung from the incoming partons. Finally this study leads to an upper limit of the production of heavy objects decaying into a W and a jet.

2. LONGITUDINAL MOTION

The longitudinal momentum of the W cannot be observed directly as the longitudinal momentum of the neutrino is principally not measured. However when one fixes the invariant mass of the lepton and neutrino system to the W mass the range of the possible Feynman x−values of the neutrino reduces to only two values. In one of them the neutrino is emitted forward in the rest frame of the W; in the other one its direction is backward. It turns out that in about a third of the events where we observe the W decay into electron and neutrino one solution is ruled out because it leads to unphysical x−values of the W ($x_W > 1$). In another third of the events only one solution is compatible with energy and momentum conservation when one takes into account the energy flow in the rest of the event. In the remaining ambiguous cases the two possible longitudinal momenta of the W are very similar. To resolve this we take the most probable solution, the one with the smallest value of X^W. This is handled in the same way in the Monte Carlo studies. For the events in which the W decays into a muon and its neutrino the momentum of the lepton is determined by magnetic deflection in the central drift chamber only, so it is measured with less resolution (typically 30%) than the electron momentum (typically 2.5%). Therefore to study the longitudinal momentum of the W in these events we restrict ourselves to the events where the transverse mass of the muon and the neutrino is less than the W mass. The longitudinal momentum of the W is evaluated in the same way as for the W's observed in the electron channel. In contrast to the electron W sample the muonic decays of the W can have rather different x−values in case of ambiguous longitudinal momenta; so we take them both into account with a weight of one half.

At $p\bar{p}$ Collider energies of 546 or 630 GeV we can assume that W's are dominantly produced by annihilating valence quarks and antiquarks in a Drell−Yan type mechanism [3]. Their structure functions are thus probed by the W resonance

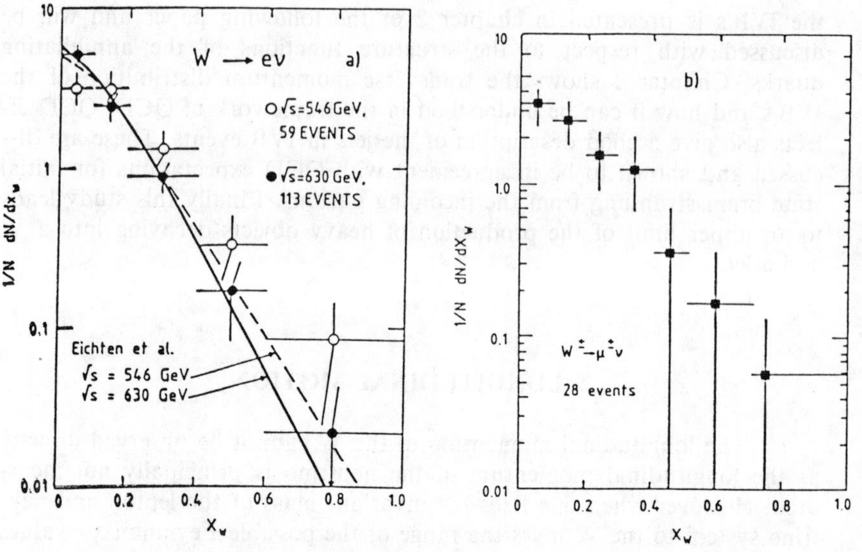

Figure 1: Longitudinal Momentum of the W

a) Feynman x – distribution for W → eν at $\sqrt{s}$ = 546 GeV/c (open circles) and $\sqrt{s}$ = 620 GeV/c (closed circles). The curve is the prediction using the model described in the text.

b) Feynman x – distribution for W → μν.

$$x_q x_{\bar{q}} = M^2{}_W/s. \tag{1}$$

Because of momentum conservation the difference of the two parton x – values determines the longitudinal momentum of the W

$$|x_q - x_{\bar{q}}| = x_W. \tag{2}$$

We would therefore expect the longitudinal momentum of the W to reflect the structure functions of the partons (Fig. 1). Choosing a parametrization of the structure functions as evaluated by Eichten et al. (Λ = 0.2 GeV) a Monte Carlo [4] prediction reproduces well the measured distribution.

Equations (1) and (2) enable us to determine the Feynman x – value of the partons annihilating in the W production process. Fig. 2 presents the parton distributions in the proton and the antiproton obtained from the W → eν sample. They are essentially identical thus providing us with another way to demonstrate the CP invariance of proton and antiproton. The distributions are in agreement with Monte Carlo expectations that use the structure functions mentioned above.

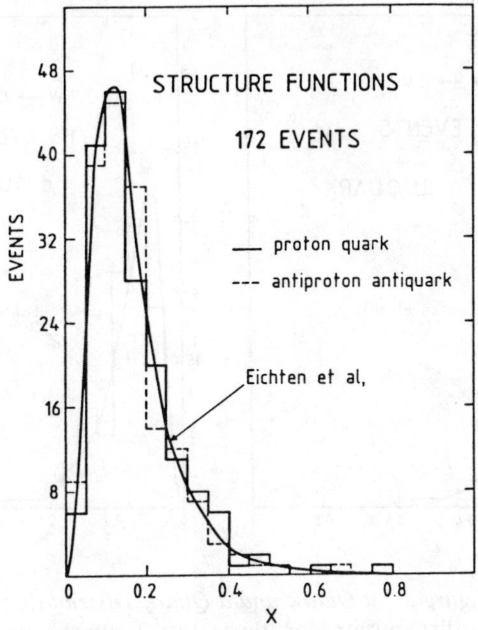

Figure 2: Parton Distributions

Feynman x−distribution for the proton (solid curve) and antiproton (dashed curve) partons making the W (W→eν sample). The curve shows the Eichten et al. prediction.

When the charge of the W is known the annihilating quark−antiquark pair can be identified as a $(u,\bar{d})$ or a $(\bar{u},d)$ pair, respectively:

$$(x_q, x_{\bar{q}}) \rightarrow (x_u, x_{\bar{d}}), (x_{\bar{u}}, x_d) .$$

By choosing the appropriate charge the $u(\bar{u})$ and $d(\bar{d})$ quark distributions (Fig. 3) can be determined for the proton(antiproton). Their different x dependances

$$xu(x) \sim (1-x)^3 \tag{3}$$
$$xd(x) \sim (1-x)^4 \tag{4}$$

are washed out as the two structure functions are sampled via the W pole. The distributions are in agreement with predictions assuming Eichten et al. structure functions.

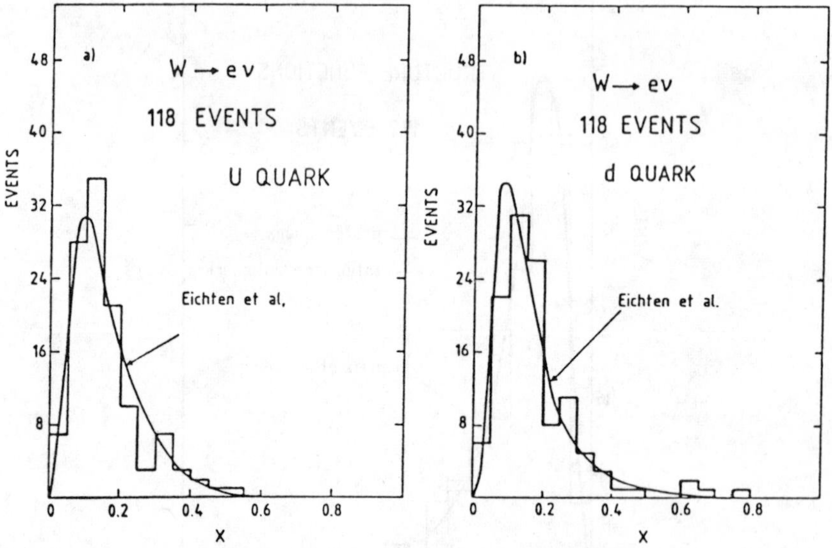

Figure 3: u Quark and d Quark Distributions
a) Feynman x–distribution for those u($\bar{u}$) quarks producing the W (W→eν sample). The curve shows the Eichten et al. prediction.
b) Feynman x–distribution for those d($\bar{d}$) quarks producing the W (W→eν sample). The curve shows the Eichten et al. prediction.

3. TRANSVERSE MOMENTUM

In the IVB production at the CERN $p\bar{p}$ Collider the incoming quarks and antiquarks carry colour and can therefore radiate gluons. In order to describe this initial state parton bremsstrahlung one has to improve the bare Drell–Yan production mechanism by including higher order QCD corrections. We would expect them to be reflected in the data by a long tail of the transverse momentum distribution of the observed IVB's and the occurrence of hadronic jets correlated with the transverse momentum of the weak bosons.

The transverse momentum of the W is determined by the vector sum of the lepton and neutrino momenta in the transverse plane. The identification of relatively low transverse energy jets is difficult due to experimental and theoretical uncertainties.

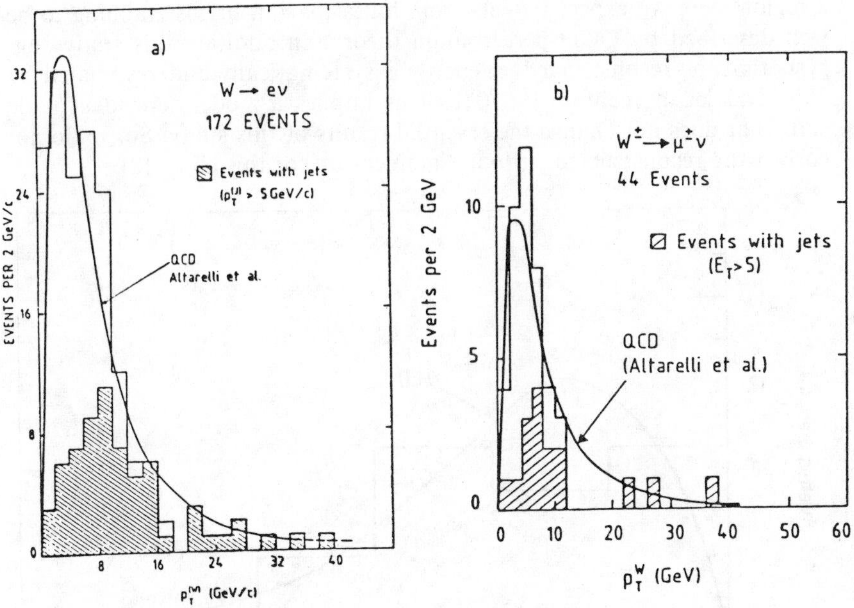

Figure 4: Transverse Momentum of the W
a)The W transverse momentum distribution (W→eν sample).
b)The W transverse momentum distribution (W→μν sample).
The curves show the QCD prediction by Altarelli et al [10]. The hatched sub-histograms show the contribution from events with at least one reconstructed jet with transverse momentum above 5 GeV/c, in the rapidity interval $|\eta| < 2.5$.

The jet detection is performed by the UA1 standard jet finding algorithm [5]. The following analysis considers jets within $|\eta| < 2.5$ that exceed a transverse energy of 5 GeV. Two jets can be resolved when they are in (η,ϕ) space further apart than one unit, their distance being $\Delta R = (\Delta\phi^2 + \Delta\eta^2)^{1/2}$, where the azimuthal angles are measured in radians. ISAJET Monte Carlo [6] events were used to study the features of the UA1 jet finding algorithm. In these events W's were generated in which a gluon balances the transverse momentum of the W. Assuming Field and Feynman fragmentation [7] the gluons were turned into hadronic jets and subsequently subjected to the jet finding algorithm using a full detector simulation. The reconstruction efficiency rises with the transverse energy from 0% and reaches 100% at about 20 GeV transverse jet energy, a region were we expect the jets from initial parton bremsstrahlung to be well described by QCD perturbation theory. In addition this study suggests that the reconstructed jet energy is systematically underestimated by 20% and jet momentum by 10%. This implies a model dependent scale shift that does not change the essential results of this study. So we do not correct the reconstructed jet four−momentum for this effect [2].

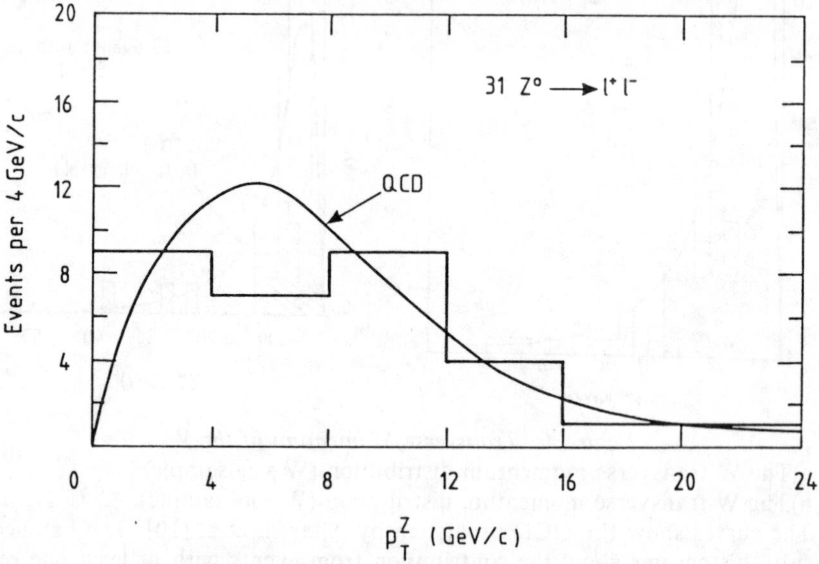

Figure 5: Transverse Momentum of the Z^0

The Z^0 transverse momentum distribution ($Z^0 \rightarrow e^+e^-$ and $Z^0 \rightarrow \mu^+\mu^-$ together). The curve shows the QCD prediction of Aurenche and Kinnunen [11].

Fig. 4 shows the distribution of the transverse momentum of the W in the muon and electron sample. The distributions peak at a value $p_t^W = 4$ GeV/c being consistent with the experimental resolution of the measurement of the missing energy in the event, and it has a tail extending to transverse momenta of 40 GeV/c. The lepton isolation requirement in the event selection causes losses in the region of high p_t W's. These losses have been studied in the W → eν sample using an event mixing technique; they rise from about 0% for the lowest transverse momentum W's to about 6% in the highest p_t region. In the W → $\mu\nu$ sample the overall event loss is higher as a back to back configuration of the muon with a jet in the transverse plane is additionnally forbidden. This jet veto protects the event sample against background from QCD events where an unidentified π/K decay gives rise to a high p_t muon. The total loss is prelimarily determined to be 12% from the analysis of the 83 data [8] only. The inefficiencies are however small enough not to be corrected for in a comparison with expectations from a QCD improved Drell–Yan modell. The shape of such a curve evaluated by Altarelli et al. [9] using the GHR structure functions [10] fits well the distribution of the data.

The transverse momentum of the Z^0 is directly obtained from the two momenta of the decay leptons in the electron sample and from the energy flow of the rest of the event in the muon sample after the substraction of the muon energy depositions in the calorimeters. Its distribution is shown in Fig. 5 . Although limited by statistics it is compatible with a QCD prediction by Aurenche and Kinnunen [11].

4. JET ACTIVITY

The events with one or more reconstructed jets in the W events are indicated by the hatched area in the transverse momentum distributions. 38% of the $W \rightarrow e\nu$ decays and 36% of the $W \rightarrow \mu\nu$ decays have additional jet activity in the event. As expected the high p_t W's are more likely to have jets in the events; all the W's with a p_t above 20 GeV/c have indeed at least one. Within the experimental resolution the jet activity in the W events balances the p_t^W. The distribution $(p_t^W - p_t^{jets})/ p_t^W$ (Fig. 6) for the $W \rightarrow e\nu$ sample is well described by a resolution curve that is obtained in a full detector simulation on ISAJET Monte Carlo generated events. The expected resolution is 25% p_t^W.

The jet transverse momentum distribution therefore essentially reflects the p_t^W distribution (Fig. 7 for the $W \rightarrow e\nu$ sample), and in the region in which we have a high reconstruction efficiency for the jet it is well decribed by the expectation from QCD perturbation theory [12] theory for jets arising from initial state parton bremsstrahlung.

This hypothesis can also be tested on the jet angular distribution. The angle Θ^* of the jet relative to the average beam direction in the rest frame of the W and jet(s) system is strongly peaked towards the beam direction as can be seen in Fig. 8 For the region with reasonably constant experimental acceptance ($\cos\Theta^* < 0.95$) a QCD bremsstrahlung spectrum which is basically $(1 - |\cos\Theta^*|)^{-1}$ fits well the shape of the distribution (solid curve in Fig. 8).

Table 1 provides a summary of the jet activity in W and Z^0 decays. In order to compare the jet rates in W and Z^0 events we define

$$R_W \equiv \sigma[W + \text{jet(s)}]/\sigma(W) \quad (5)$$
$$R_Z \equiv \sigma[Z^0 + \text{jet(s)}]/\sigma(Z^0), \quad (6)$$

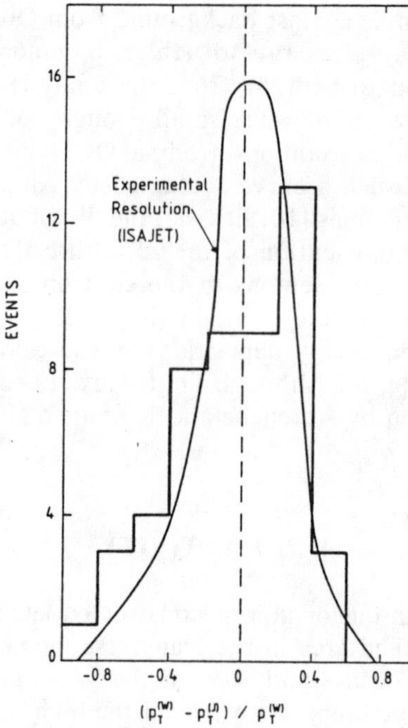

Figure 6: Transverse Balance of the W and the Jet

W – jet transverse momentum balance ($W \to e\nu$ sample). The imbalance between the W and jet transverse momenta is shown as a fraction of the W transverse momentum for those events with $p_t^W > 5$ GeV/c. The curve is the prediction using the ISAJET program for the expected experimental resolution on this quantity.

$\sigma(W)$ and $\sigma(Z)$ indicating the bare weak boson production cross sections without jets. We find $R_W = 0.60 \pm 0.01$, and from a preliminary analysis $R_Z = 0.9 \pm 0.3$. The ratio of these rates gives $R \equiv R_Z/R_W = 1.6 \pm 0.6$. In this comparison the jet rate in Z^0 events is consistent with the jet activity seen in W events.

To conclude about the jet activity in W events one has to investigate whether events with a W and one or several jets can be due to decays of heavy objects

$$p\bar{p} \to X + ...$$
$$X \to W + \text{jet}$$

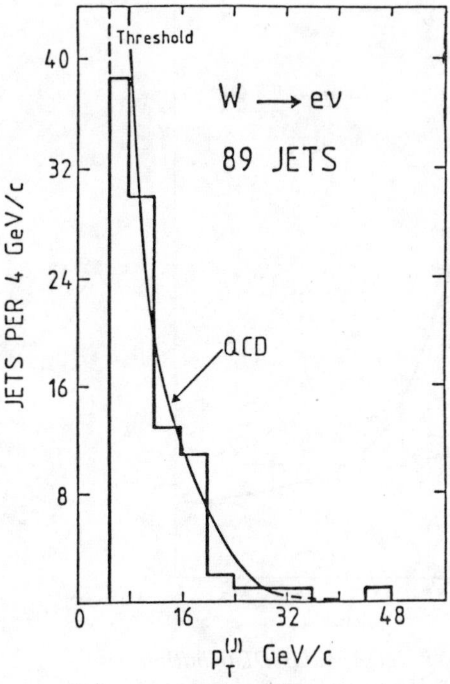

Figure 7: Jet Transverse Momentum Distribution

Jet transverse momentum distribution ($W \to e\nu$ sample) for all the reconstructed jets produced in association with the W. The curve shows the QCD prediction of ref. [12] normalized to the tail of the distribution, the region in which we expect to have good reconstruction efficiency for the jets.

In the W + jet events in which the W decays into an electron the mass distribution of the (W,jet) system has been systematically searched for a resonance (Fig. 9).

It can be compared with a predicted distribution that presents W + jet masses to be found in W decays generated by ISAJET together with one jet from initial state bremsstrahlung. The measured distribution is slightly broader than the Monte Carlo curve. However when one uses an event mixing technique in which the observed W four-vectors are randomly associated with the four-vectors of the jets in the W sample the expected distribution fits the data well. From the mass spectrum we obtain no evidence for a heavy object decay into a W and a jet. Therefore we give an

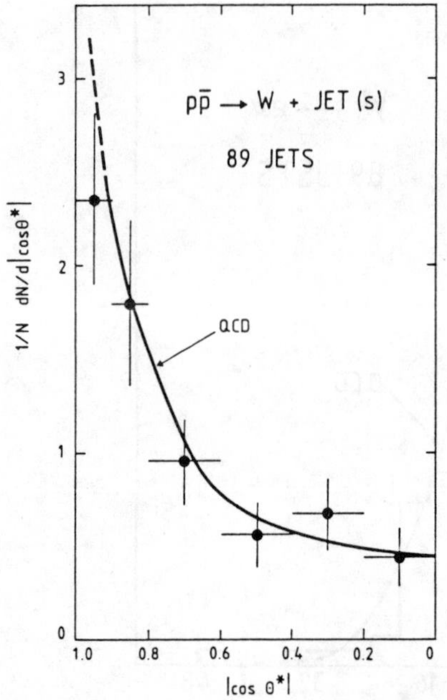

Figure 8: Jet Angular Distribution

The angular distribution for jets reconstructed in $W \rightarrow e\nu$ events. The distribution of $\cos \Theta^*$ is shown, Θ^* being the angle between the jet and the average beam direction in the rest frame of the W and the jet(s). The curve shows the QCD prediction from ref. [12].

Table 1: Jet Activity in W and Z^0 Decays

	Number of Events Total	≥ 1 jet	1 jet	2 jets	≥ 3 jets
$W \rightarrow e\nu$	172	65	50	8	7
$W \rightarrow \mu\nu$	44	16	14	1	1
$W \rightarrow l\nu$	216	81	64	9	8
$Z^0 \rightarrow e^+e^-$	22	9	4	4	1
$Z^0 \rightarrow \mu^+\mu^-$	9	6	4	0	2
$Z^0 \rightarrow l^+l^-$	31	15	8	4	3

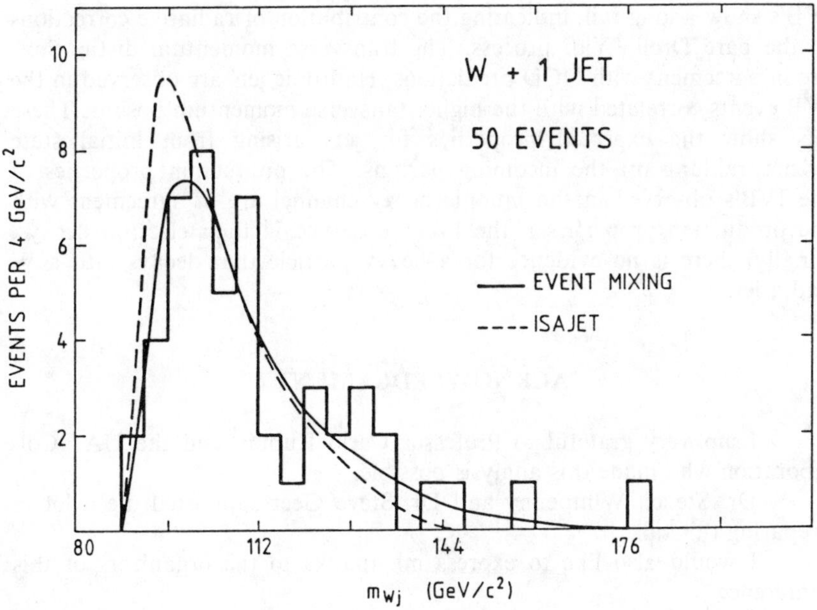

Figure 9: W + Jet Mass Spectrum

The (W + jet) − mass distribution for W → eν decays in which one jet is reconstructed. The curves show the predictions of ISAJET (dashed curve) and event mixing (solid curve).

upper limit on the cross section of heavy particles going into the above mentioned channel:

$$\sigma \cdot B(X \to W(\to e\nu) + \text{jet})/\sigma(W \to e\nu) < 0.013$$

at 90% confidence level.

5. SUMMARY

The production properties of the Intermediate Vector Bosons produced at the CERN SPS Collider are in agreement with predictions for IVB's produced by the QCD improved Drell − Yan mechanism. The longitudinal motion of the W's reflects the longitudinal momentum distributions of the incoming partons, and is well described by the structure functions of Eichten et al.. The transverse momentum distributions of the

IVB's show a long tail, indicating the contribution of radiative corrections to the bare Drell – Yan process. The transverse momentum distibutions are in agreement with QCD predictions. Hadronic jets are observed in the IVB events correlated with the higher tansverse momentum bosons. These jets show the expected properties for jets arising from initial state bremsstrahlung off the incoming partons. The production properties of the IVB's observed in the muonic decay channel are in agreement with the production properties if the IVB's observed in their electron decays. Finally, there is no evidence for a heavy particle that decays into a W and a jet.

ACKNOWLEDGEMENT

I am very grateful to Professor Carlo Rubbia and the UA1 Collaboration who made this analysis possible.

Dr. Steven Wimpenny and Dr. Steve Geer supported me a lot in preparing this talk.

I would also like to express my thanks to the organizers of this conference.

REFERENCES

[1]. G. Arnison et al. (UA1 Collaboration), Phys. Lett. **122B** 103 (1983)
G. Arnison et al. (UA1 Collaboration), Phys. Lett. **129B** 273 (1983)
G. Arnison et al. (UA1 Collaboration), Phys. Lett. **126B** 389 (1983)
G. Arnison et al. (UA1 Collaboration), Phys. Lett. **147B** 241 (1984)
M. Banner et al. (UA2 Collaboration), Phys. Lett. **122B** 476 (1983)
M. Banner et al. (UA2 Collaboration), Phys. Lett. **129B** 130 (1983)
C. Rubbia, Proc. 9th Hawaii Topical Conf. in Particle Physics, Honolulu (1983)
E. Radermacher, CERN – EP 84 – 41 (1984), to be published in Progress in Particle and Nuclear Physics, Pergamon Press

[2]. G. Arnison et al. (UA1 Collaboration), in preparation
S. Geer, CERN – EP 85 – 63 (1985)

[3]. S. Drell and T.M. Yan, Phys. Rev. **25** 1316 (1970), and Ann. Phys. **66** 1578 (1971)

[4]. E. Eichten et al., Rev. Mod. Phys. **56** 579 (1984)

[5]. G. Arnison et al. (UA1 Collaboration), Phys. Lett. **123B** 115 (1983)

[6]. F.E. Paige and S.D. Protopopescu, ISAJET program BNL 29777(1981)

[7]. R.D. Field and R.P. Feynman, Nucl. Phys. **B136** 1 (1978)

[8]. G. Arnison et al. (UA1 Collaboration), Phys. Lett. **134B** 469 (1984)

[9]. G. Altarelli, R.K. Ellis, M. Greco, and G. Martinelli, Nucl. Phys. **B246** 12 (1984)
G. Altarelli, R.K. Ellis, and G. Martinelli, CERN – TH 4015/84 (1984)

[10]. M. Glück, E. Hoffmann, and E. Reya, Z. Phys. **C13** 119 (1982)

[11]. P. Aurenche and R. Kinnunen, Phys. Lett. **135B** 493 (1984)

[12]. S. Geer and W.J. Stirling, Phys. Lett. **152B** 373 (1985)

SEARCHING FOR FOURTH GENERATION QUARKS AND LEPTONS

Howard Baer*
CERN - Geneva

ABSTRACT

Production, decay, and detection of a possible fourth generation of quarks and leptons are examined at present and future collider facilities. Interesting possibilities of new quarks include overlapping generations, where one new quark might be unusually long-lived, and also superheavy quarks which decay into real W-bosons. New charged heavy leptons with mass $m_L < m_W$ can most easily be detected through hadronic decays at $p\bar{p}$ colliders, or through leptonic decay at e^+e^- machines.

INTRODUCTION

The recent discovery of the t-quark [1] by the UA1 Collaboration has lent confirmation to the world picture of all matter being composed of three generations of quarks and leptons interacting via the exchange of vector bosons, according to the SU(3)×SU(2)×U(1) (QCD plus GWS) standard model [2]. However, a fourth generation of quarks and leptons is easily incorporated into the standard model framework, and one may hope that the new energy thresholds soon to be reached (i.e., Tevatron, SLC, LEP), or already being probed (SPS) will either unveil a new generation, or place mass limits on its existence. Indeed, within some composite models of quarks and leptons, repetition of generations is natural [3]. Also, some superstring theories which reduce to acceptable low-energy phenomenology imply the number of generations must be even [4], prompting one to look for a fourth generation. The purpose of this talk is to examine the production, decay and detection of fourth generation particles, and to single out possible features which may lead to their eventual discovery. We denote the four generations of quarks as

$$\binom{u}{d} \quad \binom{c}{s} \quad \binom{t}{b} \quad \binom{a}{v} \tag{1}$$

where, a, v are suggested by the alphabetic labelling scheme [5] a, b, c, d, ..., s, t, u, v. The four generations of leptons are

$$\binom{\nu_e}{e} \quad \binom{\nu_\mu}{\mu} \quad \binom{\nu_\tau}{\tau} \quad \binom{\nu_L}{L} \tag{2}$$

*Address after Oct. 1, 1985: Argonne Nat. Lab., Argonne, IL 60439

Possible names for the new fourth generation quarks are "virtue" and "amity". Limits from cosmology [6] and $p\bar{p}$ collider data [7] seem to imply $N_G < 4$.

a AND v QUARKS

To calculate properties of a and v quarks, we must specify their masses, mixing parameters, and couplings - the latter we take to be the same as the couplings of the first three generations. Radiative corrections to the ρ parameter constrain fourth generation doublet member's masses to be sufficiently degenerate [8]

$$(m_v-m_a)^2 + \tfrac{1}{3}(m_{\nu_L}-m_L)^2 < (310\ \text{GeV})^2; \tag{3}$$

for reliable perturbative calculations, we require

$$\begin{aligned} m_a &< 500\ \text{GeV} \\ m_L &< 900\ \text{GeV} \end{aligned} \tag{4}$$

lest weak interactions become strong[9].

Extension of the parametrization of the 6-quark KM mixing matrix [10] element's strengths to four generations can be written as

$$\left|U_{nm}\right| = (\tfrac{1}{2})^{n-m-1}\ \Theta^{2n-m-2}$$

or

$$\left|U_{nm}\right| = (\tfrac{1}{2})^{n-m-1}\ \Theta^{\frac{1}{2}(n-m)(n+m-1)} \tag{5}$$

where n(m) labels upper (lower) generational elements and $\Theta \cong 0.23$.

Production of a and v quarks through W decay ($W^- \to \bar{t}v$ or $\bar{a}v$) or Z decay ($Z \to a\bar{a}$, $v\bar{v}$) will be unimportant due to mixing suppression and/or phase space limitations. The dominant production mechanism for a and v at $p\bar{p}$ colliders will likely be through hadroproduction in gluon-gluon and quark-antiquark fusion. In Fig. 1 we plot cross-sections for heavy quark production as a function of quark mass for various collider energies, and two sets of structure functions. The SPS, Tevatron and SSC colliders could conceivably probe for quark masses of 80, 150 and 500 GeV respectively, assuming an arbitrary minimal requirement of a 10pb signal.

Possible decay modes of a and v are

$$\begin{aligned}
&a \to v + W \text{ (real or virtual)} && \left|U_{ab}\right| \sim 1; \quad m_a > m_v \\
&v \to t + W \text{ (real or virtual)} && \left|U_{vt}\right| \sim \theta^3 \\
&\text{or} && \\
&v \to c + W \text{ (virtual)} && \left|U_{vc}\right| \sim \tfrac{1}{2}\theta^4 \text{ or } \tfrac{1}{2}\theta^5
\end{aligned} \tag{6}$$

Thus we see that sufficiently high a and v masses could lead to a new source of W pair-production. If $m_v < m_t$ (overlapping generations), then there would likely be strong mixing suppression of v weak

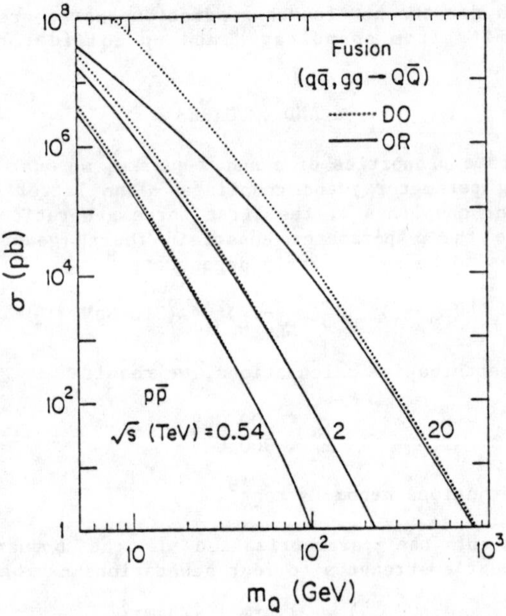

Fig. 1: Total cross-sections for hadroproduction of heavy quarks for two choices of parton distributions.

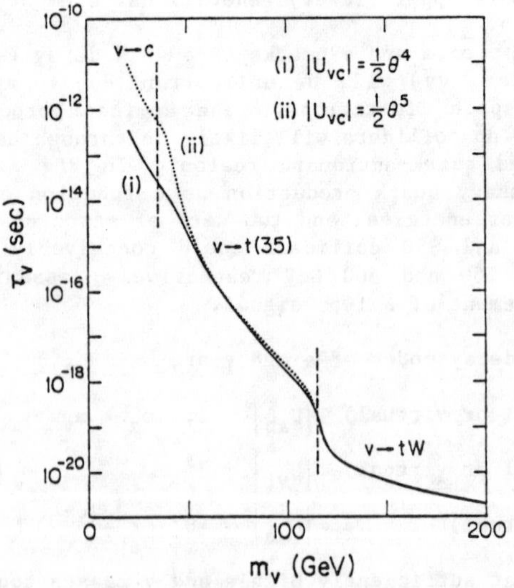

Fig. 2: v-quark lifetime vs. v-quark mass for two extensions of KM matrix to four-generations.

decays, leading to a long v lifetime, comparable to that of the b or c-quark: see Fig. 2. This case could lead to a confusion of v and t quark signals in current collider data; resolution of v and t quarks in this case might be made through use of a vertex detector, allowing one to separate the long-lived v quarks from short-lived t quarks.

Three possibilities for detecting a and v quarks at hadron colliders are i) triggering on very energetic isolated leptons from a or v semileptonic decay ii) looking for multileptons from a or v cascade decays, or iii) finding W pair events along with jets. Lepton isolation requirements in lepton + jet events have aided in the t-quark discovery, to separate t events from those due to other heavy flavours (b,c)[11]. In case $M_W + m_t > m_v > m_t$, t-quark events will be the dominant background for v quark signals. In this case, triggering on very high p_T leptons (e.g., $p_T > 20$-40 GeV) may allow one to reject the t-backgrounds by using isolation cuts since the opening angle between heavy quark decay debris increases with quark mass, and decreases with increasing trigger requirements [12]. When v-quark decay debris is identifiable, transverse mass techniques may allow determination of the v quark mass.

CHARGED HEAVY LEPTON L

In $p\bar{p}$ collisions, the heavy lepton L can be produced in the following reactions:

$$q\bar{Q} \to W^- \to L\,\bar{\nu}_L \tag{7a}$$

or

$$q\bar{q} \to Z \to L\bar{L} \tag{7b}$$

The L can decay via

$$\begin{aligned} L \to \nu_L &+ e\,\bar{\nu}_e \\ &+ \mu\,\bar{\nu}_\mu \\ &+ \tau\,\bar{\nu}_\tau \\ &+ 3d\bar{u} \\ &+ 3s\bar{c} \\ &(+\,3b\bar{t}) \end{aligned} \tag{8}$$

Reaction (7a) gives an observable rate when $m_L < m_W$ [13] at the CERN $p\bar{p}$ collider; reaction (7b) gives small cross-section after branching fractions, even for $m_L < m_{Z/2}$ [14]. However $L\bar{L}$ production via gluon-gluon fusion may be large at SSC or LHC energies[15].

First we consider the case where $W \to L \to e$ or $\mu + \not{p}_T$ (missing p_T), at the $Sp\bar{p}S$ collider. Large backgrounds to this signal come from direct $W \to e$ or $\mu + \not{p}_T$ and from $W \to \tau \to e$ or $\mu + \not{p}_T$. In order to maximize the L signal to background ratio, the best place to look is where $8 < p_{e,\mu_T} < 16$ GeV (Fig. 3), and $-0.9 < \cos\theta(\bar{p},e^+ \text{ or } \mu^+) < 0.2$. This "window" to the L leptonic signal is the best that can be done - even in this region the L signal never exceeds background. Many events would be required to establish existence of L this way, with little hope of a mass determination. (If L couples via V+A interactions, the leptonic signal becomes more visible[16].)

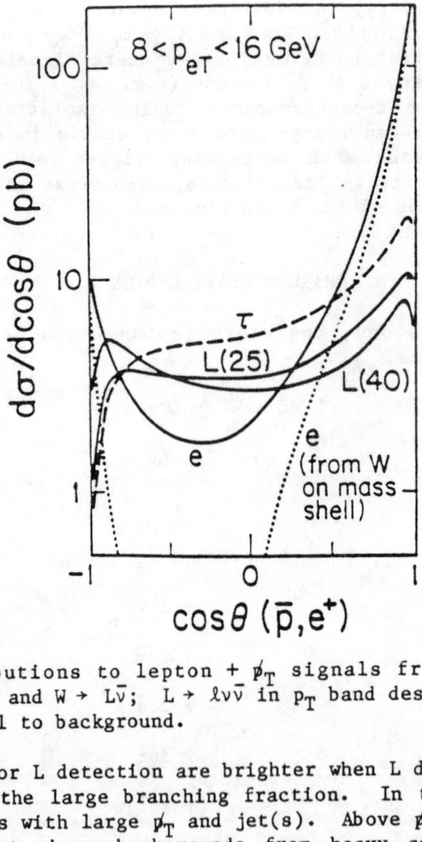

Fig. 3: Contributions to lepton + $\not{p}_T$ signals from $W \to \bar{\nu}\ell$, $W \to \tau\bar{\nu}$; $\tau \to \ell\nu\bar{\nu}$ and $W \to L\bar{\nu}$; $L \to \ell\nu\bar{\nu}$ in p_T band designed to maximize L signal to background.

Prospects for L detection are brighter when L decays to hadrons, due in part to the large branching fraction. In this case one may trigger on events with large $\not{p}_T$ and jet(s). Above $\not{p}_T = 20$ GeV, the L signal stands out above backgrounds from heavy quarks. The only competing standard model background is from $W \to \tau \to$ jets. However, the distinction between L and τ should be clear when considering jet multiplicity, broadness, and the presence of accompanying minijets in monojet events. Figure 4 is a plot of expected monojet and dijet cross-sections from $W \to L$, for $\not{p}_T > \max(15 \text{ GeV}, 4\sigma)$, $P_{Tj_1} > 25$ GeV, $p_{Tj_2} > 12$ GeV, incorporating experimental resolution and Q_{TW} smearing and using UA1 jet finding algorithm (See Ref. 17 for details). Monojet events dominate over dijet events for all values of m_L, and the signal exceeds 10 pb for $m_L < 60$ GeV (we have included a QCD motivated K-factor of 2 here).

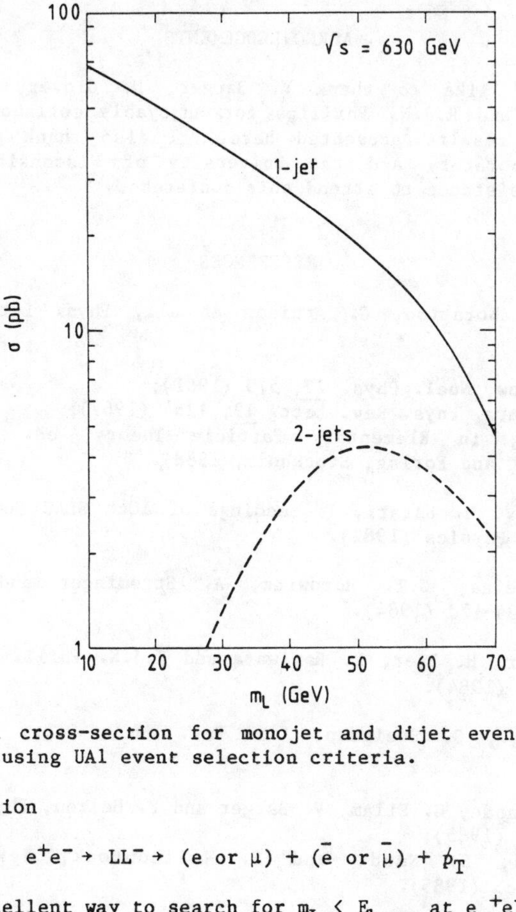

Fig. 4: Total cross-section for monojet and dijet events from $W \to L\nu_L$, using UA1 event selection criteria.

The reaction

$$e^+e^- \to LL^- \to (e \text{ or } \mu) + (\bar{e} \text{ or } \bar{\mu}) + \not{p}_T$$

will be an excellent way to search for $m_L < E_{beam}$ at e^+e^- colliders. The signature in this case is striking - if one requires the two trigger leptons to be acolinear, there is no SM background[14].

CONCLUSION

Potential properties of fourth generation quarks and leptons have been reviewed. Exciting possibilities include overlapping quark generations, where the new v quark is long-lived, and superheavy quarks which decay into real W-bosons. The CERN Sp$\bar{p}$S collider is sensitive to heavy leptons with $m_L < 60$ GeV, by looking for monojet events with $\not{p}_T < 40$ GeV; the L leptonic signal is plagued by serious background problems.

ACKNOWLEDGEMENTS

I would like to thank V. Barger, N. Glover, K. Hagiwara, A.D. Martin and R.J.N. Phillips for enjoyable collaborations which led to the results presented here. I also thank CERN, Argonne National Laboratory and the University of Wisconsin-Madison for financial assistance to attend this conference.

REFERENCES

1. UA1 Collaboration, G. Arnison et al., Phys. Lett. 147B, 493 (1984).

2. S. Glashow, Nucl. Phys. 22, 579 (1961);
S. Weinberg, Phys. Rev. Lett. 19, 1264 (1967);
A. Salam, in Elementary Particle Theory, ed. N. Svartholm (Almquist and Forlag, Stockholm, 1968).

3. See, e.g., H. Harari, Proceedings of 10th SLAC Summer Inst. on Particle Physics (1982).

4. P. Candelas, G.T. Horowitz, A. Strominger and E. Witten, NSF-ITP-84-170 (1984).

5. V. Barger, H. Baer, K. Hagiwara and R.J.N. Phillips, Phys. Rev. D30, 947 (1984).

6. See, e.g., G. Steigman, Ann. Rev. Nucl. Part. Sci. 29, 313 (1979).

7. N. Deshpande, G. Eilam, V. Barger and F. Halzen, Phys. Rev. Lett. 54, 1757 (1985);
D. Dicus, S. Nandi and S. Wollenbrock, Texas preprint UTTG-08-85 (1985).

8. M. Veltman, Nucl. Phys. B123, 89 (1977);
M.B. Einhorn, D.R.T. Jones and M. Veltman, ibid. B191, 146 (1981).

9. M.S. Chanowitz, M.A. Furman and I. Hinchliffe, Phys. Lett. 78B, 285 (1978); Nucl. Phys. B153, 402 (1979).

10. L. Wolftenstein, Phys. Rev. Lett. 51, 1945 (1983),

11. V. Barger, H. Baer, A.D. Martin and R.J.N. Phillips, Phys. Rev. D29, 887 (1984);
V. Barger, H. Baer, K. Hagiwara, A.D. Martin and R.J.N. Phillips, Phys. Rev. D29, 1923 (1984).

12. K. Hagiwara and W.F. Long, Phys. Lett. 132B, 202 (1983).

13. V. Barger, H. Baer, A.D. Martin, E. Glover and R.J.N. Phillips, Phys. Lett. 133B, 449 (1983); Phys. Rev. D29, 2020 (1984);
D. Cline and C. Rubbia, Phys. Lett. 127B, 277 (1983);
S. Gottlieb and T. Weiler, Phys. Rev. D29, 2005 (1984).

14. H. Baer, V. Barger and R.J.N. Phillips, MAD/PH/213 (1985), Phys. Rev. D (in press).

15. S. Willenbrock and D. Dicus, Texas preprint UTTG-03-85 (1985).

16. S. Gottlieb and T. Weiler, UCSD-10P10-244 (1985).

17. H. Baer, J. Ellis, G. Gelmini, D.V. Nanopoulos and X. Tata, CERN preprint TH.4194/85 (1985).

THE FOURTH GENERATION

Sandip Pakvasa
Department of Physics and Astronomy
University of Hawaii at Manoa, Honolulu, Hawaii 96822, U.S.A.
and
KEK, National Laboratory for High Energy Physics
Oho-machi, Tsukuba-gun, Ibaraki-ken 305, Japan

ABSTRACT

Constraints on masses of the fourth generation fermions in the standard as well as extended models are reviewed. Constraints on the mixing of the fourth generation quarks are summarized. Possible low energy signatures of the fourth generation such as: a lengthened t-lifetime, enhanced mixing and CP-violation in D^0-$\bar{D}^0$, B^0-$\bar{B}^0$ and a suppressed ε'/ε are pointed out. The desirability of finding the fourth generation is stressed.

The new particles I shall be talking about belong to the fourth generation of quarks and leptons. Assuming them to be sequential, they are $\begin{pmatrix} t' \\ b' \end{pmatrix}_L$, t'_R, b'_R $\begin{pmatrix} \nu_L \\ L \end{pmatrix}_L$, L_R . Some of what I say will overlap with the previous speaker[1].

Some constraints on the properties (masses and mixings) of the fourth generation fermions can be deduced by considering the effects on the "old" particles namely u, d, s, c, t, b and also W and Z.

Let me first consider constraints on masses:

(a) Fermion loops make ρ (the ratio of neutral to charged current coupling strengths) deviate from one. As given by Veltman[2] this deviation is given by:

$$\delta\rho = \frac{G_F}{8\pi^2\sqrt{2}}\left[3(m_{t'}^2 + m_{b'}^2) - 3\,\frac{2m_{t'}^2 m_{b'}^2}{m_{t'}^2 - m_{b'}^2}\ln m_{t'}^2/m_{b'}^2 \right.$$

$$\left. + (m_L^2 + m_{\nu_L}^2) - \frac{2m_L^2 m_{\nu_L}^2}{m_{\nu_L}^2 - m_L^2}\ln m_{\nu_L}^2/m_L^2 \right] \tag{1}$$

where we have dropped the contribution from the first three generation as being too small. The right hand side of Eq.(1) vanishes in the limit of isospin degeneracy i.e. $m_{t'} = m_{b'}$, $m_L = m_{\nu_L}$. If one takes the current accuracy[3] of $\delta\rho$ to be $\delta\rho \lesssim 0.04$ then (ignoring m_L and $m_{b'}$) one has the limit on the t'-mass: $m_{t'} \lesssim 300$ GeV. For $\delta\rho$

≤ 0.02 this is improved to $m_{t'} \lesssim 215$ GeV.
(b) Chanowitz, Furman and Hinchliffe[4] deduced bounds on fermion masses by imposing partial wave unitarity on S-wave $F\bar{F}$ scattering amplitudes:

$$m_Q^2 \leq \frac{4\sqrt{2}\pi}{5G_F\sqrt{N_D}}$$

$$m_L^2 \leq \frac{4\sqrt{2}\pi}{G_F\sqrt{N_D}} \tag{2}$$

where $N_D \geq 2$ is the number of degenerate families. For four generations this gives

$$m_Q \leq 470 \text{ GeV}. \tag{3}$$

(C) The one-loop correction to the Higgs potential becomes unbounded from below and hence the vacuum is unstable if fermions in the loop are too massive. Hence one can obtain an upper bound on fermion masses by requiring a stable vacuum at one loop level, as first observed by Hung[5] and by Politzer and Wolfram[6]. The bound is weakened by two effects[7]. One is the two loop contributions and the other is allowing vacuum to be not absolutely stable but only sufficiently long-lived ($\tau \gtrsim 10^{10}$ yrs). Then the bound on a fourth generation quark mass varies from 188 GeV to 215 GeV as the Higgs mass varies from 5 GeV to 200 GeV, in any case:

$$m_Q < 215 \text{ GeV} \tag{4}$$

(d) Assuming that the triviality of $\lambda\phi^4$ theory is established and the principle of "gauge dominance" i.e. that gauge couplings should dominate non-gauge couplings, bounds on Higgs as well as fermion masses can be derived[8]. For the fourth generation quarks, the bounds are:

$$m_{t'} < 213 \text{ GeV}$$

$$m_{b'} < 208 \text{ GeV} \tag{5}$$

(e) It has long been an attractive notion that the fermion masses (at least the largest ones) may be determined[9] by the Yukawa couplings reaching a fixed point. The most thorough analysis of this idea has been given recently by Bagger, Dimopoulos and Masso[10]. They find, under this assumption of infrared fixed point, for N_H heavy generations:

$$m_Q \leq 250/\sqrt{N_H} \text{ GeV}$$

$$m_L \leq 235/\sqrt{N_H} \text{ GeV} \tag{6}$$

for four generations $N_H = 1$ and one simply has

$$m_Q \leq 250 \text{ GeV}. \tag{7}$$

In fact in order to reach the fixed point it is necessary to have at least one heavy generation with masses satisfying these bounds.

A similar fixed point analysis in N = 1 supergravity improves[11] these bounds by about a factor of 2.

$$m_Q \lesssim 130 \text{ GeV}. \tag{8}$$

One can also ask if there are constraints on the number of new sequential generations? If some extra assumptions are made it is possible to obtain such constraints. For example, if the new generations have essentially massless ($m < O(\text{MeV})$), stable, neutrinos then the observed Helium abundunce constrains the number of such new generations. The argument[12] goes as follows: The observed Helium abundance by weight is about 26 % and is expected to be almost all primordial. If this was formed at $T > 100$ KeV then almost all neutrinos combined to form ^{4}He and the ^{4}He abundance is given by $f = \frac{2\alpha}{1 + \alpha}$ where α is the neutron to proton ratio. The actual temparature T^* at which $n \leftrightarrow p$ weak transition stops is found to be about $T^* = 0.7$ MeV. Hence $\alpha = e^{-\Delta/T^*}$ (where $\Delta = m_n - m_p$) is about 1/7 and $f = 1/4$. This is for 3 species of neutrinos. It turns out that (for a fixed baryon density) the change in f for every new flavor of neutrinos is $\delta f \sim 1/54$ or f increases by about 2 % for every new flavor. So this argument tells us that at most one (plus minus one) more "standard" generation is allowed.

Another number that depends on the number of generations is m_b/m_τ in the minimal SU(5) GUTS. At one loop level, for $\Lambda_{\overline{ms}}$ between 100 and 200 MeV, m_b is about 5.2 GeV and each new generation shifts m_b upward by about 10 to 15 %. So again it suggests that one extra generation is allowed but perhaps no more[13]. At two loop level[14] the situation is not so cleat.

In the very near future we should know the ratio of widths of W and Z well enough to learn the total number of light neutrino species.

Next I would like to summarize briefly some models/theories in which four generations are predicted or needed.

(1) A model proposed by Davidson, Wali and Mannheim[15] is based on the gauge group $SO(10)_V \times SO(10)_H$. The first SO(10) is ordinary GUTS and the second contains a technicolor-like group. Fermions are assigned to $(\underline{16}, \underline{10}) + (\underline{10}, \underline{16})$. The $SO(10)_H$ breaks to $SO(6) \times SO(4)_{TC}$ under which $\underline{16}$ transforms as $(\underline{4}, \underline{2}) + (\underline{4}^*, \underline{2}')$ and $\underline{10}$ as $(\underline{6}, \underline{1}) + (\underline{1}, \underline{4})$. Hence only four of the original ten $\underline{16}$'s remain light and none of the $\underline{10}$'s. So four "light" sequential generations are expected.

(2) Sugawara[16] proposed a theory in which all gauge symmetries are broken dynamically by fermion condensates. Gauge bosons as well as fermions get their masses dynamically and there are no scalars. The full gauge group for which a phenomenologically viable model could be constructed was found to be $SU(5)_U \times SU(5)_G \times SU(5)_{Tu} \times SU(5)_{Tg}$. Here $SU(5)_U$ is the ordinary GUT unifying group, $SU(5)_G$ is a generation group, $SU(5)_{T_u}$ is a techni-color-like group which breaks

$SU(5)_u \times SU(5)_G$ at 10^{15} GeV and $SU(5)_{Tg}$ is also a TC-like group which breaks generation symmetry completely. The low-lying fermions are assigned to (10, 10, 1, 1) + ($\bar{5}$, 5, 1, 1) + ($\bar{5}$, $\bar{5}$, 1, 1) and so one starts with 10 generation of quarks and leptons. However, the symmetry breaking pattern ensures that 6 generations become massive and only 4 "light" generations survive. Hence in this theory one expects a fourth sequential generation of quarks and leptons.

(3) It has been realized for some time that orthogonal groups[17], especially O(18), are good candidates for flavor unification. For example, O(18) which is anomaly free has a 256 dimensional spinor representation which is complex and under $O(18) \to O(10) \times O(8)$ transforms as ($\underline{16}$, $\underline{8}$) + ($\underline{16}$, $\underline{8}'$). Then one has 8 left-handed and 8 right-handed light sequential generations. This creates several problems; one is the degeneracy of left-handed and right-handed generations and the other is the quick demise ($\sim$ 100 TeV) of asymptotic freedom. Bagger and Dimopoulos[18] have suggested an ingenious solution to these problems. Their proposal is that O(8) breaks at GUT scale but not completely and some fermions remain light but some become heavy. (This differs from an earlier suggestion[17] in which some remained confined). Possibilities are O(8) to break to $SU(3) \times U(1)$, $SU(2) \times U(1)$, U(1), U'(1)..... All of these except U(1) (or a subgroup Z_5 of U(1)) give 6 or more light generations. A specific U(1) can be found such that four left and four right generations are real under it ahd pair off to get very large masses. Whereas four left and four right generations are complex and remain light. Hence in this theory one expects a fourth sequential family. The right-handed families can be ensured to be sufficiently heavy and the fourth left-handed neutrino may be massive (2 GeV $< M < 40$ GeV). The fourth lepton as well as the four right-handed neutrinos should be below 40 GeV if the infra-red fixed point is reached.

(4) Very recently, several candidates have emerged as mathematically consistent theories of quantum gravity which might be phenomenonologically realistic as well. These are the superstring theories in ten dimensions. with $E_8 \times E_8$ or O(32) as gauge groups. Candelas et al. studied the possible vacuum configurations requiring that (a) the geometry be of the form $M_4 \times K$(with M_4 maximally symmetric), and (b) there be an unbroken N = 1 supersymmetry in four dimensions. Then one of the E_8 factors is unbroken and one breaks to $SU(3) \times E_6$ where the E_6 can be identified with the usual GUTS with one generation of fermions assigned to a $\underline{27}$. The number of generations depends on the choice of K (it is equal to half the Euler characteristic of K). It is possible to find a choice of K that leads to unbroken N = 1 supersymmetry and four generations.

None of the above may seem compelling, all the same it is very suggestive that number four crops up in so many different contexts. To me, it suggests that there are at least four generations.

Before turning to my next topic, viz. constraints on the mixing of the fourth generation, I should summarize the mixing and CP phenomenology with three generations[20]. The charged weak current is

$$J_\mu = \overline{(u\ c\ t)}_L \gamma_\mu U_{KM} \begin{pmatrix} d \\ s \\ b \end{pmatrix}_L \qquad (9)$$

The 3 × 3 matrix U_{KM} can be parameterised[20] as

$$U_{KM} = \begin{pmatrix} c_1 & -s_1c_3 & -s_1s_3 \\ s_1c_2 & c_1c_2c_3 + s_2s_3e^{i\delta} & c_1c_2c_3 + s_2c_3e^{i\delta} \\ s_1s_2 & c_1s_2c_3 + c_2s_3e^{i\delta} & c_1s_2s_3 - c_2c_3e^{i\delta} \end{pmatrix} \qquad (10)$$

where $c_i = \cos\theta_i$, $s_i = \sin\theta_i$ and θ_i can be chosen to lie in $0 \leq \theta_i \leq \pi/2$ and the phase δ can be in $0 \leq \delta \leq 2$. The current experimental knowledge of the elements of U_{KM} can be summarised by[21]

$$U_{KM} = \begin{pmatrix} |U_{ud}| = 0.9734 & 0.223 \leq |U_{us}| \leq 0.234 & |U_{ub}| \leq 0.01 \\ 0.21 \leq |U_{cd}| \leq 0.27 & 0.8 \leq |U_{cs}| \leq 0.98 & 0.04 \leq |U_{cb}| \leq 0.07 \end{pmatrix} \qquad (11)$$

where the information on $|U_{cd}|$ comes from nuclear β-decay, $|U_{us}|$ from K_{e_3} and $Y \to NeV$, $|U_{cd}|$ from $\nu N \to \mu^- \mu^+ X$, $|U_{cs}|$ from $D \to KeV$, $|U_{cb}|$ from τ_B and $|U_{ub}|$ from $(b \to u)/(b \to c)$. Should τ_B turn out to be somewhat shorter than 10^{-12}s, say $\sim 3\cdot 10^{-13}$s, then $|U_{cb}|$ is in the range 0.08 to 0.12.

The observed CP-violation[23] in $K_L \to \pi\pi$ is then given by ε which is calculated from the box diagram. ε depends on m_t, s_i, δ and a parameter B which is the ratio of $\langle K^\circ | \bar{S}_L \gamma_\mu d_L \bar{S}_L \gamma_\mu d_L | \bar{K}^\circ \rangle$ to $|\langle K^\circ | \bar{s}_L \gamma_\mu d_L | o \rangle|^2$. The precise value of B is unknown but is probably[23] between 1/3 and 1. In order to reproduce[24] the known value of ε; (a) δ has to lie between $\pi/2$ and π, fairly close to $\pi/2$, (b) m_t can not be too small, (c) τ_B can not be too large and (d) $\Gamma(b \to u)/\Gamma(b \to c)$ can not be too small. For reasonable values of these parameters say $m_t \sim 40$ GeV, $\tau_B \sim 10^{-12}$S, $\Gamma(b \to u)/\Gamma(b \to c) \sim 0.03$, the expected mixings and asymmetries in other neutral meson systems[25] D^0, B^0 and B_s are as listed in the Table.

The description of CP-violation in $K_L \to 2\pi$ is in trouble[24] if (a) m_t is smaller than 30 GeV and $\tau_B > 1\cdot 2\cdot 10^{-12}$s, or if (b) $|\varepsilon'/\varepsilon|$ is less than 10^{-3}, or if (c) $\Gamma(b \to u)/\Gamma(b \to c)$ is very small. We take the point of view that (c) is correct and a fourth generation is needed to give large enough contribution to reproduce ε correctly.

The work to be described here was done with Xiao-Gang He[26] at Honolulu. Very similar results have been obtained by several groups simultaneously: U. Türke et al.[27] in Dortmund, C-K. Chou et al.[28] in Beijing, T. Hayashi et al.[29] in Hiroshima, and A. Anselm et al.[30] in Leningrad. Slightly earlier some related work was done by I.I. Bigi[31] in Aachen and some results on mixings were also obtained by M. Gronau and J. Schechter[32] in Syracuse. An early, pioneering study on the effects of the fourth generation on the mixing of D^0, B^0 and B_s^0 systems is by S. K. Bose and E. A. Paschos.[33]

First, we need to parametrize the 4 × 4 unitary matrix (generalized KM matrix) which can be written in terms of six rotation angles and three phases. We use the parameterization[34] favored locally:

$$U = \begin{pmatrix} c_1 & s_1c_3 & s_1s_3c_5 & s_1s_3s_5 \\ -s_1c_2 & c_1c_2c_3 + s_2s_3c_6e^{i\delta_1} & \begin{matrix} c_1c_2s_3c_5 - s_2c_3c_5c_6e^{i\delta_1} \\ + s_2s_5s_6e^{i(\delta_1+\delta_3)} \end{matrix} & \begin{matrix} c_1c_2s_3s_5 - s_2c_3s_5c_6e^{i\delta_1} \\ - s_2c_5s_6e^{i(\delta_1+\delta_3)} \end{matrix} \\ -s_1s_2c_4 & \begin{matrix} c_1s_2c_3c_4 - c_2s_3c_4c_6e^{i\delta_1} \\ - s_3s_4s_6e^{i\delta_2} \end{matrix} & \begin{matrix} c_1s_2s_3c_4c_5 + c_2c_3c_4c_5c_6e^{i\delta_1} \\ - c_2c_4s_5s_6e^{i(\delta_1+\delta_3)} \\ + c_3s_4c_5s_6e^{i\delta_2} \\ + s_4s_5c_6e^{i(\delta_2+\delta_3)} \end{matrix} & \begin{matrix} c_1s_2s_3c_4s_5 + c_2c_3c_4s_5c_6e^{i\delta_1} \\ + c_2c_4c_5s_6e^{i(\delta_1+\delta_3)} \\ + c_3s_4s_5s_6e^{i\delta_2} \\ - s_4c_5c_6e^{i(\delta_2+\delta_3)} \end{matrix} \\ -s_1s_2s_4 & \begin{matrix} c_1s_2c_3s_4 - c_2s_3s_4c_6e^{i\delta_1} \\ + s_3c_4s_6e^{i\delta_2} \end{matrix} & \begin{matrix} c_1s_2s_3s_4c_5 + c_2c_3s_4c_5c_6e^{i\delta_1} \\ - c_2s_4s_5s_6e^{i(\delta_1+\delta_3)} \\ - c_3c_4c_5s_6e^{i\delta_2} \\ - c_4s_5c_6e^{i(\delta_2+\delta_3)} \end{matrix} & \begin{matrix} c_1s_2s_3s_4s_5 + c_2c_3s_4s_5c_6e^{i\delta_1} \\ + c_2s_4c_5s_6e^{i(\delta_1+\delta_3)} \\ - c_3c_4s_5s_6e^{i\delta_2} \\ + c_4c_5c_6e^{i(\delta_2+\delta_3)} \end{matrix} \end{pmatrix} \tag{12}$$

where $c_i = \cos\theta_i$, $s_i = \sin\theta_i$ and θ_i can be chosen to be $0 \le \theta_i \le \theta_i \le \pi/2$ and δ_i can be $0 \le \delta_i \le 2\pi$. We use $|U_{ud}| = 0.9734$ and $|U_{us}| = 0.224$, allow $|U_{cd}|$ to lie between 0.21 and 0.23 (we find no solutions for $|U_{cd}| > 0.23$) and allow U_{cb} to lie between 0.04 and 0.07[21]. For simplicity we take $|U_{cb}| > 0$ which means $c_5 = 0$, $s_5 = 1$, and furthermore, the phase δ_3 can be removed from the U_{KM}. Most of our results do not depend on the precise value of U_{ub}. To fix the remaining parameters we use the information from $K_L \to \mu^+\mu^-$ and $K_L \to \pi^+\pi^-$. The difference between the observed rate for $K_L \to \mu^+\mu^-$ and the absorptive $K_L \to \gamma\gamma \to \mu^+\mu^-$ piece is given by the sum of the weak interaction contribution and the dispersive $K_L \to \gamma\gamma \to \mu^+\mu^-$ piece. We make the reasonable, conservative ansatz following Barger et al.[35] that the ratio of dispersive to absorptive contribution is identical for $K_L \to \gamma\gamma \to \mu^+\mu^-$ and for $\eta \to \gamma\gamma \to \mu^+\mu^-$. Then, using the experimental values for $\eta \to \mu^+\mu^-$, $K_L \to \mu^+\mu^-$, $\eta \to \gamma\gamma$ and $K_L \to \gamma\gamma$ rates one has k restricted to the ranges $-1.14 \times 10^{-2} \le k \le 1.14 \times 10^{-2}$, $2 \times 10^{-2} \le k \le 3.84 \times 10^{-2}$, and $-2 \times 10^{-2} \ge k \ge -3.84 \times 10^{-2}$ where k is defined by

$$k = \sum_i F(\alpha_i)\, \mathrm{Re}\, \lambda_i \eta_i \;/|U_{us}|, \tag{13a}$$

$$F(\alpha_i) = \frac{-2\alpha_i}{1-\alpha_i} + \frac{1}{2}\frac{\alpha_i^2}{1-\alpha_i} - \frac{3}{2}\frac{\alpha_i^2 \ln \alpha_i}{(1-\alpha_i)^2} \quad . \tag{13b}$$

In (13) i runs over u, c, t, t', $\lambda_i = U_{id}U^*_{is}$ and η_i is the QCD correction of order 1. The expression is valid for arbitary quark masses. Combining the above equations with λ_u and λ_c as inputs and the unitarity of U one can obtain Reλ_t and Re$\lambda_{t'}$.

Next we use the parameters of the $K^0 - \bar{K}^0$ system to further restrict the U matrix elements. Since the mass difference δm_{L-S} is dominated by the charm contribution and furthermore has large long-distance effects[37] we content ourselves to reproducing the observed value of ε. We have, from the box diagram for arbitary quark masses:

$$\begin{aligned}\varepsilon = &-(BG_F^2 f_K^2 m_K m_W^2/\sqrt{2}\; 12\pi^2 \delta m)\exp(i\pi/4)\\ &\times \mathrm{Im}[\lambda_c^2 B_{cc}\eta_{cc} + \lambda_t^2 B_{tt}\eta_{tt} + 2\lambda_c\lambda_{t'}B_{ct}\eta_{ct} + 2\lambda_c\lambda_{t'}B_{ct'}\eta_{ct'}\\ &+ 2\lambda_t\lambda_{t'}B_{tt'}\eta_{tt'} + \lambda_{t'}^2 B_{t't'}\eta_{t't'}] \quad ,\end{aligned} \tag{14a}$$

where

$$B_{ij} = B(\alpha_u, \alpha_u) - B(\alpha_i, \alpha_u) + B(\alpha_i, \alpha_j) - B(\alpha_j, \alpha_u)$$

$$\begin{aligned}B(\alpha_i, \alpha_j) = &(1 + \tfrac{1}{4}\alpha_i\alpha_j)\Big[\frac{1}{(1-\alpha_i)(1-\alpha_j)}\\ &+ \frac{1}{\alpha_i - \alpha_j}\Big(\frac{\alpha_i^2 \ln \alpha_i}{(1-\alpha_i)^2} - \frac{\alpha_j^2 \ln \alpha_j}{(1-\alpha_j)^2}\Big)\Big]\\ &- 2\alpha_i\alpha_j\Big[\frac{1}{(1-\alpha_i)(1-\alpha_j)} + \frac{1}{\alpha_i - \alpha_j}\Big(\frac{\alpha_i^2 \ln \alpha_i}{(1-\alpha_i)^2} - \frac{\alpha_i \ln \alpha_j}{(1-\alpha_j)^2}\Big)\end{aligned} \tag{14b}$$

and i, j run over c, t and t'. In the above expression $\alpha_i = m_i^2/m_W^2$, λ_i as defined earlier and the η_{ij} are the leading-log short-distance QCD corrections $\eta_{cc} \approx 0.7$, $\eta_{tt} \approx 0.6$, $\eta_{ct} \approx 0.4$ have been calculated before[38] and we estimate $\eta_{t't'} \approx 0.5$, $\eta_{ct'} \approx 0.4$ and $\eta_{tt'} \approx 0.4$. Of course all η_{ij} depend weakiy on the number of generations and quark masses[39]. We use $f_K = 0.1722$ GeV, $m_K = 0.497$ GeV, $m_c = 1.4$ GeV, $m_t \approx 40$ GeV, $m_W = 80$ GeV, $m_{L-S} = -0.352 \times 10^{-13}$ GeV, $\varepsilon = 2.274 \times 10^{-3} \exp(i\pi/4)$. We will use $B \doteq 0.33$ and allow it to vary up to

B = 0.6.

We calculate the minimum value of $m_{t'}$ needed to reproduce the correct value of ε for various values of other parameters. In Fig. 1, we show $m_{t'min}$ as a function of τ_B for $|U_{cd}| = 0.225$ for three values of k. It is seen that $m_{t'min} > 100$ GeV. We also find that for $|U_{cd}| > 0.299$ there are no solutions i.e., it is impossible to reproduce ε. Also, if $|U_{cd}|$ is small enough i.e., < 0.22 and $k > 0$, then it is possible to tolerate $m_{t'}$ as low as 50 Gev. This is seen clearly in Fig.2. The effect of increasing B is to lower the curves and the η_{ij} have very little effect.

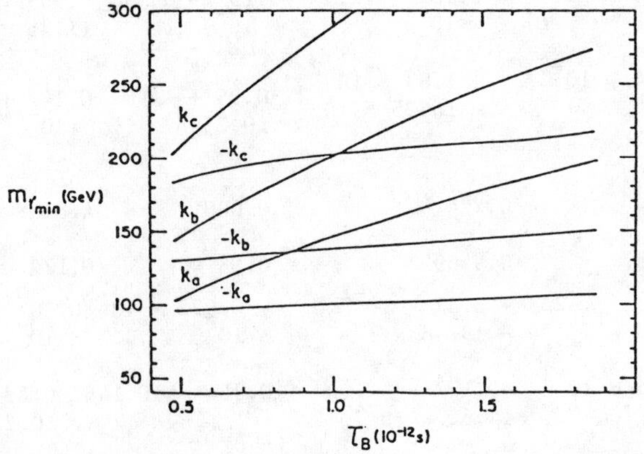

Fig. 1. $m_{t'min}$ plotted as a function of τ_B for $U_{ud} = 0.9734$, $U_{us} = 0.224$, $U_{cd} = 0.225$, $B = 0.5$ and for different values of k, $= 1.14 \times 10^{-2}$, $k_b = 2.1 \times 10^{-2}$, and $k_c = 3.84 \times 10^{-2}$.

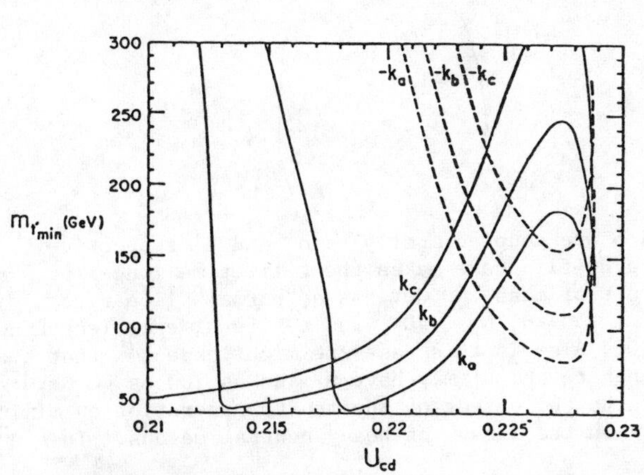

Fig. 2. $m_{t'min}$ is plotted as a function of U_{cd} at $U_{ud} = 0.9734$, $U_{us} = 0.224$, $U_{cb} = 0.06$, $B = 0.5$ and different values of k; $k_a = 1.14 \times 10^{-2}$, $k_b = 2.10 \times 10^{-2}$ and $k_c = 3.84 \times 10^{-2}$.

Next, for a given choice of $m_{t'}$, we calculate the allowed range of the angles and phases in the U. The full detailed presentation will be given elsewhere but below are two sample solutions for the 4×4 matrices. These correspond to taking $m_{t'} = 200$ GeV, B = 0.33, and k given by (a) 2.76×10^{-2} and (b) 1.14×10^{-2} respectively.

$$U_a = \begin{pmatrix} 0.9734 & 0.224 & 0.00 & 4.8 \times 10^{-2} \\ -0.21 & 0.9573 & 0.05 & 0.1915 \\ -8.96 \times 10^{-2} & 0.182 + i2.9 \times 10^{-4} & -0.157 - i1.1 \times 10^{-2} & 0.9655 - i1.35 \times 10^{-2} \\ -3.74 \times 10^{-3} & -1.83 \times 10^{-2} - i6.95 \times 10^{-2} & 0.95 + i0.2568 & 0.161 + i3.23 \times 10^{-2} \end{pmatrix} \quad (15a)$$

$$U_b = \begin{pmatrix} 0.9734 & 0.224 & 0.000 & 4.81 \times 10^{-2} \\ -0.2287 & 0.9569 & 0.05 & 0.172 \\ -1.35 \times 10^{-2} & -2.37 \times 10^{-2} - i9.48 \times 10^{-3} & -0.922 + i3 \times 10^{-3} & 0.383 + i4.4 \times 10^{-2} \\ -2.765 \times 10^{-3} & -0.177 + i4.62 \times 10^{-2} & 0.354 - i0.146 & 0.88 - i0.215 \end{pmatrix} \quad (15b)$$

The interesting thing to notice about these solutions is the emergence of the flavor-flip case U_a as an allowed solution. That is, while U_b is "normal", i.e., large along diagonal; U_a is of the form

$$\begin{pmatrix} 1 & & & \\ & 1 & & \\ & & \varepsilon & 1 \\ & & 1 & \varepsilon \end{pmatrix},$$

i.e., t and b are coupled mostly to b' and t' respectively.

The flavor-flip case makes the t-lifetime longer by a factor of ε^{-2}; e.g., a t of mass ~40 GeV has a "normal" life time $\tau_t^0 \sim 10^{-20}$ s and for $|U_{tb}| \sim 0.16$, $\tau_t \gtrsim 40\tau_t^0$ and t-life time is lengthened by a factor of 40. Also in this case one might expect[40] that t and b' to be nearly degenerate and b' may have a mass as low as 40 GeV!

We turn to the effect of the fourth generation on mixing and CP-asymmetry in the decays of heavy neutral mesons. In $e^+e^- \to M^0 + \bar{M}^0 \to \ell^+ + \ell^- + X$ let N^{+-}, N^{++}, N^{--} be the numbers of events with opposite sign, same-sign positive and same-sign negative leptons, respectively. Then the mixing parameter is

$$r = (N^{++} + N^{--})/N^{+-}, \quad (16)$$

and the CP-violating asymmetry is

$$a = (N^{--} - N^{++})/(N^{--} + N^{++}) \quad (17)$$

Experimentally all that is known[41] for the heavy quark systems is that $r_{D^0} < 0.08$ and $r_{B^0} < 0.3$. The main thing to notice is the allowed range of values compared to the values expected with three generations. For $D^0 - \bar{D}^0$ e.g., we find r_{D^0} can be as large as 3×10^{-2} whereas with three generations the expected value is less than 10^{-6}! Similarly for $B^0 - \bar{B}^0$ we find r_{B^0} can be as large as 0.3 to be compared to $r_{B^0} < 0.1$ with three generations. We also find $(ar)_{B^0}$ can be as large as 3×10^{-3} again much larger than the value 10^{-5} with three generations. For $B_s^0 - \bar{B}_s^0$ the mixing $r_{B_s^0}$ is close to 1 with three generations and we find it can vary from 1 to 10^{-6} whereas $(ar)_{B_s^0}$ can now be as large as 6×10^{-2} which is much larger than the value 10^{-4} for three generations.

Finally we turn to ε'/ε. With three generations a conservative estimate[24] is 2×10^{-3}. What is similar conservative estimate for the four-generation case? ε'/ε is given by

$$\varepsilon'/\varepsilon = (15.55)(\mathrm{Im}\, A_2 - \mathrm{Im}\, A_0/\mathrm{Re}\, A_0) \quad . \quad (18)$$

Here we have used $\mathrm{Re}\, A_2/\mathrm{Re}\, A_0 \approx 1/20$, $\phi_{\varepsilon'} \approx \pi/4$,

$$\varepsilon = 2.274 \times 10^{-3} \exp(i\pi/4).$$

It has been suggested that the main contribution to ε'/ε comes from the so-called "penguin" diagram[42]. With four generations, since the fourth generation's mass may be larger than m_W, the exact "penguin" hamiltonians must be used which are given by

$$H_{st}^{\text{"penguin"}} = -\frac{G_F \alpha_s}{12\sqrt{2}} \sum_i U_{id} U_{is}^* F_1(\alpha_i) \bar{d}\gamma_\mu \lambda^a (1 - \gamma_5)$$

$$s(\bar{u}\gamma_\mu \lambda^a u + \bar{d}\gamma_\mu \lambda^a d + \bar{s}\gamma_\mu \lambda^a s) \quad (19a)$$

$$H_{em}^{\text{"penguin"}} = -\frac{2\sqrt{2}}{3\pi} G_F \alpha_{em} \sum_i U_{id} U_{is}^* F'_1(\alpha_i) \bar{d}\gamma_\mu{}^a (1 - \gamma_5)$$

$$s(\bar{u}\gamma_\mu u - \frac{1}{2}\bar{d}\gamma_\mu d - \frac{1}{2}\bar{s}\gamma_\mu s), \quad (19b)$$

where $H_{st}^{\text{"penguin"}}$ means the strong interaction "penguin" hamiltonian and $H_{em}^{\text{"penguin"}}$ means the electromagnetic interaction "penguin" hamiltonian. $F_1(\alpha_i)$ and $F'_1(\alpha_i)$ are given by

$$F_1(\alpha_i) = \ell n\, \alpha_i - D(\alpha_i) \tag{20a}$$

$$D(\alpha_i) = \frac{1}{8}\alpha_i(1-\alpha_i)^{-3}[18 - 11\alpha_i - \alpha_i^2 + 2\alpha_i(1-\alpha_i)^{-1} (15 - 16\,\alpha_i + 4\alpha_i^2)\ln\alpha_i], \tag{20b}$$

$$F_1'(\alpha_i) = -\frac{1}{3}\left(\frac{1 - 21\alpha_i + 14\alpha_i^2}{24(1-\alpha_i)^3} + \frac{4 - 16\alpha_i + 9\alpha_i^2}{12(1-\alpha_i)^4}\ln\alpha_i\right) - \frac{1}{4(1-\alpha_i)} + \frac{4 - 8\alpha_i + \alpha_i^2}{12(1-\alpha_i)^2}\ln\alpha_i, \tag{20c}$$

where i = c, t, t'.

Taking the leading-log QCD correction into account and writing the above formulas in the way which has been used in the three-generation model we finally have:

$$|\varepsilon'/\varepsilon|_{st} = 6\,(C_6/0.1)[\langle\pi\pi(I+0)|O_6|K^0\rangle/1.0\text{GeV}^3] \times \{\text{Im}\lambda_t[\ln(m_t^2/m_c^2) - D(\alpha_t)] + \eta\text{Im}\lambda_{t'}[\ln(m_{t'}^2/m_c^2) - D(\alpha_{t'})]\}|/s_1 \ln 30^2/m_c^2. \tag{21a}$$

$$|\varepsilon'/\varepsilon|_{em} = 540\alpha_{em}|[\langle\pi\pi(I=2)\,\tilde{C}_7O_7 + \tilde{C}_8O_8|K^0\rangle/0.1\text{GeV}^3] \times \{\text{Im}\lambda_t[F_1(\alpha_t) - F_1(\alpha_c)] + \eta'\text{Im}\lambda_{t'}[F_1(\alpha_{t'}) - F_1(\alpha_c)] /s_1 \ln 30^2/m_c^2\,. \tag{21b}$$

Where η and η' are QCD corrections due to the fourth generation, they are of order 1. It has been estimated[44] that: $\tilde{C}_6 = -0.1$, $B_k' = \langle\pi\pi(I=0)|O_6|K^0\rangle/1\ \text{GeV}^3 \approx 0.2 - 0.8$[26] and, $B_k'' = \langle\pi\pi(I=0)|\tilde{C}_7 O_7 + \bar{C}_8 O_8\ K^0\ /0.1\ \text{GeV}^3$ may be of order 1[43].

We use B = 0.33. $U_{ud} = 0.9734$, $U_{us} = 0.224$, $\delta_1 = 0$ and let U_{cd} and U_{cb} varying in the entire allowed range. We found that all values of ε'/ε are consistent with the experimental value[45]. $|\varepsilon'/\varepsilon|$ can be as low as 10^{-4} and can even change sign. We also find that $(|\varepsilon'/\varepsilon|_{em}/B_k'')/(|\varepsilon'/\varepsilon|_{st}/B_k') \sim 1/3$. If $B_k''/B_k' > 1.0$ then the electromagnetic "penguin" contribution can be significant, otherwise it can be neglected. As B is increased $|\varepsilon'/\varepsilon|$ decreases.

In conclusion; a sequential fourth generation of quarks and lepton is most likely to exist. It can not be too heavy: all masses are probably below 300 GeV. It may mix very little with the three light generations, in which case CP violation has to be understood within the three generation K-M formalism and all the ensuing expectations. The more interesting possibility is that the fourth generation mixes substantially and makes significant contribution to CP violation in K-decay. In this case there can be large deviation from the expectations in the three generation case, as summarized in the table below.

quantity	KM	4 Generations
τ_t	τ_t (10^{-19}_s)	can be $40\tau_t$
r_D	10^{-7}	can be 0.03
r_B	10^{-2}	can be 0.3
$(ar)_B$	10^{-5}	can be 10^{-3}
r_{B_s}	0(1)	can be $1\text{-}10^{-6}$
$(ar)_{B_s}$	10^{-3}	can be 0.06
ε'/ε	$+ 10^{-3}$	can be $\pm 10^{-4}$

Table I.

In a nut-shell, with a fourth generation, it is possible to muddy the predictions of KM model considerably. One feature to emerge from this analysis is that it is very important to tighten the errors on the U_{km} elements such as $|U_{cd}|$, $|U_{cb}|$ and even $|U_{us}|$ because the parameters for the fourth generation depend sensitively on them.

Of course, eventually one would like to actually see the fourth generation fermions directly (as discussed here by Howie Baer[1]) rather than only see the indirect effects. I feel that it is very important to find the fourth generation. Why? Exactly one hundred years ago Balmer published[46] his paper in which he proposed the "Balmer formula" for the Hydrogen spectrum. The rest is history! Rydberg generalized the formula, Ritz gave the combination principle, the Bohr model followed and finally came quantum mechanics. The reason to bring this up is that Balmer had just four (no more) lines of hydrogen to decifer the pattern. WE NEED FOUR "LINES"! Please find the fourth generation[47].

ACKNOWLEDGEMENTS

Much of the work reported here is done in collaboration with Xiao-Gang He. I would like to thank the Theory Group at KEK for their warm hospitality while this was completed. I thank M. Kobayashi and M. Fukugita for comments on the manuscript. This work was supported in part by Japan Society for Promotion of Science and in part by U.S. Department of Energy under DOE contract DE-AM03-76SF00235.

REFERENCES AND FOOTNOTES

1. H. Baer, These proceedings.
2. M. Veltman, Nucl. Phys. B123, 89 (1979).
3. W. J. Marciano, BNL-36147.
4. M.S. Chanowitz, M. Furman and I. Hinchliffe, Phys. Lett. 78B, 285 (1978).
5. P. Q. Hung, Phys. Rev. Lett. 42, 873 (1979).
6. H. D. Politzer and S. Wolfram, Phys. Lett. 82B, 242, E421 (1979).
7. R. A. Flores and M. Sher, UCI Technical Report 82-89.
8. M. A. Bèg, C. Pangiotakapoulas and A. Sirlin, Phys. Rev. Lett. 52, 883 (1984); K. S. Babu and E. Ma, UH-511-545-84.
9. C. T. Hill, Phys. Rev. D24, 691 (1981); M. Pendleton and G. G. Ross, Phys. Lett. 98B, 291 (1981); M. Machacek and M. Vaughn, Nucl. Phys. B236, 221 (1984).
10. J. Bagger, S. Dimopoulos and E. Masso, Nucl. Phys. B253, 397 (1985); see also J. W. Halley, E. A. Paschos and H. Usler, Phys. Lett. 155B, 107 (1985).
11. H. Goldberg, these proceedings; K Tabata, I. Uemura and K. Yamamoto, Phys. Lett. 129B, 80 (1983); M. Cvetic and C. R. Preitshoff, SLAC-PUB-3685; J. Bagger, S. Dimopoulos and E. Masso, SLAC-PUB-3693.
12. V. Shvartsmann, JETP Lett. 9, 184 (1969); A. Boesgaard and G. Steigman, Ann. Rev. Astro. and Astrophys. (to be published).
13. A. Buras, J. Ellis, M. K. Gaillard and D. Nanopoulos, Nucl. Phys. B135, 66 (1978).
14. D. V. Nanopoulos and D. A. Ross, Nucl. Phys. B157, 273 (1979); Phys. Lett. B108, 351 (1982).
15. A. Davidson, K. C. Wali and P. Mannheim, Phys. Rev. Lett. 45, 1135 (1980).
16. H. Sugawara, Phys. Rev. D30, 2396 (1984).
17. M. Gell-Man, P. Ramond and R. Slansky, Supergravity, ed. by P. van Nieurenhuizen and D. Z. Freedman, N. Holland (1979) p.315; F. Wilczek and A. Zee, Phys. Rev. D25, 533 (1982); R. Mohapatra and B. Sakita, Phys. Rev. D21, 1062 (1980).
18. J. Bagger and S. Dimopoulos, Nucl. Phys. B244, 247 (1984); J. Bagger et al. SLAC-PUB-3441.
19. P. Candelas et al. NSF-ITP-84-170; see also M. Dine et al. IAS Preprint (1985); A. Strominger and E. Witten, IAS Preprint (1985).
20. M. Kobayashi and T. Maskawa, Prop. Thoret. Phys. 49, 652 (1973).
21. K. Kleinknecht, Comm. Nucl. Part Phys. 13, 219 (1984).

22. J. H. Christenson et al. Phys. Rev. Lett. 13, 138 (1984).
23. J. H. Donoghue, E. Golowich and B. R. Holstein, Phys. Lett. 119B, 412 (1982).
24. B. Holstein, UMHEP-203 (1984); L. Wolfenstein, Comm. Nucl. Part Phys. 19, 135 (1985); A. Buras, MPI-PAE/P Th/ 46/84; I. Bigi and A. Sanda, Comm. Nucl. Part. Phys. 14, 149 (1984); P. Langacker, UPR-0276T (1985).
25. A. Buras et al. Nucl. Phys. B238, 529 (1984); L-L.Chau and W-Y. Keung, Phys. Rev. D29, 529 (1984); F. J. Gilman and J. Hagelin, Phys. Lett. 113B, 443 (1983); E. Paschos and U. Turke, Nucl. Phys. B243, 29 (1984); M. Shin, HUTP-84/A024; T. Brown and S. Pakvasa, Phys. Rev. D31, 1661 (1985).
26. X-G. He and S. Pakvasa, Phys. Lett. 156B, 236 (1985).
27. U. Turke et al. DO-TH 84/26.
28. C-K Chou et al. AS-ITP-84-029.
29. T. Hayashi et al. HUPD-8505.
30. A. Auselm et al, Phys. Lett. 156B, 102 (1985).
31. I. I. Bigi, Zeit. fur Phys. C27, 303 (1985)
32. M. Gronau and J. Schechter, Phys. Rev. D31, 1668 (1985).
33. S. K. Bose and E. Paschos, Nucl. Phys. B169, 384 (1980).
34. B. Barger et al, Phys. Rev. D23, 2773 (1981); R. J. Oakes, Phys. Rev. D26, 1128 (1982).
35. V. Barger et al., Phys. Rev. D25, 1860 (1982).
36. T. Inami and C. S. Lim, Prop. Theor. Phys. 65, 297 (1981). J. S. Hagelin, Phys. Rev. D23, 119 (1981); L-L.Chau; Phys. Rep. 95, 1 (1983).
37. L. Wolfstein, Nucl. Phys. B160, 50 (1979); C. T. Hill, Phys. Lett. 97B, 275 (1980); I. I. Bigi and A. I. Sanda, Phys. Rev. D29, 1393 (1984).
38. F. Gilman and M. Wise, Phys. Rev. D27, 1128 (1983).
39. It has been pointed out that it is not always possible to factor out η_{ii} as assumed here. R. Sekhar Chivkula and J. M. Flynn, HUTP-85/A036.
40. S. Pakvasa, H. Sugawara and S. F. Tuan, Zeit. für Phys. C4, 53 (1980). V. Barger et al., Phys. Rev. D30, 947 (1984).
41. A. Bodek et al., Phys. Lett. 113B, 82 (1982). A. Avery et al., Phys. Rev. Lett. 53, 1300 (1984).
42. F. Gilman and M. Wise, Phys. Lett. 82B, 83 (1979).
43. E. Ma. and A. Pramudita, Phys. Rev. D24, 1410 (1981); L-L. Chau, H-Y. Cheng and W-Y. Keung, BNL-35163; J. Bijnens and M. Wise, Phys. Lett. 137B, 245 (1984).
44. F. Gilman and M. Wise, Phys. Rev. D9, 2392 (1979).
45. R. H. Bernstein et al., Phys. Rev. Lett. 54, 1631 (1985); J. K. Black et al., Phys. Rev. Lett. 54, 1628 (1985).
46. J. J. Balmer, Ann. d. Phys. 25, 80 (1885).
47. If the fourth generation quarks carry integral charges as suggested recently (S. Pakvasa and H. Sugawara, to be published) then they will not mix with the other three and the KM matrix remains three by three. The discussion of direct signatures also has to be modified.

RECENT RESULTS ON NEW PARTICLE SEARCHES AT PEP

Richard Prepost
Department of Physics, University of Wisconsin, Madison, Wisconsin, U.S.A.

ABSTRACT

The status of the search for new particles at the PEP storage ring is reviewed. The results of searches for supersymmetric particles by the MAC and MARK II groups are presented and mass limits are given. The HRS, MAC, and MARK II limits for monojet production are given and the results are interpreted in terms of limits on Higgs particle and heavy neutral lepton production.

INTRODUCTION

The status of particle searches at the PEP storage ring is reviewed. The searches that will be discussed are the MAC and MARK II searches for supersymmetric particles and the HRS, MAC, and MARK II searches for monojet events. The searches for supersymmetric particle production are of two kinds. Both MAC and MARK II have made a search based on single electron spectra which is a search for real selectron production. Results based on these searches have been previously published and the most recent results are presented. The MAC search for photino pair production based on single photon spectra is reviewed and the most recent results are presented. Photino pair production proceeds through selectron exchange and therefore the results are sensitive to selectron as well as photino masses. The new ASP detector now running at PEP is designed specifically for these searches and will be described separately by R. Hollebeek.

New searches for monojet production by the HRS, MAC, and MARK II groups are then described. The wide interest in the UA1 monojet events has led to various speculations for an explanation. It has been proposed that the monojet events may arise from Z^0 decay into a pair of Higgs particles, one of which decays into a jet while the other lighter one escapes undetected. The electron positron colliders at present energies while not yet capable of making real Z^0 particles nevertheless can have reactions that are energetically allowed proceed through virtual Z^0 production. Since the CERN monojet events have a relatively light mass, the PEP and PETRA colliders are sensitive to such Z^0 processes. The PEP results of this search for monojet like events is described, and the results are interpreted in terms of limits on Higgs particle masses and also, for the MAC case, as a limit on the production of heavy neutral leptons.

SEARCHES FOR SUPERSYMMETRIC PARTICLES

1) Introduction

The searches for supersymmetric particle production at PEP which will be reported here are based on the study of single electron and single photon spectra. The reactions which have been studied are: (1) $e^+e^- \rightarrow e^{\pm}\tilde{e}^{\mp}\tilde{\gamma} \rightarrow e^{\mp}$ and (2) $e^+e^- \rightarrow \tilde{\gamma}\tilde{\gamma}\gamma \rightarrow \gamma$. The searches involve triggering the detector on either a single electron or a single photon. Reaction 1 is a process for the production of a real single selectron and hence is limited to $m_{\tilde{e}} \leq \sqrt{s}$. Reaction 2, on the other hand, is the radiative production of a real photino pair via selectron exchange and hence sets bounds on combinations of the selectron and photino masses. However, for the special case of massless photinos, the limit on the selectron mass is limited only by backgrounds and luminosity and not by the beam energy.

2) Single $\tilde{e}$ Production

Limits on the selectron mass from reaction 1 result from an analysis which assumes that the $\tilde{\gamma}$ is stable and not seen in the detector. Calculations for this process[1] also show that the $e^{\pm}$ which accompanies the $\tilde{e}$ in the final state tends to escape undetected down the beam pipe. The only observed final state particle is then the $e^{\mp}$ from the $\tilde{e}^{\mp}$ decay. This electron has high energy $\approx m_{\tilde{e}}/2$ and an almost flat angular distribution. The reaction is sensitive to $m_{\tilde{e}} \leq 2E_{beam}$ depending on the $\tilde{\gamma}$ mass. MAC[2] and MARK II,[3] using this technique, have previously reported lower limits on the $\tilde{e}$ mass of 22.4 and 22.2 GeV/c^2 respectively, at the 95% confidence level. This report updates the MAC search to a data sample three times larger than previously reported.

Background single electron events can come from $ee\gamma$ final states where only one of the electrons is detected. If the detector is inefficient at detecting particles or has dead regions, this background can be several orders of magnitude larger than the expected signal. However, if the undetected particles are constrained to be at small angles relative to the beam axis then momentun conservation limits the energy distribution of the observed electron. A search region for the single electrons can then be defined so that the $ee\gamma$ background is neglible.

More serious backgrounds result from decays of $\tau\tau$, $\tau\tau\gamma$, and $ee\tau\tau$ events in which most of the energy is taken by the neutrinos. If one τ decays to a visible electron and two neutrinos, and the other decays to neutrinos and a soft electron or pion which escapes down the beam pipe, then this event is indistinguishable from the SUSY process. This background has been calculated by Monte Carlo technique.

The results of these searches are interpreted as limits on the mass of the selectron assuming a massless photino, or as correlated limits on the photino and selectron masses. Fig. 1 shows the MARK II result as a contour in photino versus selectron mass for the two cases of degenerate and nondegenerate left and righthanded selectron masses. The 90% confidence level limit for the case of zero photino mass and degenerate right and left handed selectrons is 22 GeV.

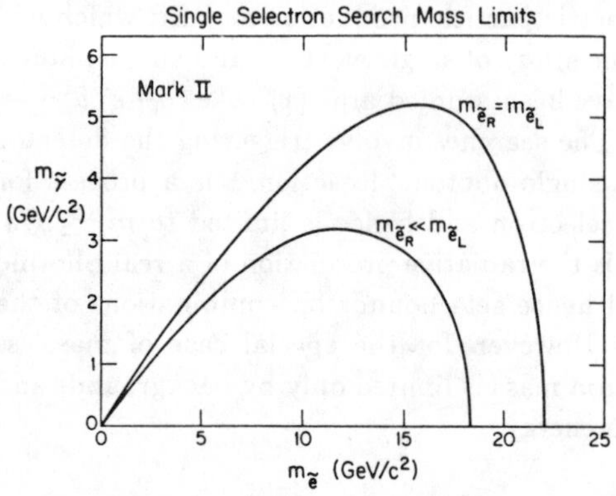

Fig. 1. The MARK II result for selectron and photino mass limits based on the single electron search described in the text. The contours represent 95% confidence level lower limits.

The MAC updated result based on a sample of 110 pb^{-1} gives an upper limit on the single electron cross section in the search region of <0.017 pb at the 90% confidence level. The corresponding mass limit is $m_{\tilde{e}} > 25$ GeV/c^2 assuming $m_{\tilde{e}_R} = m_{\tilde{e}_L}$. If $m_{\tilde{e}_L} \gg m_{\tilde{e}_R}$ then the lower limit on the lighter $\tilde{e}$ mass is 24 GeV/c^2. Future increases in the MAC data sample will only marginally improve the $\tilde{e}$ mass limit. Further improvements on the $\tilde{e}$ mass limit will necessarily be made at higher beam energies or with different reactions.

3.) $\tilde{\gamma}$ Pair Production

Reaction 2, $e^+e^- \rightarrow \tilde{\gamma}\tilde{\gamma}\gamma$ involving the detection of a single photon,[4] requires only that the photino is non interacting in the detector. This reaction also permits $m_{\tilde{\gamma}} = 0$ as a possibility, but in general the mass limits set by the experiment will be a contour with $m_{\tilde{e}}$ and $m_{\tilde{\gamma}}$ as variables.

There are potential backgrounds to the above process from radiative electromagnetic processes where charged particles are produced at angles smaller than the detector acceptance but where the photon is detected. These backgrounds include radiative Bhabha scattering and radiative tau pair production. The process $e^+e^- \rightarrow \gamma\gamma\gamma$ is a potential background if only one of the photons is emitted into the detector acceptance. There can also be single photons resulting from beam gas interactions and beam spill. Finally, the radiative neutrino pair production process $e^+e^- \rightarrow \gamma\nu\bar{\nu}$ is indistinguishable from the $\gamma\tilde{\gamma}\tilde{\gamma}$ process. However it is very desirable to also measure the cross section for this process. The cross section for radiative photino pair production has been calculated by several authors.[5] The cross section for the MAC acceptance is shown in Fig. 2 for several values of the selectron mass. The radiative neutrino pair production cross section is also shown for comparison.[6] At PEP energies, photino production is the dominant process for selectron masses less than about 50 GeV/c². The experiment is accomplished by defining acceptance criteria for the detected photon and demanding no other activity in the detector. Since the detector acceptance goes to zero below some minimum angle, this condition corresponds to setting a minimum veto angle which in turn corresponds to a minimum $E_\perp$ for the detected photon.

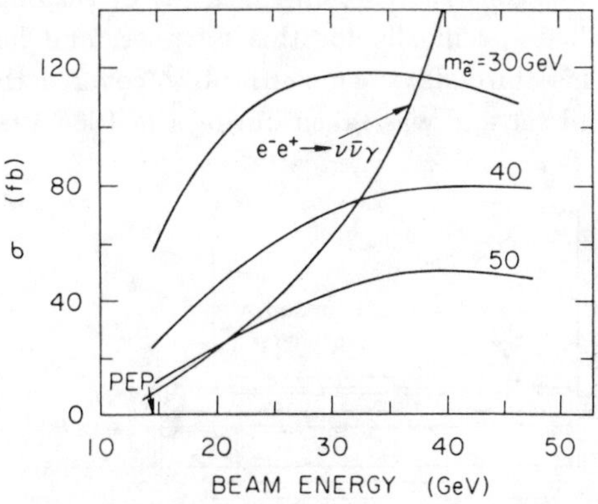

Fig. 2. The theoretical cross section for radiative photino pair production for the MAC acceptance as a function of beam energy for several values of the selectron mass. The radiative neutrino pair cross section is also shown.

The MAC analysis cuts require an electromagnetic shower with $|\cos\theta| < 0.77$ and an energy greater than 1 GeV. Below about 2 GeV the trigger efficiency for single photons begins to fall off. In addition, there can be no charged tracks in the central drift chamber. Further cuts on the electromagnetic shower profile and vertex constraints are also made. These cuts are all tuned experimentally using single electrons and tagged photons from radiative Bhabha scattering.

Three data samples were used for the analysis. For the first data set of 36 pb^{-1} the luminosity and the veto conditions were the same as used for the single $\tilde{e}$ search. The second data set of 80 pb^{-1} was taken after the installation of a special small angle tagging system which covers the region $5^\circ \leq \theta \leq 10^\circ$ with lead-proportional chamber shower counter and lead-scintillator shower counter arrays installed specifically for this experiment. The location of this veto package relative to the main detector is shown in Fig. 3. During the summer of 1984, a MAC vertex chamber was installed, and the associated geometry change required the installation of a new additional small angle tagging system in order to continue the SUSY search. This new small angle tagging system was constructed using an array of Bismuth Germanate(BGO) crystals installed specifically for this purpose, and has performed very satifactorily. This third data set of 61 pb^{-1} covered the same angular region as the second set and was taken during the 1984-1985 running period.

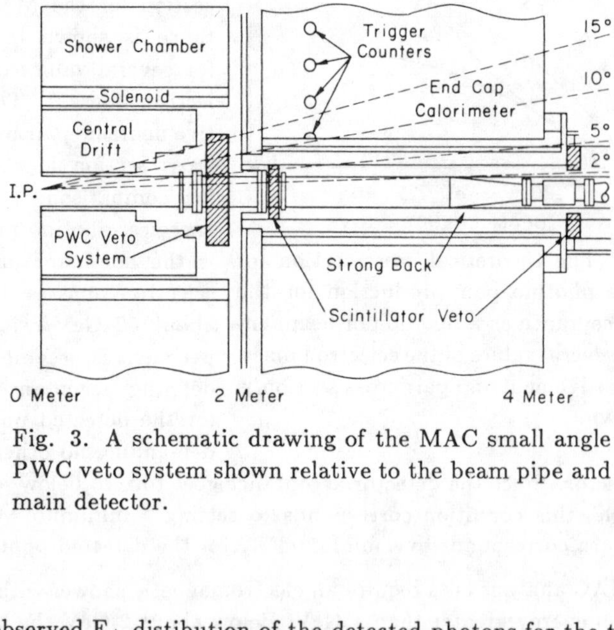

Fig. 3. A schematic drawing of the MAC small angle PWC veto system shown relative to the beam pipe and main detector.

The observed $E_\perp$ distibution of the detected photons for the three data samples is shown in Fig. 4 together with the calculated yield from radiative Bhabha scattering. The search regions were taken to be $E_\perp > 4.3$ GeV for the first data set and $E_\perp > 3.0$ GeV for the second and third data sets. The overall trigger and analysis efficiencies for the three data samples were approximately 65%. The small angle veto inefficiency was determined to be $\approx 10^{-4}$.

The most important backgrounds were calculated to be: $\nu\bar{\nu}\gamma \approx 0.8$ event, $\tau\bar{\tau}\gamma \approx 0.05$ event, and $ee\gamma \approx 0.1$ event. One event from the second data set is observed in the combined search regions at $E_\perp = 5.3$ GeV. The observed event, regardless of interpretation, limits the single photon production cross section in the detector acceptance to < 37 fb at the 90% confidence level. This corresponds to a limit of $N_\nu < 25$ for the reaction $e^+e^- \to \gamma\nu\bar{\nu}$ The calculated cross section for radiative photino pair production has been used to obtain limits for the $\tilde{e}$ and $\tilde{\gamma}$ masses. The result at the 90% confidence level is shown in Fig. 5. For $m_{\tilde{\gamma}} = 0$ and $m_{\tilde{e}_L} = m_{\tilde{e}_R}$, the limit is $m_{\tilde{e}} > 43$ GeV/c^2. For $m_{\tilde{e}_L} \gg m_{\tilde{e}_R}$, the limit is $m_{\tilde{e}_R} > 35$ GeV/c^2. These limits are significantly higher than those obtained from searches for either single $\tilde{e}$ production[2,3] or $\tilde{e}^+\tilde{e}^-$ pair production.[7] The calculation by Ware and Machacek[8] of the radiative supersymmetric neutrino pair production cross section is used to obtain a limit for the $\tilde{\nu}$ mass. For the range of $\widetilde{W}$ masses assumed in this calculation, $20 < m_{\widetilde{W}} < 29$ GeV/c^2, the limit $m_{\tilde{\nu}} > 11$ GeV/c^2 is obtained at the 90% confidence level. Results based on the first two data sets of Fig. 4 have been previously published.[9]

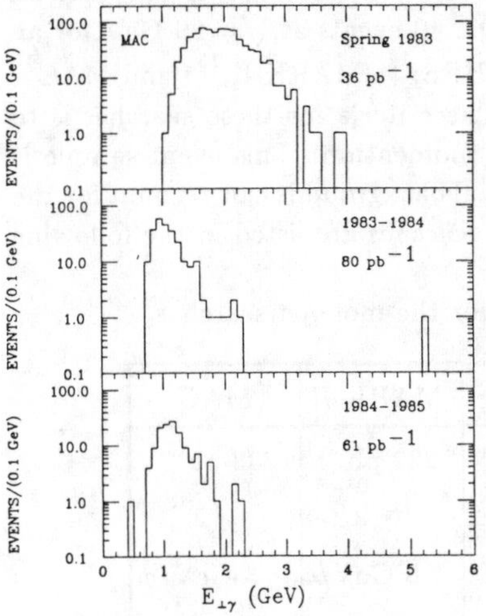

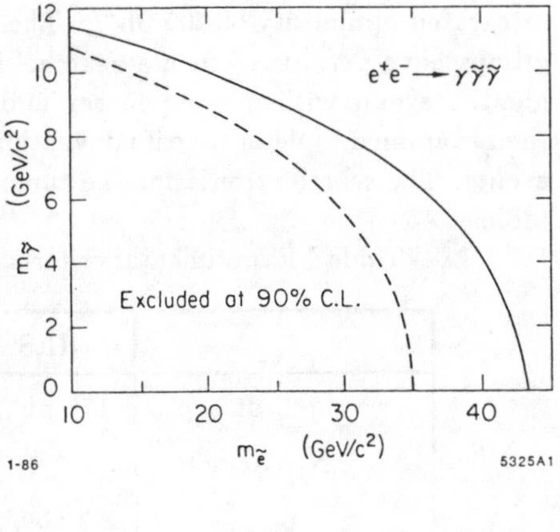

Fig. 4. The observed $E_{\perp\gamma}$ spectra for the three MAC data sets as described in the text.

Fig. 5. The lower limit for $m_{\tilde{e}}$ as a function of $m_{\tilde{\gamma}}$. The solid curve is for $m_{\tilde{e}_L} = m_{\tilde{e}_R}$. The dashed curve is for $m_{\tilde{e}_L} \gg m_{\tilde{e}_R}$. The limits are at the 90% confidence level.

SEARCH FOR MONOJET PRODUCTION

1) Introduction

The standard electroweak theory has proven very successful in describing present high energy experiments. The discovery of the intermediate vector bosons at the CERN $\bar{p}p$ collider has placed the theory on even more solid ground. On the other hand, the collider experiments have found some classes of anomalous events[10] which may not fit in the standard model and which have stimulated considerable theoretical speculation.[11] Monojet events are one such class of events. There have been proposed interpretations speculating that the monojets are direct products of Z^0 decay into either new neutral lepton pairs[12] or light Higgs particle pairs.[13] These models also predict an observable rate for monojet production in the currently operating e^+e^-storage rings via virtual Z^0 production.

2) The Experimental Searches

The search for monojets as a consequence of Z^0 decays is viable at existing e^+e^-energies since the branching ratio for Z^0 decay into light Higgs particles is about 3%. This implies a yield of about 40 events at $\sqrt{s}$=29 GeV for an integrated luminosity of 200 pb^{-1}. The HRS,[14] MARK II,[15] and MAC[16] groups have performed such a search. The criteria for these searches is to identify events with missing energy and momentum. This event sample is then examined to determine if conventional backgrounds can account for the events. The search criteria for the three searches are listed in the following table.

Table I Experimental criteria for the monojet searches

	HRS	MARK II	MAC
$\int \mathcal{L}\, dt$	176 pb^{-1}	222 pb^{-1}	238 pb^{-1}
$N_{charged}$	≥ 4	≥ 2	≥ 2
$p_\perp$	≥ 7 GeV/c	≥ 8 GeV/c	≥ 3 GeV/c
$\cos\theta$	≤ 0.5	≤ 0.67	≤ 0.8
opposite requirement	no tracks	no tracks	no tracks
efficiency	$\approx .3$	$\approx .3 - .4$	$\approx .6$
Events	1	2	11

Here the quantities $p_\perp$ and $\cos\theta$ are defined relative to the thrust axis of the jet and the selection criteria in the above table define the jet. The observed events that result from these selection criteria are then examined further. The $E_\perp$ of the jets must be sufficiently high to exclude jets that result from the two photon annihilation process $e^+e^- \to e^+e^-X$. It is still then possible to observe monojet-like events from tau pair production where one tau decays into a hard jet while the other tau decays such that the neutrino takes away almost all of the visible energy. The experiments must also correct for any lack of a hermetic seal in the direction approximately opposite the direction of the observed jet. The MAC experiment has sufficient sensitivity due to the larger $p_\perp$ acceptance to expect a background from tau decays. Fig. 6 shows the $p_\perp$ distribution of the measured events as well as the calculated distribution for tau events for two search regions corresponding to data taken with and without the small angle veto tagging system which was also used for the single photon search. The observed events are seen to be completely consistent with the tau decay hypothesis. The obseved events also have the characteristic low multiplicity of tau decays, specifically no events were found with five or more charged particles.

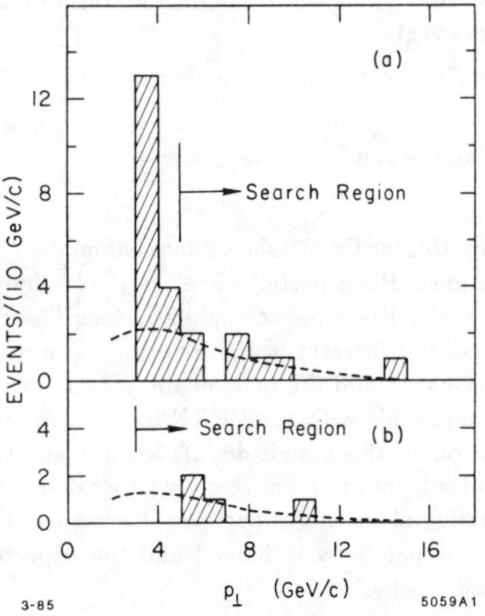

Fig. 6. The observed $P_\perp$ spectra for the MAC monojet search for two data samples corresponding to 10° and 5° veto angles respectively. The calculated tau yield is also shown in the figures as a dashed line.

Upper limits at the 90% confidence level for monojet production have been calculated from the combined data samples by subtracting the calculated tau backgrounds. The detection efficiencies used in the calculation were determined for the detector and trigger configurations for each data sample assuming the two different monojet production models described below. In order to combine the data samples, an effective detection efficiency was obtained by taking an average of these efficiencies weighted by the integrated luminosity of each data sample.

An interpretation of this cross section limit may be made in terms of virtual Z^0 production and decay. The total cross section $e^+e^- \to m_1 m_2$ may be written:

$$\sigma_{m_1 m_2} = \frac{G_F M_Z \Gamma_Z s}{\sqrt{2}((s - M_Z{}^2)^2 + M_Z^2 \Gamma_Z^2)}(1 - 4\sin^2\theta_W + 8\sin^4\theta_W) \cdot BR_{m_1 m_2}$$

where G_F is the Fermi coupling constant, M_Z=92 GeV/c^2 and Γ_Z=3.0 GeV are the Z^0 mass and width, and $BR_{m_1 m_2}$ is the Z^0 branching ratio of the decay responsible for the monojet events.

The upper limit for the branching ratio $BR_{m_1 m_2}$ has been calculated for two cases:

1) m_1=0.2 GeV/c^2 and m_2 variable, and

2) m_1=m_2 variable.

Case 1 has mass conditions suitable for the Glashow-Manohar model[13] as a monojet source, namely production of Higgs particles $e^+e^- \to \chi^0\lambda^0$ followed by $\chi^0 \to \tau^+\tau^-$ or $c\bar{c}$ decay. The scalar Higgs particle λ^0 has a long lifetime because of its small mass and does not interact in the detector. The mass of the λ^0 was assumed to be 0.2 GeV/c^2 and the mass of the χ^0 was varied from 4 to 10 GeV/c^2, a range compatible with the CERN monojet events. The production angular distribution for this case is $d\sigma/d\Omega \propto \sin^2\theta$ and the χ^0 was assumed to decay preferentially into $\tau^+\tau^-$ or $c\bar{c}$ giving a jet detection efficiency of about 60%. If no mixing of the Higgs particles is assumed, the Z^0 decay rate into a Higgs particle pair is well defined and the expected branching ratio for $Z^0 \to \chi^0\lambda^0$ is given by:[17]

$$BR_{\chi^0\lambda^0} = \frac{G_F M_Z^3}{24\sqrt{2}\pi\Gamma_Z} \cdot [1 - \frac{2}{s}(m_\chi^2 + m_\lambda^2) + \frac{1}{s^2}(m_\chi^2 - m_\lambda^2)^2]^{\frac{3}{2}}$$

where m_χ and m_λ are the χ^0 and λ^0 masses respectively.

Fig. 7 shows the measured MAC result for the 90% confidence level upper limit for the monojet cross section and Z^0 branching ratio as a function of the mass of the jet. Limits are given both for the case of the Glashow-Manohar model and for the case of neutral heavy lepton production where only one of the heavy neutral leptons gives visible energy.[18] It can be seen that both models predict significantly more events with jet mass above 4 GeV/c^2 than are observed. The Mark II and HRS experiments give a similar result with less sensitivity due to the smaller acceptance of the detectors for this type of measurement. The MARK II experiment expects 14 events based on the Glashow-Manohar model while only 2 candidate events were observed for a jet parent mass of 5 GeV/c^2. The HRS experiment bases their limits on no observed events with jet masses below 3.6 GeV/c^2 and at most one candidate at higher masses. The limits thus obtained are similar to those of the MARK II group and correspond to a monojet cross section limit varying from 50 to 120 fb over the mass range 2-10 GeV/c^2.

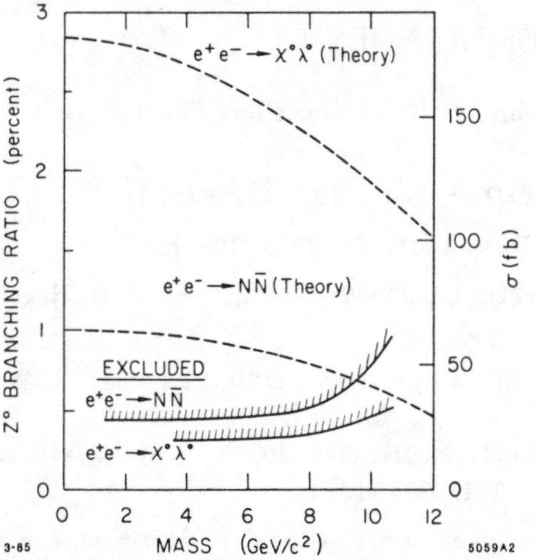

Fig. 7. The MAC upper limits at the 90% confidence level limit for the monojet production cross section and and Z^0 decay branching ratio into monojets for the two case described in the text. The calculated cross sections and branching ratios for these cases are also shown as a dashed line.

On the basis of the above limits, the hypothesis that the UA1 monojet events are due to Z^0 decays into light Higgs particles is ruled out for masses up to about 10 GeV/c^2. The hypothesis that the CERN events are due to Z^0 decays into heavy neutral leptons is ruled out for a branching ratio into monojets as small as 0.5% for Z^0 decays into masses up to approximately 10 GeV/c^2.

The author would like to thank H. R. Band and T. L. Lavine for their work on the updated MAC single photon analysis and G. Feldman and M. Derrick for supplying information about their respective experiments. This work was supported in part by the Department of Energy under contract number DE-AC02-76ER00881.

REFERENCES

1. M. K. Gaillard, L. Hall, and I. Hinchliffe, **Phys. Lett. 116B**, 279 (1982).
2. E. Fernandez et al., **Phys. Rev. Lett. 52**, 22 (1984).
3. L. Gladney et al., **Phys. Rev. Lett. 51**, 2253 (1983).
4. P. Fayet,**Phys. Lett. 117B**, 460 (1982); J. Ellis and J. S. Hagelin, **Phys. Lett. 122B**, 303 (1983).
5. K. Grassie and P. N. Pandita,**Phys. Rev. D30**, 22 (1984); J. Ware and
M. E. Machacek,**Phys. Lett. 142B**, 300 (1984); T. Kobayashi and M. Kuroda, **Phys. Lett. 139B**, 208 (1984).
6. E. Ma and J. Okada, **Phys. Rev. Lett. 41**, 287 (1978); K. J. F. Gaemers et al., **Phys. Rev. D19**, 1605 (1979).
7. H. J. Behrend *et al.* (CELLO Collaboration), **Phys. Lett. 114B**, 287 (1982) excluded $2 < m_{\tilde{e}} < 16.8$ GeV/c^2 at the 95% confidence level.
8. J. D. Ware and M. E. Machacek, **Phys. Lett. 147B**, 415 (1984).
9. E. Fernandez et al., **Phys. Rev. Lett. 54**, 1118 (1985).
10. G. Arnison *etal.*, **Phys. Lett. 139B**, 115 (1984).
11. For example some supersymmetric explanations for the CERN events are: E. Reya and D. P. Roy, **Phys. Lett. 141B**, 442 (1984); **Phys. Rev. Lett. 53**, 881 (1984); V. Barger, K. Hagiwara, J. Woodside, and W.-Y, Keung, **Phys. Rev. Lett. 53**, 641 (1984); J. Ellis and H. Kowalski, Phys. Lett. **142B**, 441 (1984); **Nucl. Phys. B246**, 189 (1984).
12. L. M. Krauss, **Phys. Lett. 143B**, 248 (1984); M. Gronau and J. L. Rosner, **Phys. Lett. 147B**, 217 (1984).
13. S. Glashow and A. Manohar, **Phys. Rev. Lett. 54**, 526 (1985); similar models are proposed by S. F. King, **Phys. Rev. Lett. 54**, 528 (1985); H. Georgi, **HUTP-85/A004** (1985 unpublished).
14. C. Akerlof et al., Argonne National Laboratory report ANL-HEP-PR-85-11.
15. G. J. Feldman et al., **Phys. Rev. Lett. 54**, 2289 (1985).
16. W. W. Ash et al., **Phys. Rev. Lett. 54**, 2477 (1985).

17. G. Pòcsik and G. Zsigmond, **Z. Phys. C10**, 367 (1981); N. G. Deshpande, X. Tata, and D. A. Dicus, **Phys. Rev. D29**, 1527 (1984).

18. J. L. Rosner, CERN-TH. 4086/85 (January 1985, unpublished).

EUROJET: A QCD BASED MONTE CARLO PROGRAM INCLUDING PERTURBATIVELY CALCULATED HIGHER ORDER PROCESSES IN $P\bar{P}$ INTERACTIONS

B. Van Eijk,
NIKHEF-H, Postbox 41882, 1009 DB Amsterdam, The Netherlands and
CERN, UA1 Coll., CH-1211 Geneva 23, Switzerland.

ABSTRACT

The richness of the experimental data brought forward by experiments at the CERN $p\bar{p}$-collider, has increased the demand for detailed Monte Carlo studies in order to compare with theoretical calculations. On the other hand, design studies for future accelerators show the necessity for predictions by Monte Carlo studies based on extrapolations of the available experimental information. These predictions sometimes cover many orders of magnitude (SLC, hadron-hadron interactions at the CERN LEP tunnel).

We will present some basic ideas underlying the EUROJET MC focussing on QCD calculations for the matrix elements involved in hard parton-parton interactions, fragmentation of (heavy) quarks and gluons into jets and finally the decays of heavy flavours. A comparison with the UA1 inclusive muon and inclusive dimuon data is made.

INTRODUCTION

The success of theoretical models based on QCD calculations in predicting several properties of hadronic jets in $p\bar{p}$-interactions, has initiated more detailed analyses. These studies seem to indicate considerable contributions from higher order processes[1].

Even stronger indications stem from the analyses of W and Z^0 events by both the UA1 and UA2 collaborations, which are accompanied by sizable jets[2,3]. The signature of these jets is fully compatible with the interpretation of the jets being initiated either by gluons emitted from the incoming quark lines or by quarks left over after gluon-quark scattering and the subsequent radiation of an Intermediate Vector Boson (IVB). Although experimental efficiencies limit the reconstruction of relatively soft jets ($E_t <$ 5-8 GeV.), one is well able to determine the global p_t distribution of the IVB's. As shown by Martinelli et all.[4] lowest order and next to lowest order QCD calculations, with in addition leading log. summations, reproduce the experimental distribution. However one can still put some question marks on how well these calculations fit the high p_t tail of the distribution. In particular when one is interested in the properties on an event by event bases, next to lowest order calculations are not sufficient. In addition one can argue that the low p_t part of the IVB p_t spectrum is highly determined by the experimental momentum resolution (e.g. missing energy measurement in W events).

These arguments have led us to construct a Monte Carlo program which contains tree level QCD calculations up to as many orders as are available in the literature. Remarkable agreement between data and the calculations of Stirling et. al.[5] have strengthened our believe that higher order QCD calculations are valid for large phase space areas and upto relatively low jet energies[6]. In other words, divergencies occurring in the matrix elements can be kept well under control, while 'soft parts' only enter in isolation studies of leptons and/or hadrons.

For a satisfying treatment of both regions we propose a scheme where the 'hard part' of distributions is described by fixed order (QCD) calculations while soft parts are approximated by a so-called parton evolution ('shower') model[7]. Before we present some results concerning the production of heavy flavours at the CERN $p\bar{p}$-collider, we will briefly outline the philosophy behind the MC program.

THEORETICAL FRAMEWORK

The perturbative expansion of QCD at relatively high Q^2 has proven to give many valuable results in the comparison with experimental data obtained at e^+e^- and $p\bar{p}$ machines. The central idea in this approach is that one can model the hard part of the interaction in terms of parton-parton scattering. Concentrating on $p\bar{p}$ (pp) interactions one can write the following expression for the cross-section:

$$\sigma = \sum_{i,j} \int dx_1 dx_2 \, f_i(x_1,Q^2) \, f_j(x_2,Q^2) \, \sigma_{ij}$$

where f_i and f_j are the parton distributions inside the colliding hadrons, i, j run over the parton species (determined by the type of hard scattering) and σ_{ij} is the Q^2 dependent and in most of the cases flavour dependent amplitude squared for the parton-parton interaction. Besides the uncertainties due to divergencies as refered to in the introduction, determination and extrapolation of the structure functions at collider energies introduces additional uncertainties. The limited x and Q^2 range accessible from, for instance, neutrino experiments introduces model dependence in the parametrizations. However, use of the Altarelli-Parisi equations[8] shows that for increasing Q^2, differences due to uncertainties at low values of Q^2, tend to diminish[9].

In order to be able to study differences in cross-sections we have included several options:

- Glück, Hoffman and Reya[10]
- Eichten, Hinchliffe, Lane and Quigg[11] (set: 1, 2)
- Duke and Owens[12] (set: 1, 2)

which allows the user of the MC also to change the QCD scale parameter (Λ) attached to these parametrizations. Since Q^2 sets the scale for the parton distributions we treat this as the invariant quantity that directly enters into the strong coupling constant or the propagator terms used in the calculation of σ_{ij}. The partons involved in the hard scatter are 'subtracted' from the colliding hadrons, leaving two parton clusters responsible for the 'underlying event' of the hard scattering. Thus overall quantum numbers, energy and momentum are conserved.

So far, calculations for W and Z^0 production upto order α_s^2 (tree level) and Drell-Yan lepton pair production have been included in the MC[5,6], with the option of having massive final states in the decay of the IVB upto order α_s. This includes the final state emission of a hard gluon in the hadronic decays. One of the nice features of the spinor techniques, used to obtain the expressions for the matrix elements, is the possibility of studying spin effects in the decay of heavy objects originating from the IVB (top, heavy lepton)[13]. In addition, neutral Higgs (H^0) production in association with IVB's has been added with the option of having either a leptonic decay or (massive) hadronic decay of the IVB[14].

Concerning the strong interactions, two and three jet production matrix elements have been included for both massive[15] (c, b, t quarks) and 'light'[16] partons (u, d, s quarks and gluons).

In order to do detailed comparisons with experimental data, calculation of global parton properties are not sufficient. Finally, a Monte Carlo event generator should produce physical particles with well defined four momenta and masses. Precise detector simulation then enables one to simulate experimental biases introduced by the apparatus and the trigger system. This means in practice that one has to simulate the conversion of the parton into (stable) particles after which one can treat the simulated event in an identical way as the real data.

We will summarize here some of the most important features of our hadronization model, a detailed description is in preparation[17]. We assume that at collider energies quarks and gluons fragment rather independent, since relative energies involved, tend to be large. However, long range correlations between individual jets are implemented in a natural way, in order to preserve quantum number conservation. In practice this means that light flavours are fragmented in a Field-Feynman type scheme[18], whereas the longitudinal fragmentation of heavy flavours is described by the Peterson et. al. parametrization[19]. Each quark-antiquark or diquark-antidiquark pair created from the vacuum obtains a relative transverse momentum in such a way that the overall p_t of the jet with respect to the jet (initial parton) axis is conserved. Relative p_t, pseudo scalar to vector meson ratio, meson to baryon ratio, different flavour contents of the vacuum etc. are freely adjustable parameters. Some of these parameters will highly depent on the jet energy region one is interested in and lack of experimental feedback.

The stopping criterium for the fragmentation is obtained as follows. The initial quark (or anti-diquark) has all it's momentum along one of the axis in a local frame. Part of it's energy and momentum will be transfered to the hadron after the creation of for instance a $q\bar{q}$ pair from the vacuum. The left over quark is treated now in a similar way as the original quark, a new hadron is created provided that enough energy and momentum are available. Building up a chain of hadrons, finally one is left with a parton unable to fragment. In order to prevent the last hadron to have a negative z-value with respect to the local axis, the last parton is thrown away and the hadron is broken into partons again. The original fragmenting parton is kept while the left-over parton recombines with left-over partons from other fragmentation chains (hard scattering and underlying event), introducing long range correlations between jets. Surprisingly, this model gives good agreements with respect to for instance charged multiplicity distributions obtained in e^+e^- experiments. Detailed comparisons will be published elsewhere[17].

This scheme has unfortunately one disadvantage, although all quantum numbers are nicely conserved, energy-momentum conservation is violated during the fragmentation process. However, these violations in general appear to be small and can be adjusted in a simple way by rescaling the four-momenta of all particles produced in the fragmentation process. Rescaling effects are typically less than 1% and decrease with increasing multiplicity in the event.

The unstable particles present in the jets are treated next and enter a decay chain, giving rise to a tree with stable mesons and baryons at it's end points. Instead of modelling all these decays, we have made extensive use of experimental information[20]. However, where necessary, appropriate matrix elements are implemented in order to describe specific decay properties. For the weak decay of heavy flavours, the V-A matrix elements have been implemented, while the phase space limits are completely determined by the specific decay mode (effects due to mass differences between various mesons and baryons containing the same heavy flavour, are properly taken into account).

INCLUSIVE Pt SPECTRUM FOR MUONS AND DIMUONS

Assuming that the bulk of the events, containing leptons with considerable transverse momenta, are either originating from heavy flavour decays or the decays of IVB's, one is able to predict the cross-section, and the shape of differential cross-sections as a function of several variables. Fig. 1

shows the inclusive muon and dimuon p_t spectra resulting from the order α_s^2 QCD heavy flavour production. The contributions from $b\bar{b}$ $c\bar{c}$ are indicated separately, while the shaded area shows the contributions from t-quark decays varying the top mass ($25 \leq m_t \leq 40$). For the single inclusive muon spectrum, we demand for each muon: $p_t > 6$ GeV/c and a limited pseudo rapidity range: $|\eta| < 2$. For the dimuons the same rapidity cut was made while at least one muon has to fulfill: $|\eta| < 1.3$. Each muon has to pass the cut: $p_t > 3$ GeV/c and in addition we require a minimum invariant mass for the dimuon system: $M_{\mu\mu} > 6$ GeV/c^2.

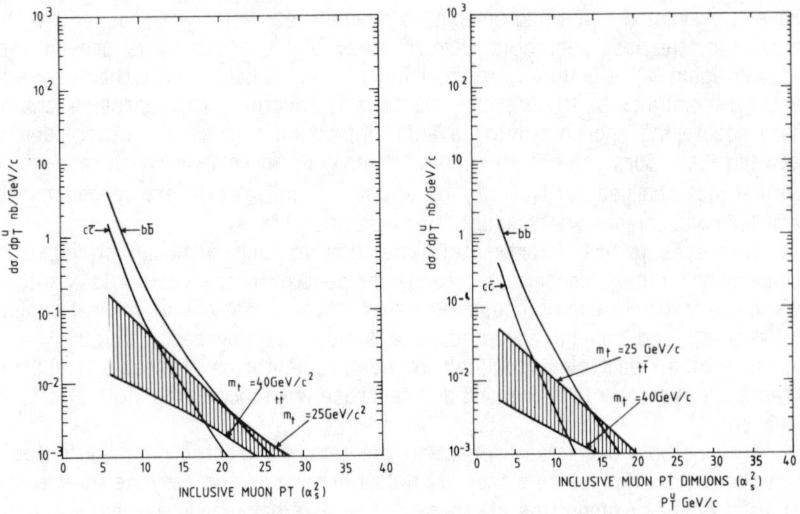

Fig. 1.

Similar distributions for order α_s^3 calculations are shown in fig. 2. For all distributions we used the structure functions of ref. 12, set 1 (Λ=0.2 GeV.), branching ratio for carmed mesons and baryons (average) into muons ~11%, branching ratio for bottom hadrons 12% and finally for top we assumed 11% for all leptonic decay modes.

The Peterson et al parametrization[19] for the charm and bottom fragmentation was tuned to fit the data available from e^+e^- experiments, while the top fragmentation is assumed to be fairly hard ($\langle z \rangle > 0.9$). The relative ratio between the charm and bottom curves in the dimuon spectra strongly depends on the branching ratio's and introduces larger uncertainties than variations in the fragmentation functions.

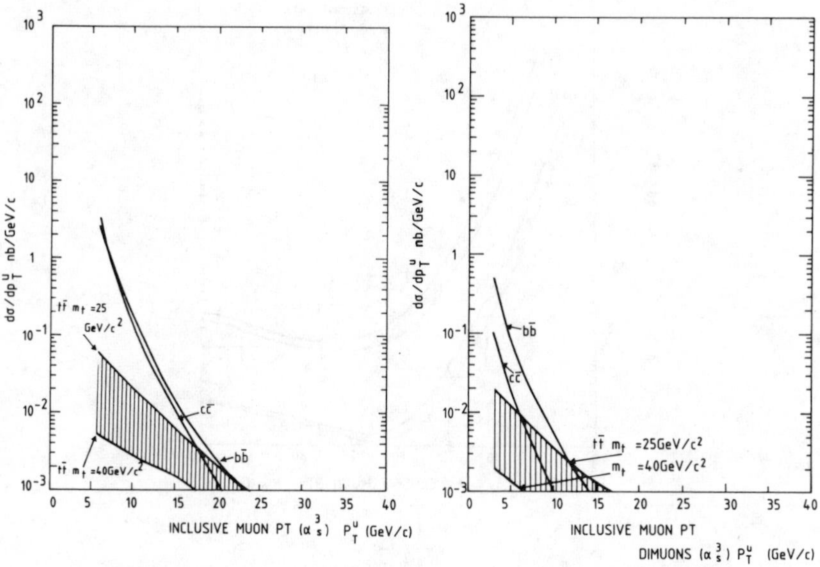

Fig. 2.

Finally we present a comparison with the preliminary UA1 data. We have included the muonic decays of W and Z^0 in both the data and the MC results. In addition heavy flavour decays of the IVB's are included in the MC calculations, however excluding the top quark from this analyses. The remarkable agreement in shape for both the muon and dimuon distributions is shown in fig. 3. The band on the theoretical curves includes statistical errors and theoretical uncertainties as mentioned above. Even more remarkable is the prediction for the absolute dimuon cross-section, whereas the theoretical curve for the inclusive single muon spectrum seems to be below the experimental one. Both the experimental curves are corrected for muonic pion and kaon decays. The subtraction of technical background sources is also included, but this analyses is not yet completely finished[22].

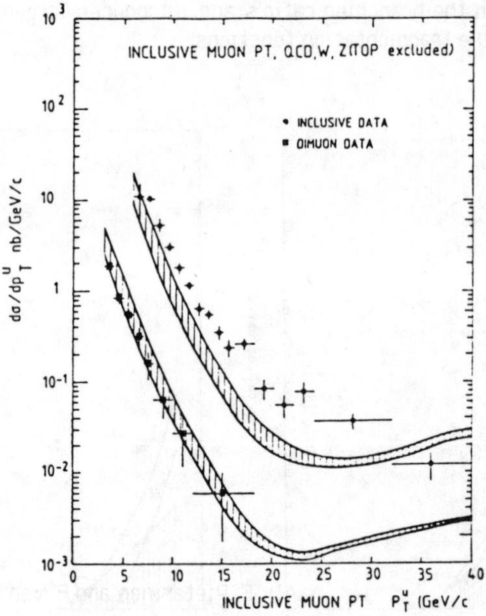

Fig. 3.

CONCLUSIONS

I have presented here a global comparison between EUROJET calculations and inclusive muon data obtained with the UA1 detector at the CERN Sp$\bar{p}$S collider. Detailed comparisons can be made as soon as the model has passed basic tests with respect to for instance the simulation of the underlying event. This iterative process of comparing data and MC calculations is in progress. However, the distributions presented here and at other occasions[21, 22] indicate the predictive power of this MC event generator.

ACKNOWLEDGEMENTS

I would like to thank my colleagues in the UA1 and EUROJET Collaboration for many useful discussions. Special thanks goes to Ingrid ten Have for her help in preparing the plots. I thank the organizers for an enjoyable conference.

REFERENCES

1. G. Arnison et al., UA1 Collaboration, Phys. Lett. 158B, 494 (1985).
2. G. Arnison et al., UA1 Collaboration, Nuovo Cimento 44, 1 (1985).

3. P. Bagnaia et al., UA2 Collaboration, presented by L. Mapelli, see ref. 4.
4. G. Martinelli, 5th Topical Workshop on Proton-Antiproton Collider Physics, Ed. M. Greco, Saint-Vincent, febr. 1985 (World Scientific Publishing Co.)
5. S.D. Ellis, R. Kleiss and W.J. Stirling, CERN Preprint TH.4096/85.
6. S.D. Ellis, R. Kleiss and W.J. Stirling, CERN Preprint TH.4185/85.
7. T. Sjöstrand, FERMILAB-Pub-85/23-T (1985).
8. G. Altarelli and G. Parisi, Nucl. Phys. B126, 298 (1977).
9. R.K. Ellis, FERMILAB-Conf-84/96-T (1984).
10. M. Glück, E. Hoffman and E. Reya, Z. Phys. C13, 119 (1982).
11. E. Eichten, I. Hinchliffe, K. Lane and C. Quigg, FERMILAB-Pub-84/17-T (1984).
12. D.W. Duke and J.F. Owens, Phys. Rev. D30, 49 (1984).
13. R. Kleiss and W.J. Stirling, private communication.
14. R. Kleiss and B. van Eijk, in preparation.
15. B.L. Combridge, Nucl. Phys. B151, 429 (1979)
Z. Kunszt and E. Pietarinen, private communication.
16. Z. Kunszt and E. Pietarinen, Nucl. Phys. B164, 45 (1980).
I. Schmitt, L.M. Seghal, H.D. Troll and P.H. Zerwas, Phys. Lett. 139B, 99 (1984).
17. The EUROJET MC program, A. Ali, E. Pietarinen and B. van Eijk,
B. van Eijk, PhD Thesis, University of Amsterdam, in preparation.
18. R.D. Field and R.P. Feynman, Nucl. Phys. B136, 1 (1978).
19. C. Peterson et al., Phys. Rev. D27, 105 (1983).
20. Particle Data Group, Rev. Mod. Phys. 56, No. 2, Part 2 (1985).
21. A. Ali, see ref. 4., B. van Eijk, idem (1985).
A. Ali, CERN Preprint TH.4207/85 (1985).
A. Ali, DESY Preprint 85-107 (1985).
22. K. Eggert, this conference.

UNEXPECTED OBSERVATIONS OF MUONS FROM CYGNUS X-3

J. W. Elbert
Physics Department, University of Utah, Salt Lake City, UT 84112

ABSTRACT

One surface experiment (Kiel) and two underground experiments (Soudan and Mt. Blanc) have detected unexpectedly large fluxes of cosmic ray muons from the approximate direction of Cygnus X-3, with signals showing the precise period of the system. The muon signals cannot be produced by any known type of elementary particle unless unexpected processes are involved.

INTRODUCTION

Recent muon observations of Cygnus X-3 are of interest to particle physicists because they do not appear to be consistent with other cosmic ray observations unless new physics is involved. A brief summary of the problem will be given in the following paragraphs. This will be followed by an introduction to Cygnus X-3, a description of > 10^{12} eV observations of Cygnus X-3, and a discussion of the muon observations.

The basic problem is that no known particles have the right properties to transmit the observed periodic signal to the earth from Cygnus X-3. The source is at a distance of > 11 kpc (>36000 lt yr). The signal is observed as a lack of uniformity of the event rates as a function of the phase within the 4.8 hour period of the source. If the muon signals originate from charged cosmic rays, the 2.5 μgauss interstellar magnetic field[1] would deflect the particles severely, so that the cosmic ray signal would not arrive from the direction of Cygnus X-3. The path lengths of different particles reaching the earth would vary greatly as well, so that no periodicity would remain in the arrival time distribution of the events. Consequently, the parent particles which produce the muon signal must be neutral.

Because of the great distance of Cygnus X-3, neutrons would decay before reaching the earth. For time dilation effects to be sufficient to keep the neutrons from decaying, the required energy is at least 10^{18} eV. But in this case the flux of 10^{18} eV particles from Cygnus X-3 would have to be much greater than the flux of all cosmic rays above 10^{18} eV.[2] If the parent particles were neutrinos, an enormous flux would be involved. In addition, the showers detected at Kiel appeared to be produced high in the atmosphere[3] and the angular distributions of muons at Soudan[2] and Mt. Blanc imply that the parent flux is attenuated strongly by large rock depths. These characteristics do not suggest that neutrinos are the source of the muon signal.

The remaining choice for the primary particles is γ-rays. But there are severe problems with this possibility. At Kiel, the average muon density in atmospheric cascades in the signal was only a little lower than the muon density for the background cosmic ray cascades. The background atmospheric cascades (or air showers) are produced by primary protons and other nuclei. However, the expected numbers of muons in air showers from γ-rays are smaller than the expected numbers from cosmic ray nuclei by at least an order of magnitude.[4,5] Similarly, the Soudan and Mt. Blanc muon fluxes in the signal are much larger than expected based on the observed very high energy "γ-ray" fluxes detected from Cygnus X-3. So, the muon signal does not appear to be produced by γ-rays from Cygnus X-3.

In the previous paragraphs, all the known particles have been rejected as likely sources of the muon signals. The remaining choices seem to be that new particles or new processes are involved or that the experimental results are wrong.

INTRODUCTION TO CYGNUS X-3

The name Cygnus X-3 implies that it was the third X-ray source discovered in the northern hemisphere constellation Cygnus. It was discovered by Giacconi et al.[6] using rocket-borne proportional counters in 1966. In 1972 it was discovered that the X-ray emission shows a 4.8 hour variation in intensity.[7] The 4.8 hour period is thought to be the orbital period of a binary system.

A radio detection of Cygnus X-3 allowed its direction to be measured more precisely than in previous X-ray observations.[8] The more accurate position allowed an infrared signal to be found by Becklin et al.[9] Although the radio signal does not exhibit the 4.8 hour period, the infrared signal does show it. The radio signal from Cygnus X-3 is extremely variable. The flux has been observed to vary by a factor of 100 on a time scale of days. At times it is one of the brightest radio objects in the sky. In 1982, high resolution radio techniques allowed "jet-like behavior" to be detected following a strong radio outburst of the system.[10] The "jet-like behavior" consisted of an expanding elliptical radio image.

Cygnus X-3 has not been observed in the visible part of the spectrum. This is attributed to the fact that it is located in the plane of the galaxy, as is our solar system, so that the line of sight to Cygnus X-3 passes through a great deal of dust. This obscuration by dust is not uncommon for stars located more than a few kiloparsecs away in the plane of the galaxy.

Since the distance to Cygnus X-3 was important in the argument that neutrons are not producing the muon signal, I will describe the distance estimation process. During a radio outburst of Cygnus X-3 in 1972, two groups studied the continuum emission near the 21 cm line of neutral hydrogen. The radiation was absorbed by hydrogen clouds in galactic arms at different distances. The gas in different spiral arms has different velocities with respect to the solar system and each arm produces absorption at a Doppler shift corresponding to the velocity of the arm. Chu and Bieging[11] detected evidence of absorption in arms at distances of ~3, 8.3, and 10.4 kpc. They concluded that Cygnus X-3 is at a distance of at least 10.4 kpc.

The second group, Lauque et al.[12] did not initially detect the 10.4 kpc absorption, and concluded that the Cygnus X-3 distance, D, is in the range 8-11 kpc. Later, however, they detected the 10.4 kpc absorption feature, and concluded that $D > 11$ kpc. During a radio outburst in 1982, Dickey[13] used the Very Large Array to obtain high sensitivity, high directional resolution data on 21 cm absorption from Cygnus X-3. He concluded that $D > 11.6$ kpc. No upper limit could be reliably obtained from the data.

Besides the 4.8 hour period, other longer periods of the system have been reported. Holt et al.[14] reported a 17.75 day period. Molteni et al.[15] reported a 34.1 day period. If the 4.8 hour period is the orbital period of a known system, a shorter period due to the rotation of the X-ray pulsar may exist. Such a period has not been detected. The 4.8 hour period is known very precisely. Van der Klis and Bonnet-Bidaud[16] give the value $0.1996830 \pm .0000004$ days, with $\dot{p} = (1.18 \pm 0.14) \times 10^{-9}$. Because of the lack of detection of a pulsar period or an optical spectrum, orbital velocity Doppler shifts that might prove that Cygnus X-3 is a binary system have not been obtained.

In the γ-ray spectrum above 40 MeV, a 1972 balloon experiment found a 3.6 σ excess from Cygnus X-3.[17] A later balloon flight by the same group did not support this result. Another balloon experiment by another group reported a flux upper limit well below that of the positive result mentioned above.[18] In 1977, a 4.5 σ effect from Cygnus X-3 was obtained in SAS-2 data.[19] This was for γ-rays above 35 MeV. The 4.8 hour period was detected. More recently, COS-B has not detected Cygnus X-3,[20] in spite of a persistent search for a signal. The situation in the MeV γ-ray range is not simple. Although the idea may be repugnant to particle physicists, the γ-ray emission from Cygnus X-3 appears to be time dependent. Higher energy γ-ray observations will be described in the next section.

TEV AND PEV AIR SHOWER STUDIES

Before discussing the TEV and PeV (10^{15} eV) γ-ray observations of Cygnus X-3, I would like to describe the main techniques used in these studies. In this energy range, the atmospheric cascades produced by γ-rays (and cosmic ray nuclei) result in thousands to millions of charged particles traveling through the air. The two kinds of systems used to detect these particles (mostly e±, with some μ±) are the Cerenkov light detector and the air shower array.

The Cerenkov system detects light produced in the air above the detector by electrons with E>21 MeV. Most of the light is emitted within a degree or two of the direction of the primary γ-ray or cosmic ray direction. On the ground, the more intense light is spread over an area of $>10^4$ m^2. Thus, an angular resolution on the order of 1° and a collection area of $>10^4$m^2 can be obtained with a single detector consisting of a parabolic or spherical mirror and a photomultiplier tube. The light pulse widths are in the nanosecond range. Nanosecond timing of the light arrival times in multiple detectors can improve the angular resolution to a few tenths of a degree[21]. Detailed imaging of the Cerenkov flash also can reach this resolution.[22]

Background light fluctuations, photoelectron statistics, and flashes from nearby particles limit the sensitivity of the systems. By using coincidence techniques, fast electronics, and large mirrors, showers of energies as low as a few tenths of a TeV can be detected.

With its large collection area, the Cerenkov technique is particularly valuable in detecting sporadic sources. The energy threshold is lower than for a counter array, so the Cerenkov technique is the only practical technique in the 0.3-10 TeV energy range. Disadvantages of the Cerenkov technique are the need for good atmospheric conditions (no clouds) and darkness (operation at night when the moon is not bright, away from city lights). Because of these limitations, a Cerenkov system can only be operated about 5-10% of the time. Manned operation is usually necessary in order to protect the system from bad weather, curious people with flashlights, etc.

Air shower counter arrays in the PeV range usually consist of a number of ~1m^2 scintillator slabs viewed by fast photomultiplier tubes. These are arranged in an array on the ground with separations of a few tens to a hundred meters. Charged particle densities are sampled at a number of locations and the total number of particles is estimated, from which the shower energy is obtained. Fast timing of the particle arrival times allows the shower direction to be obtained. Counter arrays can be operated nearly 100% of the time and operators need not be present. High altitude locations, such as Plateau Rosa at 3500 m, allow the shower threshold energy to be as low as 30 TeV.

For showers that are well above the energy threshold of a counter array, the angular resolution can be a fraction of a degree.[23] Existing systems have usually been designed to study cosmic ray properties rather than to search for point sources of γ-rays. As a result, the existing systems tend to have angular resolutions of a few to many degrees.

For Cerenkov as well as counter array systems, the sensitivity for detection of γ-ray showers is usually limited by Poisson fluctuations in the number of ordinary cosmic ray (p or nucleus) showers. The sensitivity improves with increasing detector area, angular resolution, and running time. (For sporadic sources, however, the relevant time is not the running time but the time that the source is "on".) Ideas for improving the sensitivity of systems by rejecting ordinary cosmic ray showers have had modest success. One idea which may produce a big increase in sensitivity is to accept only muon-poor air showers. According to expectation, this should be a powerful technique. However, the Kiel results (discussed in the next section) make this assumption questionable. Muon-poor showers can be selected by adding shielded scintillators to air shower arrays so that muon densities can be measured.

Now let's look at TeV observations of Cygnus X-3. Basic information is given in Table I. All observations given there were done by the atmospheric Cerenkov method. The threshold energies, E_γ, and numbers of standard deviations of the observed signals are given. The phases of the significant peaks are given in units such that the phase goes from 0 to 1 in one 4.8 hour period. Phase 0 corresponds to the minimum of the X-ray emission. Usually, the histogram contained one main peak, but two peaks as well as no peaks were observed. The integral spectrum of Cygnus X-3 γ-rays goes roughly as E^{-1} so the quantity $E_\gamma I_\gamma(>E_\gamma)$ is expected to be nearly energy independent. It can be seen that this quantity falls in the range 18-89 eV $cm^{-2}s^{-1}$ for all the observations given here, with 6 of 10 in the interval 25-40 eV $cm^{-2}s^{-1}$. The flux used in calculating $E_\gamma I_\gamma$ was the average flux during the entire (0-1) phase interval.

TABLE I Cygnus X-3 TeV γ-Ray Detections

Location	Years	E_γ (TeV)	σ	Phase	$E_\gamma I_\gamma$ ($eVcm^{-2}s^{-1}$)	Groups*	Ref.	Comment
Crimea	72,73	1.2	3.5,5	0.1	84,29	CAO	24	
Crimea	74	1.2	4.6	0.1,0.6	72	CAO	24	Two peaks
Crimea	75	1.2	4.6	0.1	18	CAO	24	
Tien Shan	76-79	3	~4	0.2	26	CAO,LEB	25	Two peaks
Mt. Hopkins	80	2	3.5	0.65	30	D,HS	26	
Crimea	79,80	1	3.6		89	CAO	27	Sporadic no peak
Jet.Prop. Lab.	81	0.5	4.2	0.6	40	ISU,JPL, UCR	28	
Dugway	81,82	1.0	4.1	0.625	~38	Dur	29	Average flux 34 day depen.
Mt. Hopkins	83	0.8	4.4	0.63	36	D,HS,Haw, Dur	30	Peak flux variable

*
CAO=Crimean Astrophysics Obs.
D=U. of Dublin
ISU=Iowa State University
UCR=U. of California, Riverside
Haw=U. of Hawaii
LEB=Lebedev Physical Inst.
HS=Harvard Smithsonian Center for Astrophys.
JPL=Jet Propulsion Lab.
Dur=U. of Durham

The pioneering work of the Crimean group is evident in this table. It can be seen that the phase observations are concentrated near 0.1-0.2 and 0.6. A number of 4σ observations have been made. The Mt. Hopkins data show evidence of variability in the TeV flux.[30]

Data for the PeV region are given in Table II. Although the observations cover a long time span, the detection of a Cygnus X-3 PeV γ-ray signal was first reported by Samorski and Stamm, at Kiel, in 1983.[3] Shortly afterwards, the Leeds group confirmed this observation by finding a signal from Cygnus X-3 in the Haverah Park data.[31] In Table II, the "Method" column contains "CA" for counter array experiments and "Ch" for atmospheric Cerenkov light experiments. The quoted phase values are roughly the centers of the intervals of phase peaks. I have attempted to give phase values according to the quadratic ($\dot{p}\neq 0$) ephemeris of van der Klis and Bonnet-Bidaud.[16] The phases occur near 0.25 and 0.6, something like in the TeV case. The values of $E_\gamma I_\gamma$ vary over a very broad range. There must be major effects of systematics between the different experiments. It would seem to be stretching things to claim that these systematic variations produce more than an order of magnitude discrepancy between different experiments. The Dugway[32] and the Gulmarg[33] experiments did not detect fluxes from Cygnus X-3 during August-September 1984. The data seem to be telling us that Cygnus X-3 is an extremely variable source in the PeV region. The data from Bhat et al[33] were obtained at Gulmarg using only two photomultiplier tubes without mirrors! These data were only recently searched for evidence of a 4.8 hour period, which appears to be very strong. A very weak signal was obtained at Akeno.[34] This result was obtained by making a severe cut which required the showers to be muon-poor. Even much weaker cuts seem to allow only about a factor of 2 higher fluxes, however.

Table II Cygnus X-3 PeV γ-Ray Detections

Location	Years	Method	Ethr (PeV)	σ	Phase	$E_\gamma I_\gamma$ ($eVcm^{-2}s^{-1}$)	Groups	Comment
Kiel	76-80	CA	2 10	4.4	0.25	148 110	Samorski & Stamm	Ref. 16 phase, muons
Plateau Rosa	80-83	CA	0.03	2.8	0.63	126	Turin	Poss. correlation with 34.1 day period
Dugway	83	Ch	1.0	3.5	0.25	320	Utah	1984→upper limit
Gulmarg	76,77	Ch	0.5	5	0.60	1300	Srinagar	1984→upper limit
Akeno	81-84	CA	1.0	3.1	0.50	11	Tokyo, Kyoto	Muon-poor showers Ref. 16 phase

MUONS AND CYGNUS X-3: THE KIEL DATA

Samorski and Stamm[3] made the first PeV detection of Cygnus X-3. In Ref. 35, they also described the measurement of muon densities in the shower sample in which Cygnus X-3 was detected. The air showers were detected with an array of 28 scintillation counters. The counters were 1 m^2 in area and were separated by distances up to 100 m from the center of the array. By 1 ns accuracy timing, the array achieved angular resolution of 1°. The array was located near sea level at Kiel. Shower energies were above 2×10^{15} eV.

Showers were accepted if the densest part, the core, fell within 30 meters of the center of the array. An effective running time of 16,775 hours was accumulated between March 1976 and January 1980. Data were only accepted at zenith angles less than 30°, leaving 3,838 hours of observation time for Cygnus X-3. The expected number of showers in the vicinity of Cygnus X-3 was based on the rates in background regions with the same declination as Cygnus X-3. It was expected that air showers from γ-rays would have broad lateral distributions and consequently would have larger observed values of the so-called age parameter, s. A cut was made requiring $s \geqslant 1.1$. With this cut, a 4.4σ excess of showers (16.6 above a background of 14.4 ± 0.4) was seen in a 3° square directional bin centered on Cygnus X-3.

Next, the shower arrival times were converted to phases in the 4.8 hour period using X-ray results for the exact period and a time at which the phase was zero. The result is shown in Figure 1. The Poisson probability of the peak occurring in one of the 10 bins was 5×10^{-8}. This established that a signal was present from Cygnus X-3. The flux above 2×10^{15} eV was $7.4 \pm 3.2 \times 10^{-14} cm^{-2} s^{-1}$. Above 10^{16} eV it was $1.1 \pm 0.6 \times 10^{-14} cm^{-2} s^{-1}$.

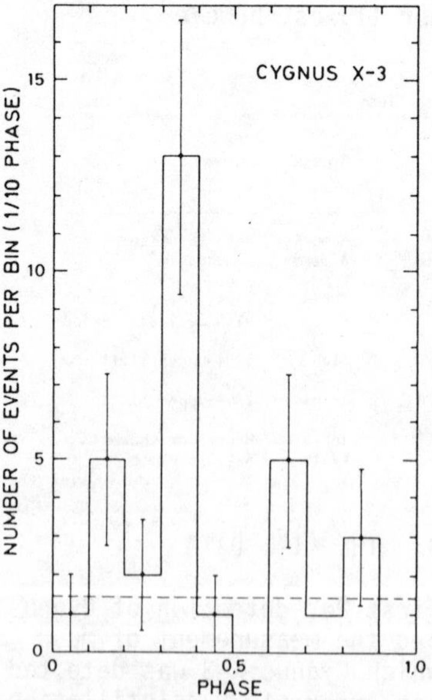

Fig. 1 Phase histogram of the arrival times of extensive air showers of size $N_e \geqslant 10^5$ particles and age parameter $s \geqslant 1.1$ from the direction of Cygnus X-3. The dashed line corresponds to the average off-source rate of 1.44 ± 0.04 showers per bin.

Muons were detected in a shielded flash tube hodoscope of physical area 64.6 m^2 and effective area 21.4 m^2. The shielding consisted of 880 g cm^{-2} of concrete. (See Figure 2). The muon threshold energy was 2 GeV. The hodoscope consisted of 367,500 spherical flash tubes. A number of steps were taken in the analysis procedure. First, regions with "bursts" were excluded. These are cascades produced by muons or hadrons in

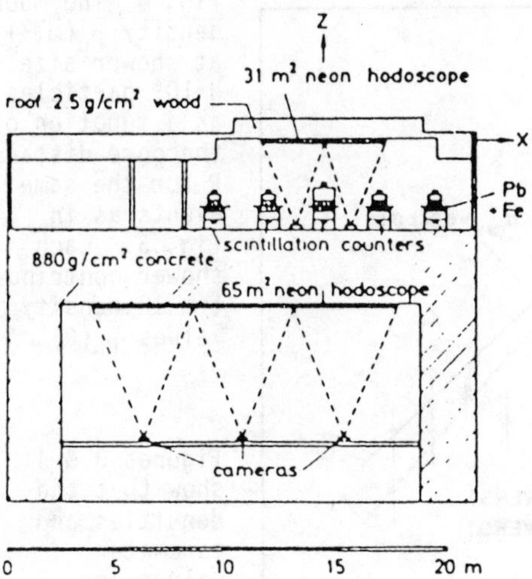

Fig. 2 The Muon Detector at Kiel

the concrete. Using all the showers, small regions in which more tubes fired than expected from Poisson statistics were identified and rejected. A residual background flash rate was subtracted. Then the detector was divided into 12 regions and the muon density was measured in each region. These densities were then fitted to the following function.

$$\rho_\mu(R)=C(10^6/N)^\gamma R^{-\alpha} \quad (1)$$

Then $\rho_\mu(R=10\text{ m})$ was used to characterize the muon content of the shower.

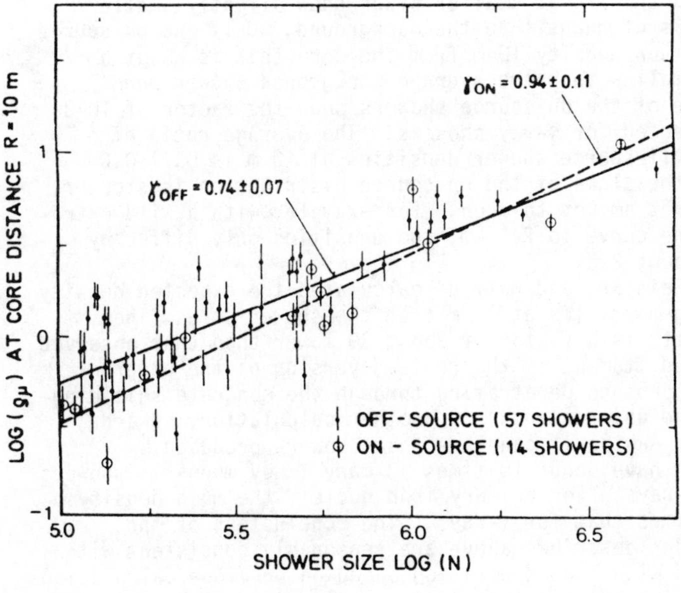

Fig. 3. The muon density ρ_μ (m^{-2}) at core distance R = 10 m as a function of the shower size N for 14 on-source showers from Cyg X-3 and 57 off-source showers. (Kiel data)

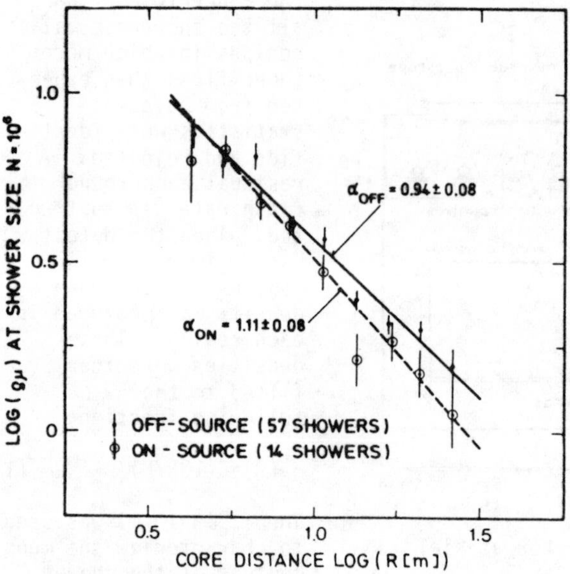

Fig. 4 The muon density $\rho_\mu(m^{-2})$ at shower size $N=10^6$ particles as a function of the core distance R for the same events as in Fig. 3. Each shower contributes 12 density values $\rho_\mu(R)$.

Figures 3 & 4 show that the densities and parameter values are similar for the showers from the signal (on-source) and the background showers (off-source). In Figure 3, the 9 on-source showers above shower size (number of charged particles) $N=3x10^5$ seem to have muon numbers very close to those of the off-source sample. The 5 showers at smaller sizes show slightly smaller average numbers of muons than the background. Only one on-source shower has a muon density 10 m from the core that is about a factor of 4 smaller than the average background shower muon density. None of the on-source showers show the factor of 10-30 reduction expected for γ-ray showers. The average ratio of on-source to off-source shower densities at 10 m is 0.77±0.09. In Figure 4, the slope of the on-source distribution is steeper, but this doesn't amount to much. For example, with a wild extrapolation of the curve to R=1 km, the densities only differ by a factor of about 2.5.

Stanev, Gaisser, and Halzen[4] calculated the expected density of penetrating particles at 10 m from the shower core. The expected density is a factor of about 30 lower than that observed by Samorski and Stamm. In the revised version of the preprint, the effect of photons penetrating through the concrete shielding was found to be unimportant. In another calculation, Edwards, Protheroe, and Rawinski[5] conclude that showers produced by protons should have about 10 times as many 1 GeV muons as showers produced by γ-rays. For primary iron nuclei, the muon density is about 50 times that for γ-rays. The conclusions of the two calculations described above are reasonably consistent with each other and with the expectation based on previous calculations. The muon densities measured at Kiel are surprisingly high if the primaries are γ-rays.

The recent observation of Cygnus X-3 by the Akeno group[34] does not confirm the muon result obtained by Samorski and Stamm. The Akeno group was able to find a small Cygnus X-3 signal by selecting showers with muon densities less than 1/30 of those seen in average showers. Furthermore, this muon-poor component appeared to make up at least half of the total periodic signal from Cygnus X-3. Since the Cygnus X-3 flux detected at Akeno was much smaller than that seen at Kiel, and since the peak phases were different, it might be argued that the results are not necessarily contradictory because they may not have been observing the same phenomenon. A simpler conclusion is that more observations are needed.

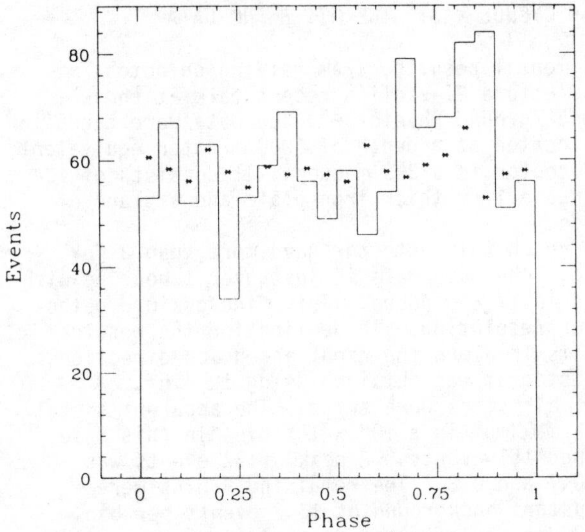

Fig. 5 Phase plot for events within 3° of the observed position of Cygnus X-3. The solid histogram shows the observed data. The points represent the expected number of events for a constant, isotropic source. (Soudan data)

MUONS AND CYGNUS X-3: THE SOUDAN DATA

The Soudan results will be covered in Professor Marshak's report and have recently been published,[2] so I will not give many details here. The detector is 2.9 m x 2.9 m x 1.9 m and is located underground at a depth of 1800 m water equivalent. It is operated by a Minnesota-Argonne collaboration. It is a multi-layered nucleon decay detector and yields an angular resolution of about 1.5° for muon tracks.

Muons require about 600 GeV or more energy at the surface in order to reach the detector. Although a significant flux of high energy muons is not expected from Cygnus X-3 on theoretical grounds, a histogram of the phases of the muon arrival times in the 4.8 hour period was plotted. It is shown in Figure 5. The muons were from a 3° x 3° directional bin close to the Cygnus X-3 coordinates. A fairly broad enhancement is present between phases 0.65 and 0.9. It consists of 84±20 events, corresponding to a muon flux of $\sim 7 \times 10^{-11} cm^{-2} s^{-1}$. The χ^2 test yielded a probability of 2×10^{-4} that the experimental distribution was in agreement with the background.

If $E_\mu > 600$ GeV, the flux given just above gives $\Sigma_\mu = I_\mu E_\mu > 42$ eV $cm^{-2} s^{-1}$. By energy conservation, the value of Σ for the primary particles must be greater than or equal to Σ_μ. The γ-rays would need to have energies in at least the TeV range. If the muon production process by γ-rays is extremely inefficient, as expected, then $\Sigma_\gamma >> \Sigma_\mu$. But, we can see from Table I that the observed TeV Σ values are comparable with the Σ_μ values from the Soudan experiment. Therefore, such large muon fluxes from Cygnus X-3 are completely unexpected.

MUONS AND CYGNUS X-3: THE MT. BLANC DATA

For these experimental results, I am relying on notes taken from Prof. B. D'Ettore Piazzoli's recent talk at the First Symposium on Underground Physics.[36] The data were taken in the NUSEX detector, located at a depth of 5000 m water equivalent in Mt. Blanc. The detector is a 3.5 m cube. It consists of 134 layers. Each layer has a 1 cm thick iron plate and a plane of plastic streamer tubes.

The muons which reach this detector have more than 3 TeV energy at the surface. The muon rate is just over 1 $hour^{-1}$, with 20,784 muons observed in 17,856 hours. This fine-grained detector gives good angular resolution. In declination the resolution is 1/2°, and it is 1° along the right ascension direction. The 4.8 hour phase histogram was obtained using 10 bins. A number of directional bin sizes were tried. The apparent signal from Cygnus X-3 was a maximum in a 10° x 10° bin. In this case, the histogram contained 151 events. A peak of 32 events was seen between phases 0.7 and 0.8. The remaining 9 bins were consistent with a constant background of 13.2 events per bin. The χ^2 was 29.1 for 9 degrees of freedom, yielding a probability of 6×10^{-4}.

The events in the phase interval 0.7-0.8 tended to come from shallow depths, like the background events. This is consistent with the Soudan result that the muons are not produced by neutrinos. The fact that the optimal bin size was 10° x 10° is puzzling. This implies that larger angular deviations are present than expected from the angular resolution and multiple Coulomb scattering in the rock, as Dr. Stanev has pointed out. However, I do not know how strongly the data support a large bin size (as opposed to smaller sizes), so I shouldn't overemphasize this point.

ACKNOWLEDGEMENTS

I would like to thank Dick Steinberg and Vic Stenger for providing me with their notes on the Mt. Blanc data. I would also like to acknowledge helpful conversation with Marvin Marshak, John Learned, and Trevor Weekes.

REFERENCES

1. M.S. Longair, High Energy Astrophysics (Cambridge University Press, Cambridge, 1981), p.258.
2. M.L. Marshak et al., Phys. Rev. Letters 54, 2079 (1985).
3. M. Samorski and W. Stamm, Ap.J. (Letters) 268, L17 (1983).
4. T. Stanev, T.K. Gaisser, and F. Halzen, Bartol Research Foundation preprint, 1985.
5. P.G. Edwards, R.J. Protheroe, and E. Rawinski, Journal of Physics G, (in press, 1985).
6. R. Giacconi et al., Ap.J. (Letters) 148, L119 (1967).
7. D.R. Parsignault et al., Nature Phys. Sci. 239, 123 (1972). P.W. Sanford and F.M. Hawkins, Nature Phys. Sci. 239, 135 (1972).
8. L.L.E. Braes and G.K. Miley, Nature 237, 506 (1972).
9. E.E. Becklin et al., Nature 245, 302 (1973).
10. B.J. Geldzahler et al., Ap.J.(Letters) 273, L65 (1983).
11. K.W. Chu and J.H. Bieging, Ap.J. (Letters) 179, L21 (1973).
12. R. Lauque, J. Lequeux, and Nguyen-Quang-Rieu, Nature Phys. Sci. 239, 118 (1972).
13. J.M. Dickey, Ap. J. (Letters) 273, L71 (1983).
14. S.S. Holt et al., Nature 260, 592 (1976).
15. D. Molteni et al., Astron. Ap. 87, 88 (1980).
16. M. van der Klis and J.M. Bonnet-Bidaud, Astron. Ap. Lett. 95, L5 (1981).
17. A.M. Galper et al., 14th ICRC Munich 1, 95 (1975).
18. S.P. McKechnie et al., Ap.J. (Letters) 207, L151 (1976).
19. R.C. Lamb et al., Ap.J. (Letters) 212, L63 (1977).
20. K. Bennet et al., Astron. Ap. 59, 273 (1977).
21. A.I. Gibson et al., 17th ICRC Paris, 8, 38 (1981).
22. T.C. Weekes, 17th ICRC Paris, 8, 34 (1981).
23. J. Linsley, U. of New Mexico preprint, 1985.
24. A.A. Stepanian, B.M. Vladimirski, and Yu.I. Neshpor, 15th ICRC Plovdiv, 1, 135 (1977).
25. J.B. Mukanov et al., 16th ICRC Kyoto 1, 143 (1979).
26. S. Danaher et al., Nature 289, 568 (1981).
27. V. P. Fomin et al., 17th ICRC Paris 1, 28 (1981).
28. R.C. Lamb et al., Nature 296, 543 (1982).
29. J.C. Dowthwaite et al., Astrom. Ap. 126, 1 (1983).
30. M.F. Cawley et al., Harvard-Smithsonian Center for Astrophysics preprint (1985)
31. J. Lloyd-Evans et al., Nature 305, 784 (1983).
32. R.M. Baltrusaitis et al., Ap.J. (in press, 1985).
33. C.L. Bhat, M.L. Sapru, and H. Razdan, Bhabha Atomic Research Center preprint, 1985).
34. T. Kifune et al., U. of Tokyo preprint, 1985.
35. M. Samorski and W. Stamm, 18th ICRC Bangalore 11, 244 (1983).
36. Private communication with R.I. Steinberg and V.J. Stenger.

OBSERVATION OF CYGNUS X-3 AT ENERGIES OF 10^{11} TO 10^{13}eV USING THE ATMOSPHERIC CHERENKOV TECHNIQUE

Trevor C. Weekes
Whipple Observatory, Harvard-Smithsonian Center for Astrophysics, Box 97, Amado, AZ 85645-0097

ABSTRACT

Five independent groups have reported the detection of radiation from Cygnus X-3 at energies of 0.1 to 10 TeV; all the observations showed evidence for phased emission and all observations were made using the atmospheric Cherenkov technnique which does not distinguish the nature of the primary. The observations are consistent with a light-curve which shows long-term variations; this would be expected from a source whose emitted power is close to the total galactic cosmic ray replenishment rate.

INTRODUCTION

Recent observations of 1 to 10 PeV emission from Cygnus X-3 (Samorski and Stamm 1983; Lloyd-Evans et al. 1983) and the possible detection of penetrating radiation (M. Marshak et al. 1985; J. Elbert, this meeting) might give the impression that the observation of very high energy emission from Cygnus X-3 is a new phenomenon. In fact Stepanian and his co-workers at the Crimean Astrophysical Observatory first detected radiation at energies of 1 TeV in September, 1972 using the atmospheric Cherenkov technique (Vladimirsky et al. 1973); although these observations were largely ignored in the West, they were in fact confirmed by the same group and at least four other groups using the same technique in subsequent years.

In this paper we will catalog some of the salient features of the Cygnus X-3 source that have been seen at other wavelengths; we will then briefly describe the atmospheric Cherenkov technique and give a chronological account of the observations of Cygnus X-3 at energies of 0.1 to 10 TeV.

Cygnus X-3: a compendium.

Cygnus X-3 was discovered as an x-ray source in a rocket-flight in 1967 (Giacconi et al. 1967); its binary nature was established in 1972 (Parsignault et al. 1972). The large radio outburst of 1972 (Hjellming 1973) led to its identification with an infra-red source. Radio jets were observed in 1982 (Geldszahler et al. 1983).

Any model of Cygnus X-3 must account for the following features:

(1) the asymmetric sine wave x-ray light-curve of Cygnus X-3 with the absence of complete eclipse; this is not a characteristic of other galactic binary sources.

(2) Absorption of low energy gamma rays by gas in the vicinity of the source (more of the absorption is due to inter-

stellar gas in the galactic plane).

(3) Infrared modulation in the H and K bands that exhibits a weak 4.8 hour periodicity.

(4) Irregular emission of strong radio bursts which appear to be associated with jets which expand at velocities $\sim$ 0.25c.

(5) No 4.8 hour periodicity is seen in the radio emission but at shorter wavelengths there is evidence of a 4.95 hour modulation which may persist for only a few cycles.

(6) The absence of any fast pulsed emission at radio, infrared or x-ray wavelengths but the possible existence of longer periodicities at 17, 19 or 34.1 days.

(7) Quasi periodic x-ray emission on time scales of 50 to 1500 seconds.

(8) Emission at 100 MeV energies which was seen by SAS-II in 1973 but not by COS-B in 1975-80.

Atmospheric Cherenkov Technique

Optical emission from cosmic ray air showers was first detected by Galbraith and Jelley in 1953. Cherenkov light is given off by the atmosphere when the relativistic particles in the air shower have velocities which exceed the velocity of light. At sea level the Cherenkov angle is 1.3^{o} and the rate of Cherenkov light production is 30 photons/meter of path length. The light is concentrated at blue wavelengths and behaves like a penetrating component of the shower. At ground level the light pool has a diameter of 200 m and a thickness of 1 m; its angular spread (determined by Coulomb scattering of the shower particles and the Cherenkov emission angle) is $\sim$1 to 2°.

Cosmic rays of TeV or greater energies have traditionally been studied using ground-based arrays of detectors which give some indication of the nature of the primary from the distribution of the secondaries. In this respect the atmosphere is the detection medium and the collection area is determined by the lateral spread of the secondary particles. The detection of Cherenkov light photons follows the same principles with the important advantage that, unlike secondary particles, photons can be brought to a focus and light detectors are remarkably sensitive and inexpensive. At energies below 100 TeV not enough secondary particles reach even mountain altitude to make detection by conventional arrays feasible; muon telescopes can operate at lower energies but the expected flux of muons in gamma-ray air showers is too small to make this approach competitive with optical detectors.

The traditional atmospheric Cherenkov detector consists of an array of ex-army searchlights operated in some form of coincidence. The Crimean Astrophysical Observatory uses two 1.5 m reflectors which scan through the direction of the source as the earth rotates. A second set operates out of phase so that the source is continuously monitored. At Tien Shan the Lebedev Institute operates an array of three 1.5 m reflectors; the University of Durham operated four sets of 1.5 m reflectors at the Dugway Proving Ground. The J.P.L. - Riverside - Iowa State collaboration used the two 11 m Solar Concentrators at

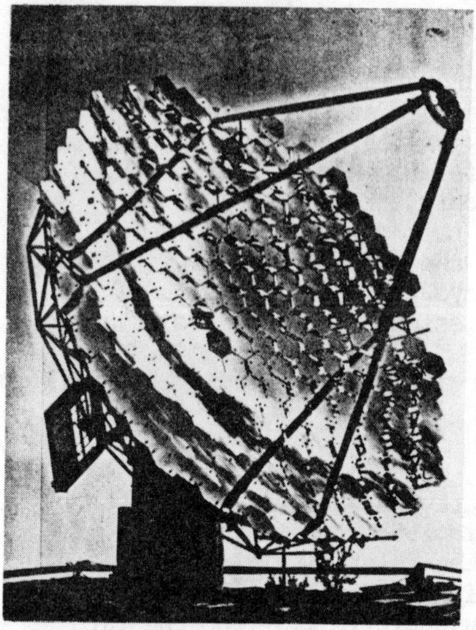

Edwards Air Force Base. A variety of reflectors including the 10 m Optical Reflector have been used to observe Cygnus X-3 from the Whipple Observatory (figure 1).

Figure 1. The Whipple Observatory 10 m Optical Reflector which was built in 1968. It is the largest purpose-built gamma-ray telescope.

To a first approximation the Cherenkov light distribution from a gamma-ray or proton initiated air shower is similar. Hence most gamma-ray astronomy experiments have concentrated on the statistical detection of a directional anisotrophy due to gamma rays amongst the isotropic background of charged cosmic rays. Many experiments have concentrated on improving the angular resolution of the detectors rather than on any attempt to differentiate the primaries. Recent studies have shown that there are differences in the light distributions which more sophisticated systems can exploit.

Observation Summary

(a) Crimean Astrophysical Observatory

The first very high energy gamma-ray observations were made just three days after the radio outburst of September 2, 1972 and were continued until November 1, 1972, a total of eleven nights of observation (Vladimirsky et al. 1973). The source was observed at the Crimean Astrophysical Observatory sequentially by two sets of detectors in the drift-scan mode. A significant excess was seen in only one of the detector sets on two consecutive nights of observation. The excess was at the 5 σ level (figure 2) and its significance was estimated by the authors as having a probability of 6.10^{-6} of arising from chance fluctuations. The authors considered the most likely explanation for the null detection in the second set of detectors as due to their energy threshold being 1.8 times that of the first

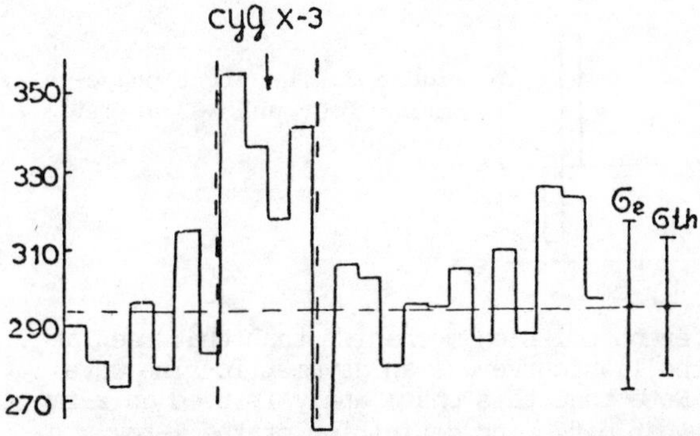

Figure 2.
Drift-scans across the position of Cygnus X-3.

set; they concluded that the source had an integral power-law spectral index in excess of 3.2. With hindsight the explanation may be that the emission was short-lived and only happened to coincide with observations by the first detector set. Later observations also point to a much smaller spectral index. The observed flux level during the outburst was 2.10^{-10} photons cm^{-2} s^{-1} with an energy output of 3.10^{36} erg s^{-1} for photons > 1 TeV.

The first announcement of the detection of very high energy gamma rays from Cygnus X-3 came in a paper in the Proceeding of the 13th International Cosmic Ray Conference at Denver, USA in August of 1973. The paper entitled "High Energy Gamma-Ray Outburst in the Direction of the X-ray Source Cygnus X-3" by B.M. Vladimirsky, A.A. Stepanian and V.P. Fomin should have attracted international attention but did not do so for a variety of reasons: (a) the paper was not presented at the conference, (b) the Crimean group had previously presented evidence for other transient galactic gamma-ray sources which had not been confirmed, (c) the observations presented in the paper were not entirely convincing, (d) the results were not included in the special issue of Nature, Physical Science, which contained other observations dealing with the September, 1972 radio outburst; they were not referenced in Hjellming's comprehensive account of the outburst (Hjellming 1973), (e) the atmospheric Cherenkov technique was not in the mainstream of high-energy astrophysics. Unfortunately this initial lack of interest in results that had a high statistical significance was to extend to the subsequent Crimean observations in the seventies.

The C.A.O. observations were continued in August, 1973 through October, 1973 when again an excess was seen from the direction of Cygnus X-3. When these observations were subjected to a phase analysis at the 4.8 hour period measured in x-rays, the result indicated that the emission was at phase 0.3 and 0.8 (figure 3). The width of the active emitting phases was 0.05 which

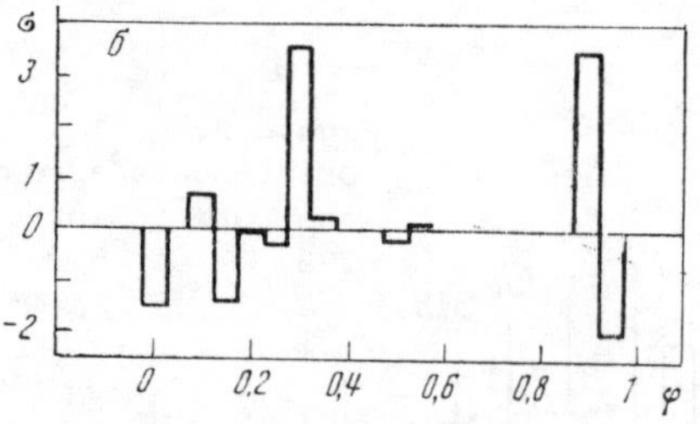

Figure 3. The first phase-ogram of Cygnus X-3 at TeV energies.

suggested a quite different emission mechanism than that seen at lower energies where the lightcurve was an asymmetric sine wave (Fomin et al. 1975). Note that this phase analysis used an x-ray emphemeris with a somewhat different definition of the x-ray minimum (phase zero) than was later employed. The later definition would put the active emission at phases 0.2 and 0.7 approximately.

In 1974, the C.A.O. group adopted a more efficient operating procedure using shorter drift-scans which increased the time on source by an order of magnitude (Vladimirsky et al. 1975). 357 drift scans between July and November, 1974 (compared with 54 in 1972-3) gave a net ON/OFF ratio of +2.7 σ (+1.8 σ in 1972 and +1.8 σ in 1973). Over the three years the net ON/OFF effect for all phases was +3.3 σ ;there was strong evidence for variability on a month-to-month basis. In August, 1974 the ON/OFF ratio was +4.4 σ . Once again the flux was found to be emitted during a narrow phase interval of the 4.8 hour period. Using the published UHURU period of 0.199622 days, the C.A.O. group noted that the active phase region in the 1972-3 data underwent an apparent phase shift in 1974. This phase shift (by about 0.2 of the phase) could be eliminated by using a period of 0.199682 days. A period close to this value was independently measured in x-rays. The ability of the gamma-ray observations to utilize the narrow phase of the emission to predict the 4.8 hour period was a remarkable confirmation of the reality of the observed emission. During the 1974 observations for the first time sporadic emission of TeV quanta was noted; this emission was not at a particular phase and persisted for a few days (A.A. Stepanian, private communication).

Continued observations in 1975 (August through November) and a new analysis of all four years of data gave the light-curve shown in figure 4 (Stepanian et al. 1977). The most significant excess (4.6σ) was seen at phase 0.130 to 0.185; there was also significant emission (3.0σ) at phase 0.740 to 0.795. It was concluded that this phased emission did <u>not</u> show any monthly variations. The sporadic emission did show monthly variations but its time averaged flux, $1.7 \cdot 10^{-11}$ photons cm^{-2} s^{-1}, was almost a factor of ten less than the flux at the active phases of the periodic emission. Also the sporadic component appeared to have a power law distribution with integral spectral exponent > 3.0 whereas the periodic component seemed to have the same distribution as the cosmic ray background (integral exponent $\sim$ 1.7).

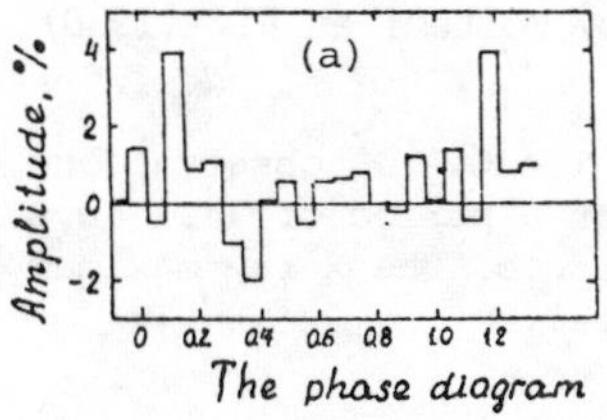

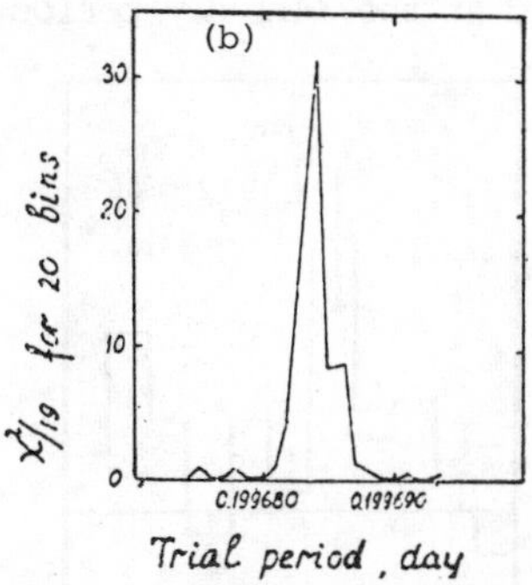

Figure 4. (a) Phaseogram of Cyg X-3. (b) Period scan.

190 hours of observations of Cygnus X-3 between 1972 and 1977 allowed the period to be determined with high accuracy. After correction for heliocentric motion, a period of 0.199683 ± 0.000001 days was determined (Neshpor et al. 1980). With another 60 hours of observations in 1978, it was possible to place a limit on the value of the $\dot{P}$ term which x-ray observations had put at $\dot{P} = 2.83.10^{-9}$ s/s but which was in dispute. The x-ray observations suffered from the uncertainty in alignment between cycles because of the width and variability of the x-ray light-curve. By contrast the gamma-ray observations, although they had a smaller signal-to-noise, had a sharper light-curve. The gamma-ray observations were compatible with a value, $0.3.10^{-9}$ s/s $< \dot{P} < 2.5.10^{-9}$ s/s.

The Crimean observations were continued in 1979 with the light-curve shown in figure 5 (Neshpor et al. 1980). It was apparent that the emission at TeV energies had diminished at

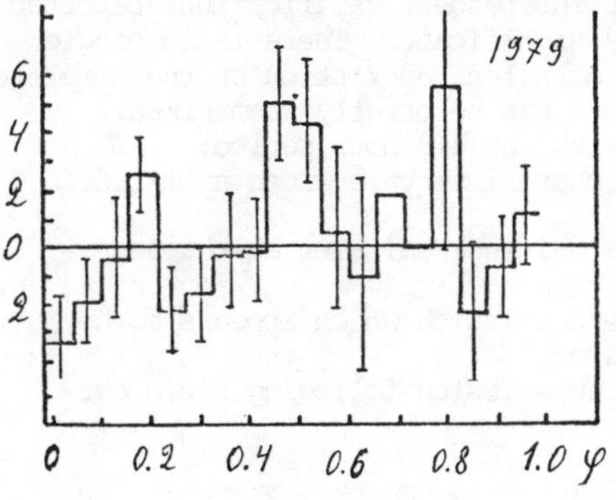

Figure 5. The phaseogram from the 1979 observations of Cygnus X-3.

phase 0.2 but there was evidence for emission at later phases.

There was also the possibility that there was another underlying periodicity which was modulating the 4.8 hour emission (Stepanian et al. 1982). Figure 6 shows the percentage flux folded at the 34.1 day period reported by Molteni et al. (1980).

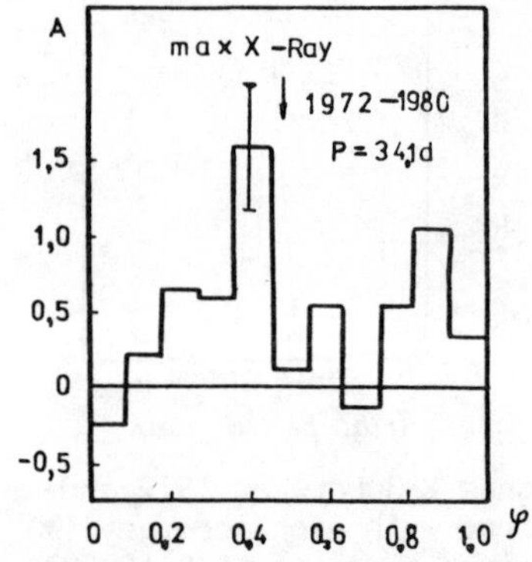

Figure 6. The C.A.O. observations folded with a period of 34.1 days. The position of the X-ray maximum is indicated.

There is some evidence that when the x-ray emission is a maximum, the periodic 4.8 hour TeV emission is also enhanced. However even in x-rays the reality of the 34.1 day modulation is in question.

Between July and October 1980 the C.A.O. made 675 drift-scans across Cygnus X-3 for a total ON time of 90 hours. Only in October was a significant excess (3.6 σ) noted. No periodic modulation was seen during this month or any other month. The October effect which apparently corresponds to the sporadic component was preceded by a radio outburst (Fomin et al. 1981).

Summary: the C.A.O. observations, taken with detectors with uniform sensitivity over an eight year period, constitute the largest and most impressive data base on 1 TeV emission from Cygnus X-3. Even without the independent verifications reported below, they are statistically significant. There is no single publication which contains a detailed description of the complete data base; the principal points can be briefly summarized:

(1) periodic emission with the 4.8 hour period;

(2) emission in narrow phase intervals whose amplitude may have varied with time;

(3) although not always evident, emission at phases separated by 0.6;

(4) sporadic or transient emission which appears to be unrelated to the 4.8 hour phase;

(5) evidence for enhanced emission following radio outbursts.

(b) Tien Shan Array

The first semi-independent confirmations of the Crimean observations were reported by Mukanov (1981). This optical array was located beside the particle array at Tien Shan (3.3 km above sea level). The threshold energy was 5 TeV. Using essentially the same technique as that employed at the Crimean Astrophysical Observatory, an excess of events was seen from Cygnus X-3 in 1977 and 1978. Figure 7(a) shows the net excess

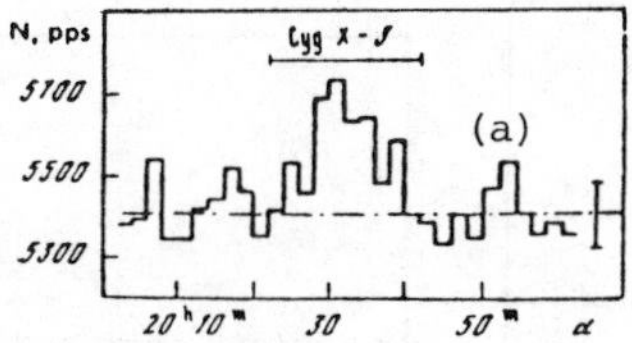

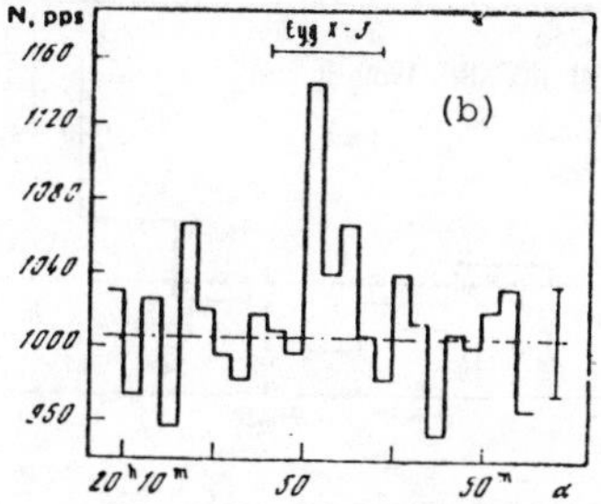

Figure 7. Cyg X-3 phaseogram. (a) 1977, all data. (b) 1978, selected phases.

at all phases observed in 1977. Figure 7(b) shows the excess in 1978 when only scans at phases 0.157 - 0.212 and 0.768 - 0.823 were included. In the first interval the excess was at the 4σ level , in the second 2σ . These observations were in close agreement with those reported by the C.A.O. group, with an apparent flux of 1.6.10^{-10} quanta cm^{-2} s^{-1}.

(c) Whipple Observatory

The first completely independent confirmation of the U.S.S.R. results came in a joint Smithsonian Astrophysical Observatory - University College, Dublin observation at the Whipple Observatory (elevation 2.3km) in 1980 (Danaher et al. 1981). Using two 1.5 m reflectors operated in coincidence, an excess was seen at phase 0.7 - 0.8 at the 3.5 σ level. This phase was determined using the ephemeris derived by the C.A.O. group from their gamma-ray observations and involved some extrapolation (figure 8(a)). Using the contemporary ephemeris derived from x-ray observations by van der Klis and Bonnet-Bidaud (1981), this emission is at phase 0.6 - 0.7. This ephemeris is now taken as the standard and is the one best used for intercomparison of observations. The measured flux during this active phase was 1.5.10^{-10} quanta cm^{-2} s^{-1} at an energy of 2 TeV.

During the period of these observations (April-June 1980), Cygnus X-3 underwent a major change in x-ray activity. A breakdown of the gamma-ray observations showed that the emission occurred at the peak of the x-ray emission (Weekes et al. 1981). Since Cygnus X-3 is not routinely monitored in x-rays, there is no other information available on the x-ray very high energy correlation (figure 8(b)).

A previous attempt to detect gamma rays from Cygnus X-3 using the Whipple Observatory 10 m optical reflector (Weekes and Helmken; 1977) was unsuccessful. These observations were made as part of a survey of 100 MeV gamma-ray sources and did not include any periodicity analysis. The upper limit deduced for the total emission was 2.1 x 10^{-11} quanta cm^{-2} s^{-1} for energies greater than 1 TeV.

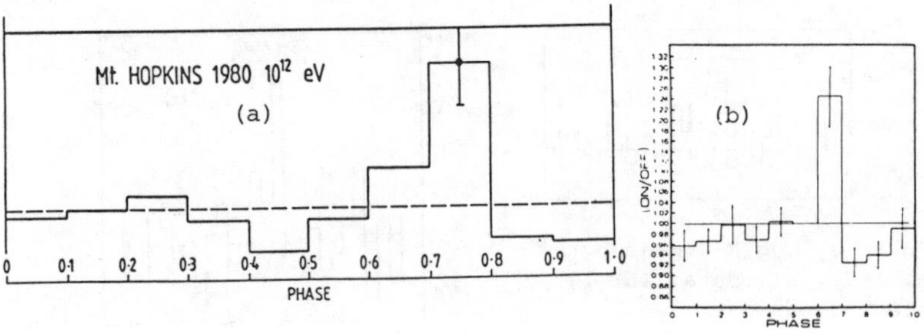

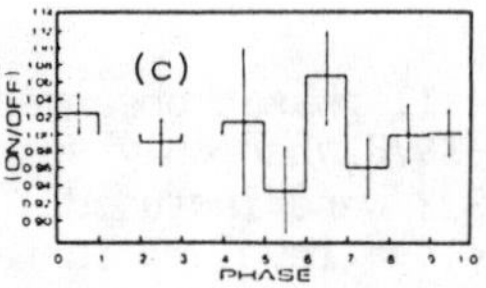

Figure 8. Phaseogram of Cygnus X-3 from Mt. Hopkins. (a) all data with C.A.O. emphemeris (b) and (c), same data but with X-ray ephemeris. (b) time of X-ray maximum (c) X-ray minimum.

Since 1981 there have been four separate detections of emission from Cygnus X-3 using the 10 m optical reflector in a variety of configurations. 54 hours of observations of Cygnus X-3 taken in 3 to 5 hour tracking runs resulted in the observation of two apparent transients (figure 9) of eight-minute duration (Weekes 1982).

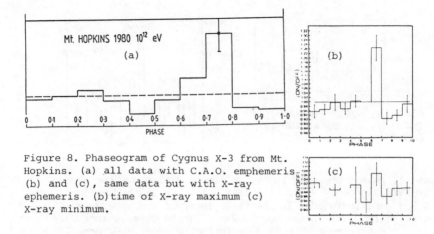

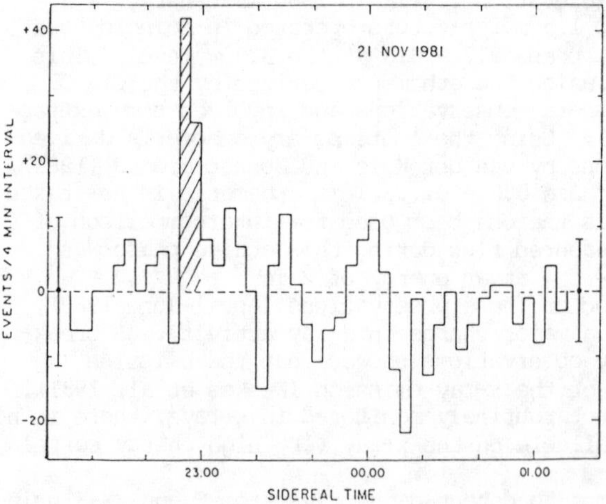

Figure 9. One of two transients observed in 1981.

These were at the 3.9σ and 4.5σ level and were at phase 0.80 and 0.87. The flux was 8.10^{-10} quanta cm^{-2} s^{-1} at energies > 0.4 TeV. In 120 hours of control data no similar transients were seen. A search for fast periodicity (in the 4 - 30 msec range) during these outbursts was unsuccessful; this emission might be expected if the Cygnus X-3 system contains a fast pulsar and would be best detected in an outburst with high data rate such as these (Fegan et al. 1982).

In 1982 the Whipple Observatory 10 m Reflector was converted to an imaging detector to improve its sensitivity to gamma-ray sources. In May 1983 an excess at the 3 level was detected from Cygnus X-3 in the 4.8 hour phase interval stretching from 0.55 to 0.70 (Clear et al. 1983).

In the Oct.-Nov. 1983 dark run the light-curve shown in figure 10 was measured (Cawley et al., 1985 a) with the full 37

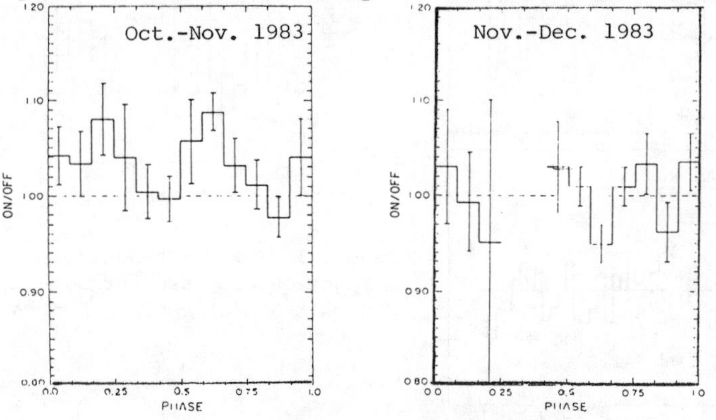

Figure 10. Cygnus X-3 phaseograms.

element camera. Note that there is some evidence for emission around phase 0.20. The effect at phase 0.58 to 0.67 is at the 4.4σ level and corresponds to a flux during this interval of $5.1 \pm 1.1 \times 10^{-10}$ quanta cm^{-2} s^{-1} at an energy of 0.8 ± 0.4 TeV. In the following month there was no evidence for emission at any phase, giving evidence for variability at the 3σ level. A large radio outburst was reported at the end of September, 1982.

During 1984 using the same detector the Whipple Observatory collaboration (University College, Dublin - University of Hawaii - Iowa State University - Smithsonian Astrophysical Observatory) failed to detect any signal from Cygnus X-3 except at the very highest energies (50 TeV). In four nights in June 1984 the reflector was used under the full moon with filters that only passed the near ultra-violet. This increased the energy threshold to 20 TeV with the detector most sensitive to showers with a penetrating (local) component. In the highest energy channel (10% of all data) an excess (4.6σ) was seen at phase 0.6 (Cawley et al. 1985 b).

(d) Edwards A.F.B.

The two 11 m Solar Concentrators was used by a joint J.P.L. - Riverside - Iowa State collaboration to successfully detect Cygnus X-3 at energies of 0.5 TeV during August - September 1981 (Lamb et al. 1982). By using fast timing between the two detectors, showers could be preferentially selected from the center of the fields of view (and hence the direction of the source during transit). An enhancement was seen when Cygnus X-3 was in the field of view (figure 11(a)). The dotted portion of recorded histogram was believed to be caused by the bright star γ Cygni

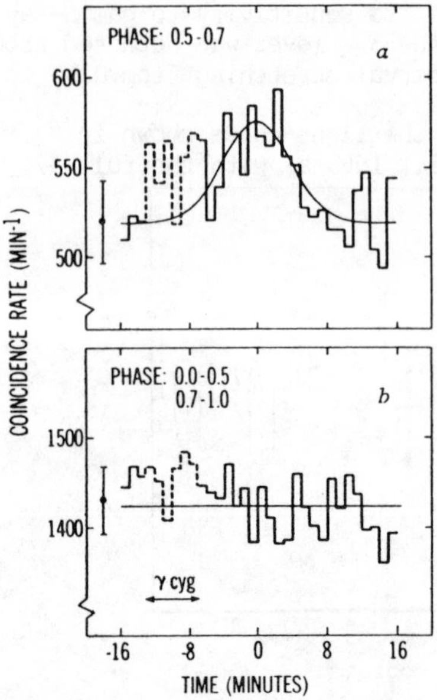

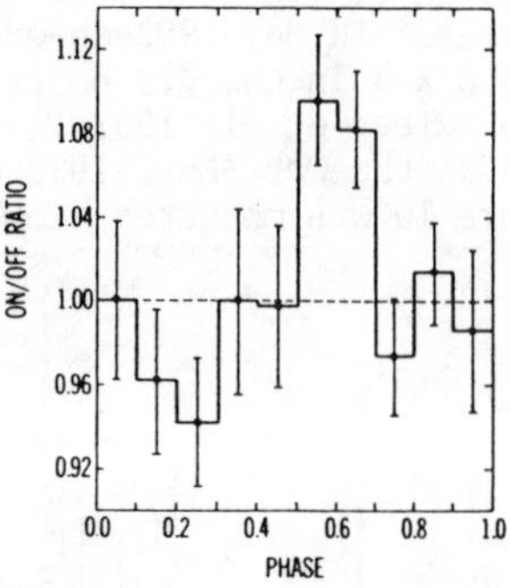

Figure 11 (a) Drift-scans as a function of phase through Cygnus X-3. (b) the 4.8 hour phaseogram.

which was in the field of view at this point; this portion was not included in the data analysis. The resulting light-curve is shown in figure 11(b). The emission region, phases 0.5 to 0.7, is broader than previous observations but is in the same general phase region. The effect is at the 4.4 σ level with the star-contaminated region rejected. The reported flux (>0.5 TeV) is 4.10^{-10} quanta cm^{-2} s^{-1} during the active phase or 8.10^{-11} quanta cm^{-2} s^{-1} averaged over the whole cycle.

(e) Dugway

The University of Durham operated a gamma-ray telescope at the Dugway Proving Ground from 1981-4 (elevation 1.4 km). The telescope consisted of four sets of three 1.5 m aperture detectors seperated by 100 m. The sets could be treated seperately or as part of an array.

Observations of Cygnus X-3 were made in 1981 and 1982 (Gibson et al. 1982). In Sept-Oct 1981 a total of 83 scans (times 4) were made across Cygnus X-3. The net ON/OFF ratio was not significant but a 2 σ excess was seen centered on phase 0.625. In 1982 the tracking mode was employed (no OFF observations) and on the assumption that the emitting phase lasted only 10 minutes, a 2.9 σ excess was seen centered on phase 0.625. Substructure on a time-scale of 3 minutes within this interval was also suggested. One of the Dugway mirror sets had a lower energy threshold than the others; when data from just this telescope is considered the net excess (over the two years of observations) is +4.0 σ corresponding to a flux of 3.10^{-10} quanta cm^{-2} s^{-1} at energies greater than 1 TeV. This suggests that the observed flux from Cygnus X-3 has a steeper spectrum than the background cosmic radiation.

Evidence for a transient outburst was also reported. There was evidence for correlation of the periodic 4.8 hour phase 0.625 emission with the x-ray maximum of the 34.1 day period (Dowthwaite et al. 1983). Because of the uncertainty about the 34.1 period, this correlation is not unambiguous.

In a later paper (Dowthwaite et al. 1985) the Durham group reported on observations of the galactic plane in the Cygnus region. These observations suggest that Cygnus X-3 lies in a "hole" in the galactic plane and that the emission at right ascension 5° on either side of it is 8 $\pm$ 3% higher than at 0°. Since these are the regions normally used as the "background" to define the signal for Cygnus X-3 this would make the Durham detections at plane 0.625 significant at the 5σ level (figure 12).

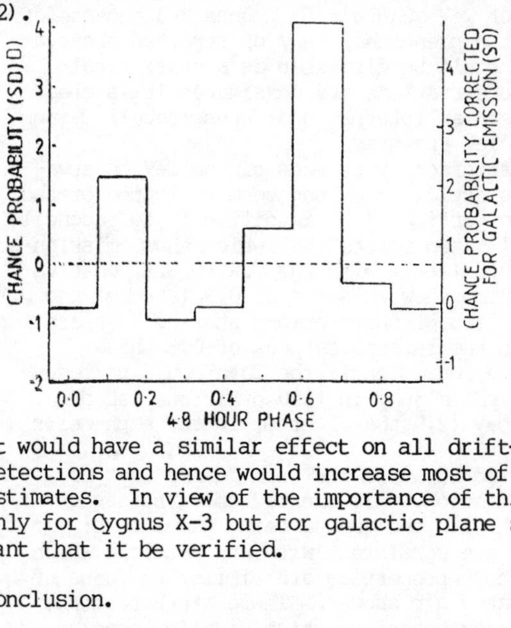

Figure 12. The Dugway phaseogram, all data. Left axis, uniform background; Right axis, with correction for gal-plane emission (K.E. Turver, private communication)

It would have a similar effect on all drift-scan or ON/OFF detections and hence would increase most of the measured flux estimates. In view of the importance of this measurement (not only for Cygnus X-3 but for galactic plane studies) it is important that it be verified.

Conclusion.

The large number of published papers and conference reports that constitute the TeV data base on Cygnus X-3 in the period 1972-84 give a rather confusing picture. It must be remembered that the atmospheric Cherenkov technique, that is the basis for these measurements, is still under development and that the field still lacks the standard candle that can be used as a reference. Other detections have been reported (e.g. PSR0531, Hercules X-3, Crab Nebula, Centaurus A, M31) but these lack the coverage that has been extended to Cygnus X-3. Since the sensitivity of the high energy gamma-ray observations is such that Cygnus X-3 can only be detected under favorable circumstances at the 3-4 σ level, it is not surprising that there is some inconsistencies amongst the reported observations. Given the large number of observations (trials) it is to be expected that some reported detections or correlations may be statistical fluctuations. Hence, any isolated observation must be treated with caution; only where some observed property is confirmed should it be given full credence.

It is clear from the observations that Cygnus X-3 is not a steady emitter of TeV quanta; there are enough null as well as positive observations to indicate that the emission is variable on a number of time-scales. It should not be surprising that a source with the power of Cygnus X-3 should be so variable. Conditions required to accelerate particles to energies of 1 TeV and above must be difficult to attain in an accreting binary; it should be expected that they often break down with the resultant loss of signal.

It is very difficult to quote an overall statistical significance for the detection of TeV quanta from Cygnus X-3 because of the transient nature of the phenomena. Any of reported observations taken in isolation could be dismissed as a statistical quirk. When all of the observations are considered it is clear that (unless there has been an international conspiracy!) Cygnus X-3 has been detected at TeV energies.

One of the most unsatisfactory aspects of the TeV observations is the inability to explain the long term evolution of the 4.8 hour light-curve. In particular it is difficult to reconcile the USSR observations all taken before 1980 which show emission at phases 0.2 (strong) and 0.8 (weak) with the U.K.-U.S.A. observations taken after 1980 which show emission at 0.6 (strong) and 0.2 (weak). Apart from both sets of observations sharing a separation between the main peak and the interpulse peak of 0.6 phase intervals, there is little consistency; the data could be reconciled if there was a 0.5 phase jump in 1980 or if one set of observations had a half-day (2.5 phase) error in the emphemeris used!

Thus far we have tried to avoid the term "gamma ray" to describe the observed radiation. All the TeV observations of Cygnus X-3 are consistent with the quanta being gamma rays; equally the observations are consistent with any particle which produces an air shower whose properties are similar to those of a hadron or photon initiated air shower. There are variations of the atmospheric Cherenkov technique which do allow some differentiation between proton and gamma-ray showers but these have not yet been used on Cygnus X-3. They include the double beam technique developed by Grindlay (1972), the use of fast timing (J. Frye, private communication), color and shape differences (Stepanian 1983). The recent suggestion by Hillas (1985) that imaging systems, such as those used at the Whipple Observatory and the Crimean Astrophysical Observatory, can uniquely identify gamma-ray showers amongst the much more numerous cosmic ray background may be the most significant development of the technique in many years. It could be that the small angular spread of the gamma-ray initiated air shower will be characteristic at TeV energies as pair production is at 100 MeV energies.

Acknowledgements

This work is supported by the U.S. Department of Energy, and the Smithsonian Scholarly Studies Fund.

References

(Abbreviations: ICRC = Proceedings of International Cosmic Ray Conference. Izv. Krym. Astro. Ob. = Izvoestiya Krymskoi Astropizicheskoi Observatorii. (in Russian) Ooty Workshop = Proceedings of International Workshop on Very High Energy Gamma Ray Astronomy, Ootacamund. Published by Smithsonian Institution and Tata Institute)

Cawley, M.F. et al., Astrophys. J. in press (1985a)
Cawley, M.F. et al., 19th I.C.R.C. San Diego, submitted (1985b)
Clear, J., et al., 18th I.C.R.C., Bangalore, 9, 53 (1983)
Danaher, S., et al. Nature, 289, 568 (1981)
Dowthwaite, J.C. et al. Astron. Astrophys. 126, 1, (1983)
Dowthwaite, J.C. et al., Astron. Astrophys. 142, 55 (1985)
Fegan, D.J., et al., Proc Ooty Workshop, 267, (1982)
Fomin, V.P., et al., Izv. Krym. Astro. Ob. 53, 59 (1975)
Fomin, V.P., et al., 17th I.C.R.C., Paris 1, 28 (1981)
Galbraith, K., Jelley, J.V., Nature 171, 349 (1953)
Geldzahler, B.J., et al., Astrophys. J. Lett. 273, L65 (1983)
Giacconi, R. et al., Astrophys. J. Lett. 148, 119 (1967)
Gibson, A.I., et al., Proc. Ooty Workshop. 97 (1982)
Grindlay, J.E., Ap. J. Lett. 174, L9 (1972)
Hillas, A.M., 19th I.C.R.C., San Diego, submitted (1985)
Hjellming, R., Science, 182, 1089 (1973)
Lamb, R.C., et al., Nature 296, 543 (1982)
Lamb, R.C., et al., Proc. Ooty Workshop 86 (1982)
Marshak, M. et al., preprint (1985)
Molteni, D. et al., Astro. Astrophys. 87, 88 (1980)
Mukanov, J.B. et al., 16th I.C.R.C., Kyoto 1, 143 (1979)
Mukanov, J.B. et al., Izv. Krym. Ast. Ob. 63, 151 (1981)
Neshpor, Yu. I. et al., Astrophys. Sp. Sci. 61, 349 (1979)
Neshpor, Yu. I. et al., Izv. Krym. Ast. Ob. 61, 61 (1980)
Neshpor, Yu. I. et al., Zyskin, Yu. L., Izv. Krym. Ast. Ob. 63, 157 (1981)
Neshpor, Yu. I. et al., Phil. Trans. R. Soc. A301, 633 (1981)
Parsignault, D.R. et al., Nature Phys. Sci. 239, 123 (1972)
Samorski, M., Stamm, W., Astrophys. J. Lett. 268, L17 (1983)
Stepanian, A.A. et al., 15th I.C.R.C., Plovdiv, 1, 135 (1977)
Stepanian, A.A. et al., Izv. Krym. Ast. Ob. 60, 80 (1979)
Stepanian, A.A. et al., Proc. Ooty Workshop 43 (1982)
Stepanian, A.A. et al., Izv Krym. Ast. Ob. 66, 234 (1983)
Van der Klis, M., Bonnet-bidaud, J.M., Astro. Astrophys. Lett. 95, L5 (1981)
Vladimirsky, B.M., et al., 13th I.C.R.C. 1, 456 (1973)
Vladimirsky, B.M., et al., Sov. Ast. Lett. 1, 57 (1975)
Vladimirsky, B.M., et al., 14th I.C.R.C., Munich, 1, 118 (1975)
Weekes, T.C. et al., Astron. Astrophys. 104, L4 (1981)
Weekes, T.C. et al., Proc. Ooty Workshop, 270 (1982)
Weekes, T.C., Helmken, H.F., Proc. 12th ESLAB Symposium, Frascati, 39 (1977)

CYGNUS X-3: IS IT UNIQUE?

R. C. Lamb
Physics Dept., Iowa State University, Ames, IA 50011

ABSTRACT

On the basis of ground-based observations spanning the range of 10^{11}eV to 10^{16}eV, Cygnus X-3 has been established as a source of very high energy radiation. However, in view of the deep underground results the nature of that radiation is not clear. This paper discusses recent observations, at very high energies, of two X-ray binary pulsars, Hercules X-1 and 4U0115+63, which may be similar to Cygnus X-3. All emit very high energy radiation, all are binaries, and if a recent possible observation of a cyclotron emission feature from Cygnus X-3 is correct, all contain spinning neutron stars.

INTRODUCTION

The possible detections of Cygnus X-3 by the Soudan I and the NUSEX (Mt. Blanc) deep underground detectors, if correct, appear to call for a type of radiation which has not been seen before. That is to say, the primary particle is neither a photon nor a neutrino. There are at least two separate issues here: Has Cygnus X-3 really been observed at very high energies? And, if so, what is the nature of the radiation (or radiations)?

From the cumulative weight of the ground-based detections from $\sim10^{11}$eV to $\sim10^{16}$eV by ten independent groups (summarized by Weekes and by Elbert, this conference), it seems clear that Cygnus X-3 has been observed. However, in light of the deep underground results, the tacit assumption that the above-ground detectors have been seeing primary photons needs to be critically evaluated.

It is not within the scope of this paper to examine this assumption. Rather, I take the point-of-view that Cygnus X-3 is a very high energy source and ask: Are there other sources like it? Recent very high energy detections of emission from spinning neutron stars in binary systems, in particular, Hercules X-1 and 4U0115+63, suggest similarities to Cygnus X-3. In the remainder of the paper I will review the ground-based observations of Hercules X-1 and 4U0115+63 and compare these sources with each other and with Cygnus X-3.

OBSERVATIONS

Hercules X-1

Hercules X-1 is a bright X-ray source, first observed by the UHURU satellite[1], which consists of a spinning neutron star in a binary orbit around a ~two solar mass companion. (See Joss and Rappaport 1984,[2] for example.) There are at least three well-established time scales for Hercules X-1, its spin period (1.24 seconds), the orbital period (1.7 days) and a 35-day X-ray modulation period caused perhaps by the precession of a tilted accretion disk. Figure 1 shows a scale drawing of the Hercules X-1 system taken from Middleditch et al., (1985).[3]

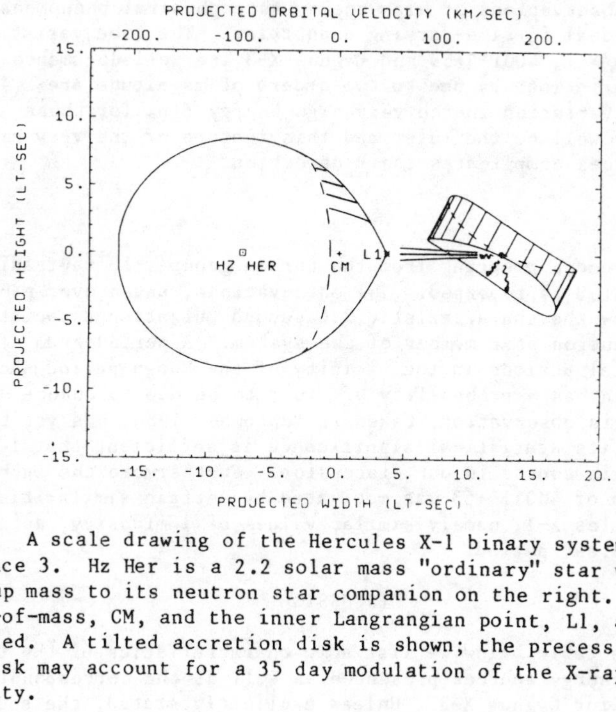

Fig. 1. A scale drawing of the Hercules X-1 binary system from reference 3. Hz Her is a 2.2 solar mass "ordinary" star which gives up mass to its neutron star companion on the right. The center-of-mass, CM, and the inner Langrangian point, L1, are indicated. A tilted accretion disk is shown; the precession of this disk may account for a 35 day modulation of the X-ray intensity.

Hercules X-1 has now been observed by three different groups at energies near 10^{12} and 10^{15}eV. The first observation at energies greater than 10^{12}eV, by the Durham group,[4] reported a three minute transient event in April, 1983, which showed the characteristic 1.24 second pulsation at a chance probability level of 4×10^{-4}, as well as some weaker evidence for pulsed emission extending throughout a subsequent ten day observing period. The Utah group, operating the Fly's Eye detector, has reported[5] a flux of pulsed emission from Hercules X-1 at energies above 500 TeV observed in July, 1983. The emission was confined to 40 minutes and had a chance probability of $2x10^{-4}$. A third observation of Hercules X-1 has now been reported by the Mt. Hopkins collaboration[6] above 0.2 TeV. Hercules X-1 was observed for approximately 30 hours on 24 separate nights. On two nights, pulsed emission was seen with an overall significance level of $2x10^{-4}$ to be chance.

A totally convincing case for Hercules X-1 as a very high energy source cannot come from any one of the three observations; however, taken together, they suggest that it is a source. Unlike high energy physics results which are based on invariant cross sections, observations of high energy astrophysical phenomena frequently deal in time-varying quantities. The time variations of Hercules X-1, 4U0115+63 and Cygnus X-3 are well documented in X-rays. Variations by one to two orders of magnitude are reported. Variation in the very high energy flux for these sources may well be the rule, and this feature of the very high energy sources complicates their detection.

4U0115+63

In a recent preprint from the Durham group[7], 10^{12}eV radiation from 4U0115+63 is reported. The observations, taken over nine nights, show the characteristic 3.6 second pulsation associated with the neutron star member of the system. A periodogram of various trial periods in the vicinity of the known period shows an effect which has a probability of $\sim10^{-5}$ to be due to chance. Although this observation, taken in September 1984, has yet to be confirmed, its statistical significance is sufficient that I have included this source in our discussion. Furthermore the Durham observation of 4U0115+63 was motivated by certain similarities of it to Hercules X-1, namely similar values of luminosity, spin-up rate and pulsar period.

DISCUSSION

In the table below we list some characteristics of the two very high energy sources presented as well as the corresponding quantities for Cygnus X-3. Unless explicitly stated, the source information is taken from the review of Bradt and McClintock (1983)[8] or, in the case of the cyclotron features, the review of Joss and Rappaport (1984)[2]. Columns 4 and 5 refer to the X-ray and very high energy (VHE) luminosities, respectively. In regard to the VHE luminosity, the assumption made hitherto by all the atmospheric Cerenkov groups is that the primary radiation is electromagnetic. If this is not true, but the radiation has a similar interaction length to the photon, then these estimates for the VHE luminosity should need little modification. The values given for the VHE luminosity are, in the case of Hercules X-1 and Cygnus X-3, the maximum observed during an orbital cycle.

Table I. Comparison of Hercules X-1, 4U0115+63, and Cygnus X-3

1 Source - distance	2 Periods	3 Cyclotron feature	4 L_x (2-11keV) erg/s	5 VHE L (1-10TeV) erg/s
Hercules X-1 ~5kpc	1.24 s, spin 1.7 d, orb. 35 d, disk precession(?)	Yes: 38 or 53 keV B=3.8 or 5.3 $\times 10^{12}$G	0.3 to 7×10^{36}	$\sim 10^{36}$
4U0115+63 ~3kpc	3.6 s, spin 26 d, orb.	Yes: 11.5 & 23 keV B=1.2×10^{12}G	< 0.1 to 8×10^{36}	$\sim 3\times 10^{35}$
Cygnus X-3 ≥11.6kpc (Dickey, 1983[9])	? spin 4.8 h, orb.	Yes(?): 85 keV B=8×10^{12}G (Ling et al., 1985[10])	0.2 to 1×10^{38}	$\sim 10^{37}$

The X-ray intensity of both Hercules X-1 and 4U0115+63 is modulated by the spin period of the neutron star member of the system. The observed spin period is further modulated by the neutron star's orbital motion. Orbital parameters are determined from analysis of this modulation, resulting in a detailed model of the binary system as illustrated, for the case of Hercules X-1, in Figure 1. The physical picture then is: a spinning neutron star in orbit around a companion star. The companion supplies mass to the neutron star. Gravitational potential energy is released as thermal energy and radiation as the matter accretes. Since each of the neutron stars is observed to be slowly gaining rotational kinetic energy via the accretion process, it is clear that the ultimate source of energy in these binary systems is gravitational energy rather than the spin energy of the neutron star itself. Precisely how the energy is transferred to the very high energy portion of the spectrum is unknown; however, mass transfer rates of 10^{-5} solar masses per year and an assumed 2% conversion efficiency of the rest mass energy will produce an overall luminosity of $\sim 10^{40}$ erg/s, more than enough to power the X-ray and very high energy radiations. Mass loss rates of as much as 10^{-5} solar masses per year are observed for certain isolated giant stars, therefore a mass transfer rate of this order, although high, may be obtained in close binaries.

What similarities do we see between Hercules X-1 and 4U0115+63? In the first place they are the only two X-ray pulsars for which cyclotron emission lines have been reported. (See Column 3 of the table.)

Also, they have similar values of X-ray luminosity (approximately a factor of 10^2 below that of the most luminous of the X-ray pulsars), neutron spin period (among the fastest observed), and rate of change of that period (spin-up rate). The spin-up rates of these two sources are the smallest observed. That is to say, the neutron stars are spinning nearly at their equilibrium spin rate. This occurs when the spin-up accretion torque balances the magnetic braking torque from a magnetic field not aligned with the rotation axis of the neutron star. In the sense that both neutron stars are spinning nearly as fast as they can, they are "fast rotators". (See, for example, Nararan et al., 1985[11].)

To what extent is Cygnus X-3 similar to the other two sources? We take the conventional point-of-view that the 4.8 hour modulation seen for Cygnus X-3 is orbital, therefore Cygnus X-3 is also a binary X-ray source which, like the other two, has been observed at very high energies. Kepler's law gives the size of the orbit as $\sim 3M^{1/3}$ light-seconds where M is the combined mass of the system in solar units. Thus, the Cygnus X-3 system may be considerably more compact than that of Hercules X-1 (See Figure 1). There is, as yet, no evidence for any spin period to be associated with the presumed neutron star member of the binary in spite of many efforts in X-rays as well as at high energies to discover one. (For a different picture of Cygnus X-3 as an accreting black-hole see, for example, Grindlay, 1982[12].) Recently Ling et al.[10] have presented evidence for an emission line near 85 keV which is consistent with the direction of Cygnus X-3. If this is indeed from Cygnus X-3 and is a cyclotron feature, then the black-hole possibility for Cygnus X-3 would be ruled out.

CONCLUDING REMARKS

1. Very high energy radiation has been seen from several X-ray binary sources. In at least two cases, the very high radiation is modulated at the spin period of a neutron star member. (There are also reports of 10^{15}eV radiation from two other X-ray pulsars, Vela X-1[13] and LMC X-4[14], as yet unconfirmed. In these cases no modulation at the spin period was reported.)

2. If Cygnus X-3 contains a spinning neutron star rather than a black hole, then on the basis of the similarities to Hercules X-1 and 4U0115+63, a spin period in the range of 1 second might be expected rather than the fast millisecond type periodicity postulated by some.[15] One might expect that a neutron star associated with Cygnus X-3 should be a fast rotator; i.e., spinning at nearly its equilibrium rate.

3. If Cygnus X-3 contains a neutron star then the correspondence between it and the two sources discussed in this paper would be strong. The answer to the question posed by the title would then appear to be that Cygnus X-3 is not unique. And

a class of VHE sources which are also X-ray pulsars would be established. In this regard it is important to note (as pointed out by Weekes, this conference) that the optical absorption in the direction of Cygnus X-3 prevents detailed optical studies. On the other hand the Hercules X-1 system is a factor of $\sim 10^4$ brighter optically and has been profitably studied in the optical region.

4. How the liberated gravitational energy is transferred to the very high energy radiation is unknown. Even the physics of the apparently simpler case of an isolated spinning neutron star is not understood. (See, for example, C. Michel's review of radio pulsars.[16])

5. The question of the nature of the very high radiation should be examined critically by all the ground-based groups. More detailed information about the longitudinal and lateral shower develpment may be crucial in establishing the identity of the radiation.

REFERENCES

1. H. Tananbaum et al., Ap. J. Lett. 174, L143 (1972).
2. P. C. Joss and S. A. Rappaport, Ann. Rev. Astron. Astrophys. 22, 537 (1984).
3. J. Middleditch, R. C. Puetter, and C. R. Pennypacker, Ap. J. 292, 267 (1985).
4. J. C. Dowthwaite et al., Nature 309, 691 (1984).
5. R. M. Baltrusaitis et al., Ap. J. Lett., in press (1985).
6. M. F. Cawley et al., 19th Int. Cosmic Ray Conf. (San Diego) paper OGG 2.2-9 (1985).
7. P. M. Chadwick et al., preprint (1985).
8. H. V. D. Bradt and J. E. McClintock, Ann. Rev. Astron. Astrophys. 21, 13 (1983).
9. J. M. Dickey, Ap. J. Lett. 273, L71 (1983).
10. J. C. Ling et al., preprint (1985).
    J. C. Ling et al., B.A.A.S. (Tucson) paper 24.07 (1985).
11. S. Nararan et al., Ap. J. 290, 487 (1985).
12. J. E. Grindlay, Workshop on VHE Gamma Ray Astronomy, Ootacamund, India, 178 (1982).
13. R. J. Protheroe, R. W. Clay, and P. R. Gerhardy, Ap. J. Lett. 280, L47 (1984).
14. R. J. Protheroe and R. W. Clay, Nature 315, 205 (1985).
15. G. F. Bignami, L. Maraschi, and A. Treves, Astr. Astrophys. 55, 155 (1977).
    M. Milgrom and D. Pines, Ap. J. 220, 272 (1978).
16. F. C. Michel, Rev. Mod. Phys. 54, 1 (1982).

EVIDENCE FOR UNDERGROUND MUONS FROM CYGNUS X-3

M. L. Marshak
School of Physics, University of Minnesota, Minneapolis, MN 55455

ABSTRACT

The observation of underground muons apparently related to the x-ray binary Cygnus X-3 is not easily understood. We describe the relevant data from the Soudan 1 detector and discuss some details of the analysis. These data suggest that flux from Cygnus X-3 has a time structure in addition to the 4.8 h modulation.

I. WHY MUONS SHOULD NOT BE OBSERVED

The recent observation by the Soudan 1 detector[1] and the NUSEX experiment[2] of underground muons related to the x-ray binary Cygnus X-3 is an unexpected result. There exists a set of arguments as to why such an effect should not occur. These hypotheses should be examined carefully. If the observations reported here are verified, one or more of the following statements must be incorrect:

a) The highest energy primaries observed via secondary muons by the Soudan 1 detector (area: ≈ 8 m^2; depth: 1800 m water equivalent) in one year of live time have energies $\approx 10^{16}$ eV. Fluxes higher than one event per year correspond to lower energies. This conclusion is based on flux comparisons with air shower data.

b) Charged primaries with energies resulting in an observable flux in the Soudan 1 detector cannot point back to a definite source at any reasonable galactic distance because of galactic magnetic fields. For example, the gyroradius of 10^{15} eV protons in a magnetic field of 10^{-6} G is about 1 pc or 10^{-4} of the galactic radius. Charged particles of such energy traversing any substantial fraction of the galactic disk will not point back to their origin.

c) The distance[3] to Cygnus X-3 is at least 10 kpc. Neutrons from Cygnus X-3 with energies of 10^{15} eV will decay before reaching the earth.

d) Neutrinos have a mean free path long compared to the ≈ 600 m of rock over the Soudan 1 detector. Most models of x-ray binaries predict ν fluxes comparable to γ fluxes or, at most, two orders of magnitude larger. At such fluxes, ν-induced events would not be observed in a detector as small as Soudan 1.

e) Photons do not often produce high energy muons. The inelastic photoproduction cross-section is about 1/300 of the pair production cross-section. Thus, a secondary muon flux comparable in magnitude to the primary photon flux cannot be generated by a conventional electromagnetic shower mechanism.

f) There are no neutral, stable particles other than neutrons, neutrinos and photons.

II. THE ANALYSIS OF THE CYGNUS X-3 DATA

The identification of underground muons with Cygnus X-3 relies on directionality and periodicity. The directional analysis is straight-forward. Each muon track is reconstructed as a line in local coordinates. Using the time-of-arrival, this local trajectory is transformed into the celestial coordinates of declination δ and right ascension α. Events pointing within a 3° half-angle cone centered on the nominal direction of Cygnus X-3 (α=307.6°, δ=40.8°) are included in the chosen sample.

The integrated background, that is, the expected number of events in the absence of a source within the selected cone, can be determined by interpolation from adjacent directions. In particular, let *A* equal the total number of events satisfying (α<304.6° *OR* α>310.6°) *AND* (δ<37.8° *OR* δ>43.8°). *B* is the number of events satisfying (α<304.6° *OR* α>310.6°) *AND* (δ>37.8° *AND* δ<43.8°). *C* is the number of events meeting the condition (α>304.6° *AND* α<310.6°) *AND* (δ<37.8° *OR* δ>43.8°). Then the expected number of events from the direction (α>304.6° *AND* α<310.6°) *AND* (δ>37.8° *AND* δ<43.8°) is *BC/A* . A small solid angle correction yields the expected number of events in the 3° half-angle cone centered on Cygnus X-3.

The periodicity analysis of the selected events is performed using the Cygnus X-3 ephemeris[4] measured with keV x-rays. The period *P* of any event at time *T* may be determined by solving the equation

$$1/2\, p_{dot} p_o P^{\,2} + p_o P \;=\; T - T_o \qquad (1)$$

where $p_{dot} = 1.18 \times 10^{-9}$, $p_o = 0.1996830$ d and T_o = JD 2440949.8986 (JD indicates Julian Date). Note that the Julian day starts at 12 h Universal Time.

Plotting the fractional part of *P* for each event yields a phase plot. There are a number of statistical tests which can be used to determine the probability that this phase plot results from the random fluctuation of a uniform background. The least sensitive is a χ^2 analysis which simply measures the bin-by-bin deviation from uniformity without any sensitivity to an organized pattern. More sensitive tests are unconstrained moment analyses which do not require the binning of the data and which are sensitive to a correlation among adjacent bins. The most sensitive tests are constrained moment analyses, which measure not just a deviation from uniformity but rather **a deviation of the type expected to result of Cygnus X-3.**

The procedure for performing the moment analysis is as follows: Assign to each muon observation a unit vector whose phase is given by the phase calculated using the Cygnus X-3 ephemeris. The nth moment is defined by the equation

$$M_n = \{\, [\, \Sigma \sin n\phi_i \,]^2 + [\, \Sigma \cos n\phi_i \,]^2 \,\}^{1/2} \qquad (2)$$
$$\text{Phase of } M_n = \tan^{-1}\{[\, \Sigma \sin n\phi_i \,]/[\, \Sigma \cos n\phi_i \,]\}$$

where the ϕ_i are the phases of the observed muons. The number of moments used in the analysis is somewhat arbitrary, but for the asymmetric, bimodal distribution expected from Cygnus X-3, the first two moments should provide the best signal-to-background discrimination.[5]

For the unconstrained moment analysis, the next step is to calculate the statistic[5]

$$Z_m^{\,2} = (2/N)\, \Sigma\, |M_i|^2 \qquad (3)$$

where N is the total number of events. $Z_m^{\,2}$ has a probability density function equal to that of a χ^2 function with 2m degrees of freedom. Thus, the probability that the observed distribution represents a random fluctuation of a uniform background can be determined by looking up the observed $Z_m^{\,2}$ in a table of χ^2 probabilities.

The constrained probability may be calculated according to the same general scheme, although its evaluation is more complicated and is perhaps best done by Monte Carlo methods. For each moment, it is necessary to select a particular phase from either air shower measurements or a simple, binary model of the source. For the data discussed below, we have chosen a first moment phase of 0.725 from Ref.6, although phases as low as 0.6 could also be selected[7]. For the second moment, a simple model of an x-ray binary predicts a phase of 0.5. The important quantity for the constrained test is not the calculated moment itself, but rather the projection of this moment along the axis chosen *a priori*.

The probability that this projected moment arises from a random fluctuation of a uniform background is then evaluated independently for each of the first two moments. If these probabilities are P and Q, respectively, then the joint probability of observing both moments is not simply the product PQ. This statement is true because a small product PQ can be obtained in a variety of ways--a small P and a small Q, a very small P and a somewhat larger Q, a very small Q and a somewhat larger P, etc. The correct confidence level R of two independent tests is given by the equation

$$R = PQ\ (1 - \ln PQ) \qquad (4)$$

III. THE SOUDAN I DATA

The Soudan 1 detector was designed for a proton decay experiment[8]. Details of this device and the data analysis procedures have been discussed elsewhere[9]. Briefly, the detector consists of an array of 3456 proportional tubes arranged in 48 layers of 72 tubes each. Adjacent layers are rotated by 90°, in order to provide two orthogonal views of each event. The tube diameters are 2.8 cm. The size of the stack is 2.9 m by 2.9 m by 1.9 m high. The mass of the array, including the heavy concrete in which the tubes are embedded, is 31 metric tons. The detector is located at a depth of 590 m (equivalent to 1800 m of water), at a latitude of 48° N, and at a longitude of 92° W. One set of tubes points 7° east of north.

We report here on data collected using the Soudan 1 detector during the period from September 1981 through November 1983. During that period, the detector live time aggregated 0.96 year. The single muon sample (those events which fit well to the hypothesis of a single, linear track) included 1.05×10^6 events. For the analysis here, muon tracks were required to have at least 8 proportional tube hits in each of the two orthogonal views in order to insure adequate pointing resolution. The application of this cut reduced the data sample to 784,000 events. Each event was reconstructed in local coordinates by a computer algorithm. The time of day and date, recorded with each event from a quartz-crystalled-controlled clock, were then used to calculate the declination δ and the right ascension α for each event.

The selection of 3° for the half-angle of the Cygnus X-3 cone was made after consideration of the resolution properties of the Soudan 1 detector. A study of the measured parallelness of multimuon events indicates that the local angular resolution is $\approx 1°$. However, there are two effects which may tend to produce a worse angular resolution in the case of a fixed object like Cygnus X-3. One is a $\pm 1°$ uncertainty in the absolute azimuth of the detector. The second is that the detector tubes are arranged on a 4 cm lattice, so that a quantization effect is imposed on angular measurements at certain angles. After consideration of the possible size of these effects, we chose 3° for the magnitude of the angle cut.

This geometric cut produced a sample of 1183 events. The expected number of events given from an isotropic background was determined as 1144 events by the interpolation method described above. Thus, there is an excess of 39 ± 34 events in the direction of Cygnus X-3, independent of phase. The phases of these 1183 events are shown in Fig. 1(a). The phase distribution of the background, determined by interpolation to be flat, is indicated by the dashed line in the figure. The peak for phases between 0.65 and 0.90 consists of 60±17 events. The results of a moment analysis of this plot are given in Table 1.

Previous air shower measurements[10] of flux from Cygnus X-3 suggest that this source is episodic. It is therefore reasonable to expect that the signal-to-background ratio in Fig. 1(a) might be improved by selecting only high flux times. On average when Cygnus X-3 is high in the sky, we observe an event within the 3° half-angle cone about every 3 hours. A high rate period might be indicated by observing 2 events within the cone within 0.5 h. Fig. 1(b) shows the phase plot for pairs of events meeting this criterion. Only one point (at the mean phase) is plotted per pair. The background for this plot has been established by interpolation from nearby directions. The major peak for phases between 0.65 and 0.9 consists of 29±6 event pairs. A possible secondary peak for phases between 0.30 and 0.35 consists of 5±3 event pairs. Similar plots for α=297.6°, δ=40.8° and α=317.6°, δ=40.8° are shown in Figs. 1(c) and 1(d). A moment analysis of Fig. 1(b),(c),(d) is also listed in Table 1.

Table 2 provides a slightly different perspective on the same data. Here we show the number of muons from the direction of Cygnus X-3, which were observed during fixed quarters of the 4.8 h period for both on-source and nearby off-source points. With the exception of the on-source phase peak, we can fit these data well with a simple Monte Carlo model which hypothesizes a uniform, random source and a detection efficiency as a function of zenith angle similar to that for all single muons. The χ^2 values for this fit are also listed in the table. The mean number of muons when the source is vertical is 0.43±.03 h^{-1}.

Fig. 1(b) and the data in Table 2 do not uniquely specify the functional form of any flux modulation which may exist in addition to the 4.8 h period. We can estimate some parameters of this variation in the context of a simple model in which the flux is enhanced only during the phase peak (namely phase from 0.65 to 0.90) of only a certain fraction of all observable 4.8 h periods. In this model, the efficiency for detecting both the signal and the background varies with zenith angle in a manner similar to that observed for all single muon events. Such a model yields a very likely fit when 7±4 percent of all possible cycles actually have flux enhancements. The total flux during the enhanced period (signal plus background) is then 1.3±0.7 muons h^{-1}.

This time variability suggests that for the purpose of comparison with air shower data we should report fluxes for each of several conditions. From the ≈8 m^2 area of the Soudan 1 detector and the 0.96 year live time, we can use the above model to estimate the following fluxes for muons with greater than 650 GeV:

(a) Average detected flux for the entire observation period: ≈2.5 x 10^{-11} cm^{-2} s^{-1}.

(b) Same as (a) if Cygnus X-3 were always directly overhead: ≈7.3 x 10^{-11} cm^{-2} s^{-1}.

(The following flux values are for the directly overhead geometry.)

(c) Average flux during all potentially active times with phase between 0.65 and 0.90: ≈2.9 x 10^{-10} cm^{-2} s^{-1}.

(d) Flux during "on" times with phase between 0.65 and 0.90 with 7 percent of cycles "on": ≈4.2 x 10^{-9} cm^{-2} s^{-1}.

(e) Flux averaged over entire 4.8 h period during 7 percent of time source is "on": ≈1.0 x 10^{-9} cm^{-2} s^{-1}.

The uncertainty in these fluxes is estimated at +50, -25 percent.

These fluxes may be compared with fluxes attributed to Cygnus X-3 by air Cerenkov experiments at Similar energies. Ref. 10 reports a peak pulsed flux (measured over 0.5 h) of (5.1+1.1) x 10^{-11} cm^{-2} s^{-1} for a threshold energy of (8+4) x 10^{11} GeV. That experiment observed no significant signal a month later, indicating that this flux corresponded to a time when the source was "on." The summary in Ref. 11 includes a report of a flux averaged over the 4.8 h cycle of $\approx 8 \times 10^{-11}$ cm^{-2} s^{-1} at a threshold energy of 500 GeV. Thus, our muon fluxes are apparently larger than fluxes reported from air Cerenkov measurements at similar energies. However, deducing a primary flux from the secondary muon flux requires a knowledge of the number of muons per primary which reach the Soudan 1 depth. Because this quantity is not known, a direct flux comparison is not possible. We also note that the air shower experiments observe somewhat narrower phase peaks.

If this analysis is correct, other experiments should observe an identical modulation in the Cygnus X-3 flux. In particular, the times at which this data sample shows the largest flux enhancements (3 or 4 muons within the 1.2 h phase peak) are (Universal Time) 29.82 December 1981, 30.78 January 1982, 4.39 June 1982, 19.98 October 1982, 27.94 October 1982, 23.87 December 1982, 3.86 January 1983, 17.50 April 1983 and 19.46 May 1983

Early x-ray measurements[12] of the Cygnus X-3 flux suggested the existence of a 34.1 d period in addition to the 4.8 h modulation. Fig. 2 shows the times listed above plotted with a 34.1 d phase. (Note that the absolute phase has been selected using these data and that it differs from the phase in Ref. 11 by almost half a period.) The points lie within half of the available phase plot. The probability of such a effect resulting from a random fluctuation is about 1 percent. In addition, Fig. 2 shows the 34.1 d phase for the three largest radio flares[13] from Cygnus X-3, which occurred during the period of our data and two air shower bursts[14] which were reported during the same period. These data are clearly anecdotal but their near-zero phase warrants further, more systematic analysis.

IV. CONCLUSIONS

With regard to the experimental data, we can summarize the current situation as follows:

(a) Continued and more sophisticated analysis yields a statistical significance for the Soudan 1 data at the 10^{-3} to 10^{-4} level.

(b) The flux appears to have a time modulation in addition to the 4.8 h period. Such modulation may have a 34.1 d period, but resolving this question requires further data.

(c) The apparent duty factor of 7 ± 4 percent implies that the flux when the source is "on" is higher than the average flux previously reported. The discrepancy between the observed secondary flux and that expected from inelastic photoproduction is thus increased.

These results worsen not improve the dilemma described in Ref. 1. A number of suggestions have been made to resolve the apparent paradox. Most involve either new mechanisms for muon production by photons or neutrinos or new neutral particles. In some hypotheses, these new neutrals are composites of known neutrals, such as the 6-quark bag including 2 u, 2 d and 2 s quarks. Other possibilities include supersymmetric particles such as photinos. Based on the current experimental data, no one hypothesis is overwhelmingly favored.

There exist a number of experimental measurements which should now be done. Some of these are

(a) More data by different detectors should be collected to decide unambiguously on the existence of the effect.

(b) Various parameters of the signal events such as zenith angle dependence, muon multiplicity, multimuon separation, stopping muon rate, etc. should be compared for signal and background events.

(c) The apparently large angular spread in the signal should be investigated.

(d) Flux variations other than the 4.8 h period should be tracked over a longer length of time and should be compared with radio and x-ray observations.

(e) A surface array and underground detector in the same location should be used to determine the characteristics of air showers associated with muons from Cygnus X-3.

The Soudan 1 experiment is the work of my colleagues John Bartelt, Hans Courant, Ken Heller, Steve Heppelmann, Terry Joyce, Earl Peterson, Keith Ruddick and Mike Shupe at the University of Minnesota and Dave Ayres, John Dawson, Tom Fields, Ed May, Larry Price and K. Sivaprasad at Argonne National Laboratory. In particular, the Monte Carlo models were first devised by Tom Fields. I am also appreciative of the assistance of the State of Minnesota, Department of Natural Resources, particularly the staff at Tower-Soudan State Park. This research has been supported by the U.S. Department of Energy and the Graduate School of the University of Minnesota.

REFERENCES

1 M. L. Marshak et al., Phys. Rev. Lett. **54**, 2079 (1985).
2 G. Battistoni et al., Phys. Lett. **155B**, 465 (1985).
3 J. Dickey, Astrophys. J. **273**, L71 (1983).
4 M. van der Klis and J. M. Bonnet-Bidaud, Astron. and Astrophys. **95**, L5 (1981).
5 R. Buccheri et al., Astron. and Astrophys. **128**, 245 (1983).
6 S. Danaher et al., Nature **289**, 568 (1981).
7 See data summarized in J. Lloyd-Evans et al., Nature **305**, 784 (1983).
8 J. Bartelt et al., Phys. Rev. Lett. **50**, 651 and 655 (1983).
9 J. Bartelt et al., Phys Rev. D, to be published.
10 M. F. Cawley, et al., submitted to Astrophys. J.
11 R. C. Lamb et al., Nature **296**, 543 (1982).
12 D. Molteni et al., Astron. and Astrophys. **87**, 88 (1980).
13 B. J. Geldzahler et al., Astrophys. J. Lett. **273**, L65 (1983) and K. Johnston, private communication.
14 G. Smith et al., Phys. Rev. Lett. **50**, 2110 (1983); T. C. Weekes, Astronomy and Astrophys. **121**, 232 (1983). The first experiment does not explicitly measure directionality but Cygnus X-3 was near overhead at the time of the burst.

Table 1: Moment analysis of Soudan 1 Data

	Fig. 1(a)	Fig. 1(b)	Fig. 1(c)	Fig. 1(d)
No. of events	1183	174	181	171
First Moment				
Magnitude	59.6	27.6	15.3	9.1
Phase	0.79	0.77	0.37	0.76
Second Moment				
Magnitude	58.8	32.9	6.8	19.0
Phase	0.52	0.56	0.41	0.17
Z_n^2	11.85	21.20	3.10	5.19
Overall Unconstrained Probability	<0.02	<0.0004	<0.55	<0.30
Magnitude of First Moment in Direction 0.725	54.7	26.5	-9.4	8.9
Magnitude of 2nd Moment in Direction 0.5	58.3	30.6	5.7	-9.1
Probability of Constrained First Moment	<0.012	<0.0024	<0.83	<0.15
Probability of Constrained 2nd Moment	<0.0085	<0.0007	<0.27	<0.85
Overall Constrained Probability	<0.0011	<0.00002	<0.56	<0.39

Table 2: Number of Cygnus X-3 cycles in which n muons are observed in 1.2 h from within 3° of Cygnus X-3

Direction	Phase	n=1	2	3	4	χ^2
on-source	0.15-0.40	206	38	2	1	3.7
	0.40-0.65	198	28	3	0	2.2
	0.65-0.90	218	49	7	2	22.5
	0.90-0.15	222	23	3	0	7.3
α=297.6°	0.15-0.40	203	45	5	1	5.5
	0.40-0.65	202	33	5	1	3.0
	0.65-0.90	218	36	5	1	4.5
	0.90-0.15	203	38	1	0	2.5
α=317.8°	0.15-0.40	166	29	6	0	8.3
	0.40-0.65	198	36	5	0	0.6
	0.65-0.90	207	32	7	1	5.7
	0.90-0.15	199	34	4	0	0.3
Fit in text		199.5	35.4	3.8	0.24	

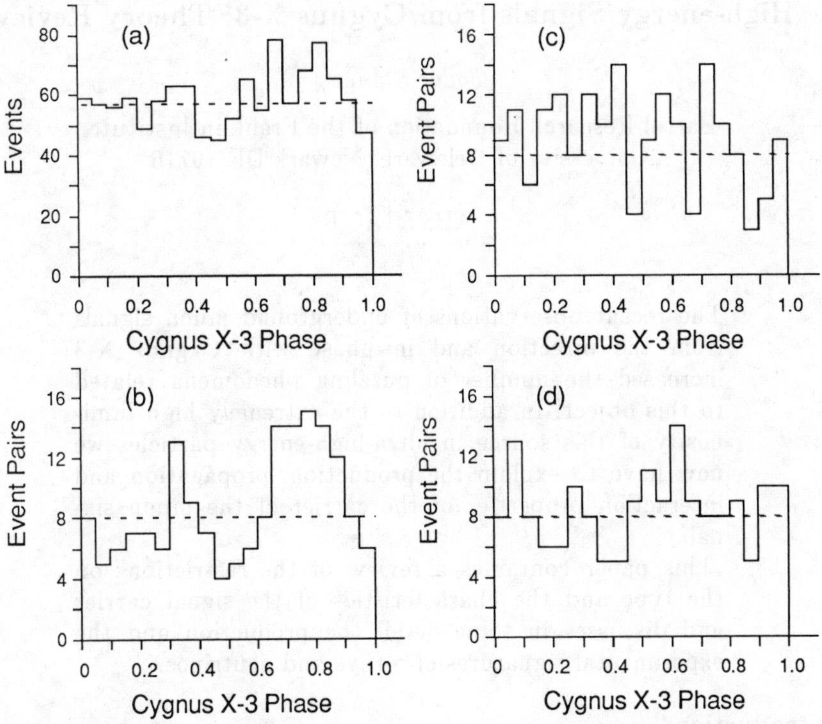

Figure 1. (a) The Cygnus X-3 phase plot for all muon events arriving within 3° of the nominal direction of Cynus X-3. (b) The same plot showing only the mean phase for pairs of events arriving within 0.5 h. (c) and (d) Similar pairs-of-events phase plots for events within a 3° half-angle cone centered at α=297.6° and α=317.6°, respectively, and the same declination as Cygnus X-3.

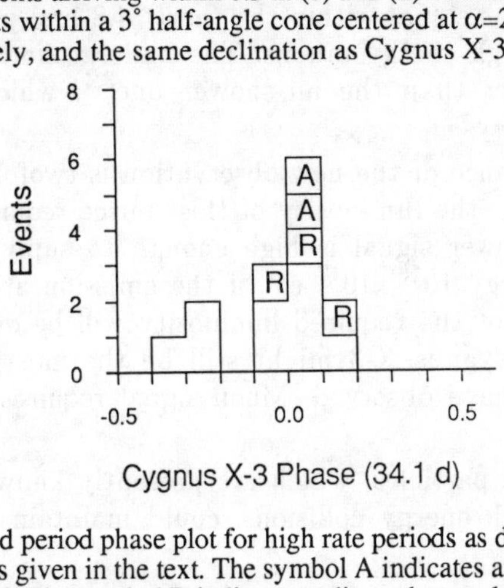

Figure 2. The 34.1 d period phase plot for high rate periods as defined in the text using the ephemeris given in the text. The symbol A indicates air shower bursts described in Ref. 14. The symbol R indicates radio outbursts described in Ref. 13.

High-energy Signals from Cygnus X-3. Theory Review.

*Todor Stanev**

Bartol Research Foundation of the Franklin Institute,
University of Delaware, Newark DE 19716

ABSTRACT

The recent observations of underground muon signals from the direction and in phase with Cygnus X-3 increased the number of puzzling phenomena related to this object. In addition to the extremely high luminosity of this source in ultra-high-energy particles we now have to explain the production, propagation and interaction properties of the carrier of the muon signal.

This paper containes a review of the restrictions on the type and the characteristics of the signal carrier and discusses in some detail the production and the experimental signatures of γ-rays and neutrinos.

Introduction

Recent observations of underground signals pointed at and in phase with Cygnus X-3[1,2] inspired a lot of thinking about the possible ways to carry the signal within the short (4.8 hrs) phase of the source and produce high energy muons. The underground signal appears to be even stronger than the air-shower one[3,4], which is traditionally atributed to γ-rays.

The importance of the new observation is twofold. As pointed out earlier by Hillas[5] the luminosity of this source required to explain the observed air-shower signal is high enough to suply our Galaxy with particles of energy 10^{16}- 10^{17} eV, if the emission at the source is isotropic. If it is not the required luminosity will be decreased by a factor of 10 and Cygnus X-3 might still be the most powerful particle accelerator we have observed. Muon signal requires still higher luminosity.

None of the particles, which are presently known to be copiously produced in high energy collisions, could maintain the direction and periodicity of Cygnus X-3 on the distance of at least 10 kpc and produce muons underground. It is likely, therefore, that we will have to involve an unknown (or even inexpected) particle as signal carrier.

Short Discussion of the Observed Signals.

The nature of the detected signals is not exactly known. The air-shower signal is presumably caused by primary γ-rays because of the following two reasons:

(a) Cygnus X-3 is already known to emit very-high-energy (10^{12} eV) γ-rays[6]

(b) The showers from the direction of this object appear to have large age parameters.

The age parameter s is a measure of the stage of development of a cascade. On the average a shower induced by 10^{15} eV γ-ray has to penetrate through 800 g/cm^2 of air to reach $s > 1.1$, exhibited by all 'in phase' showers at Kiel. Experimentally the age parameter is obtained from the lateral structure of the shower. Large age parameters, however, could also be observed in showers from primary particles whose interaction products have extremely high multiplicity and transverse momenta.

A contraindication for the γ-ray origin assumption is the large number of muons in the Kiel showers[7], since γ showers are expected to be μ-poor.

The underground signals complicated the situation even more. There is no doubt that the particles detected underground are muons. Underground muons can be produced in interactions in the atmosphere, mostly through π and K decays, or directly in the rock, which surrounds the detector. Atmospheric production requires interaction cross-section of more than 1 mb/nucleon for the primary particles. Production in the rock implies assumption of prompt muon production and correspondingly lower cross-sections, which substantially affects the estimate of the signal carrier's flux. If the cross-section were as low as 10 μb, there should be some signal carriers interacting within the large nucleon decay detectors such as IMB and Kamioka.

It is also difficult to understand the magnitude of the underground signal which appears to be much higher than the air-shower one as shown on Fig. 1. A calculation performed by T.K. Gaisser [8] shows that if the underground muons are produced through π and K decays their parent particles must have flux about 20 times higher than that of the TeV γ-rays.

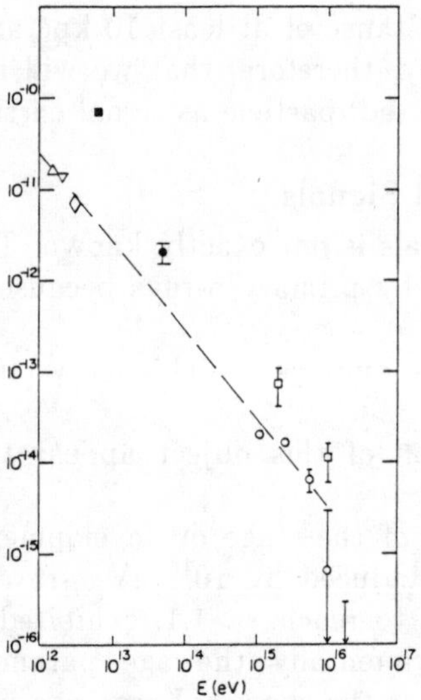

Fig.1 Integral fluxes of the detected 'in phase' signals from Cygnus X-3. Open symbols represent γ-ray and air shower measurements. The full square - Soudan and the full circle - NUSEX muon signals in the assumption that the average energy of the signal carrier is $10E_\mu$.

The two underground measurements somewhat disagree on the angular dispersion of the 'in phase' muons signal, although both detectors have angular resolutions of less than 1°. While the Soudan signal is pretty well concentrated in a cone with half opening of 3°, NUSEX events appear to randomly distributed in a 10° x 10° square centered at Cygnus X-3. If one subdivides it in 100 1° squares 72 empty squares can be counted in Fig. 3 of Ref. 2, as well as 24 and 4 containing one and two muons. A corresponding Poissonian distribution with average 0.32 gives 73, 23 and 4 squares containing 0, 1 and 2 muons.

This large scattering cannot be explained with transverse momentum acquired at production, unless the observed muons are decay products of a very heavy object. The scattering of the underground muons is dominated by the multiple Coulomb scattering of the muons in the rock, which is shown on Fig. 2 for depths of 2 and 5 km.w.e. for muons with a $E^{-3.7}$ spectrum, typical for the ordinary cosmic ray muons. A flatter muon spectrum, caused by flatter primary spectrum (like the E^{-2} spectrum derived from the air-shower signal) or/and prompt muon production will generate even narrower angular spread.

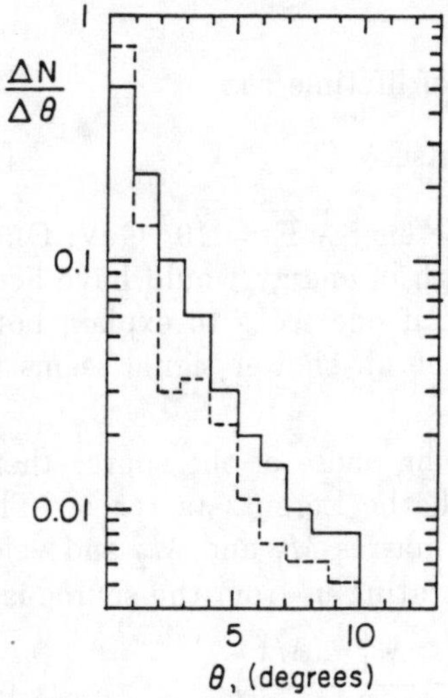

Fig. 2. Muon angular deflection due to multiple Coulomb scattering in rock at 2 and 5 (dash line) km.w.e. for $E_\mu^{-3.7}$ *muon energy spectrum.*

Restrictions on the Signal Carrier due to Propagation.

Although recent measurements estimate the distance to Cygnus X-3 to 12 - 30 kpc[9] (3.7-9.3x10^{19} cm.) we shall use in our considerations a canonical distance of 10 kpc. This distance alone puts severe restrictions on the charge, lifetime and mass of the signal carrier. A larger distance will, of cource, strengthen our arguments.

The diffusion length in the galactic magnetic field is[10]

$$L_d(cm)=3.10^{15}\beta R^{1/2} \tag{1}$$

which is only of order 10 pc for rigidity $R=10^6$ GV. Having in mind that the Soudan muons have an effective energy threshold of 0.65 TeV, i.e. the average primary energy per nucleon for production of single muons is of order of 10 TeV, not only protons, but also any conventional ionized nucleus is ruled out as signal carrier. An assumption that the signal carrier has much higher rigidity simply contradicts to the relatively well measured total cosmic ray flux. Some additional arguments considering special cases where particles are accelerated in tightly collimated beam sweeping through the direction of the magnetic field lines are given in Ref. 11. The signal carrier must be neutral as charged particles would totally disperse in both direction and time in the interstellar magnetic field. The energy estimates for the muons and parent particles at NUSEX are only less than one order of magnitude higher and the conclusion applies to that signal too.

The distance travelled by a particle of lifetime τ is

$$L(pc) = 10^{-8}\frac{E}{M}\tau(\sec) \tag{2}$$

which gives for neutrons and L = 10 kpc energy E = 10^9 GeV. Once again a significant flux of neutrons so high in energy would have been observed in the cosmic ray flux. Besides, if one seeks to explain both types of signals with the same carrier, the air-shower signal seems to vanish at 10^8 GeV.[2]

As the radiation remembers the 4.8 hr phase of the source there is also a restriction on its mass through the Lorentz factor γ. The time delay Δt between two particles of masses M_1 and M_2 and velocities β_1 and β_2 over the distance L separating us from the source is

$$\Delta t = \frac{L}{c}(1/\beta_1 - 1/\beta_2) = \frac{L}{2c}\left(\frac{M_1^2}{E_1^2} - \frac{M_2^2}{E_2^2}\right) \tag{3}$$

and assuming that $M_1 = M_2 = M$ and $E_1 E_2 = E^2$, i.e. assuming the existence of an average mass and energy of the signal carrier, the energy dispersion of the carrier at generation is

$$\frac{(E_1^2 - E_2^2)^{1/2}}{<E>} \leq <\gamma>(2c\,\Delta t/L)^{1/2} \tag{4}$$

For Δt = 2.10^3 s (which is approximately 1/10 of the phase) and L = 10 kpc we therefore obtain

$$\Delta E/E \leq 6.10^{-5}\gamma \tag{5}$$

i.e. the restriction on the energy dispersion of the signal carrier is not very strong for $\gamma > 10^5$. Notice however that for $(\Delta E/E) = 1$ Eq. (5) implies that the mass $M < 6.10^{-5}E$. The mass M must therefore be less or about a GeV for the roughly 10 TeV particles originating the underground muon signal, unless they are produced as a monoenergetic beam.

Gamma-rays and neutrinos.

The relations of the previous section still do not rule out gamma-rays and neutrinos as signal carriers. Furthermore the production of γ-rays and neutrinos at a binary source follows from the geometry of the source. Let us consider the model of Vestrand and Eichler[12], which was constructed to explain the production of 10^{12} eV γ-rays at

Cygnus X-3.

The binary system consists of a fast rotating neutron star of mass $1M_{\odot}$ orbiting around a $4M_{\odot}$ companion star. TeV γ-rays in this model are produced when the pulsar (at which the primary acceleration of protons and nuclei is performed) is behind the edge of the companion star from the point of view of the observer, i.e. just before and after the eclipse, when the accelerated beam of nucleons hits the material of the companion star, interacts with it and generates γ-rays through $\pi^o - 2\gamma$ decays. The authors also note that neutrinos will be produced on the whole duration of the eclipse.

A more precise calculation was carried out by Hillas[5]. In his model the accelerated beam energy deposition inside the star lifts off a thick layer of matter to generate a gas shroud in the environs of the companion star. The γ-rays above a few GeV are then produced by cascade processes in the gas shroud. The dominant processes of the cascading in the presence of a strong magnetic field, associated with the pulsar, are synchrotron radiation and pair production. The assumption of a roughly monoenergetic beam of 10^{17} eV protons at 10^{39} erg s^{-1} and subsequent cascading in 80 - 160 g cm^{-2} of matter accounts best for the observed γ-ray spectrum in this model.

It is not likely, however, that γ-rays are responsible for the observed underground signal or the high number of GeV muons in the Kiel showers. A recent calculation by Stanev, Gaisser and Halzen[13] confirmed older results on the low efficiency of muon production in γ-initiated showers. The dominant channel for muon production is photoproduction of hadrons and π and K decays. The photoproduction cross-section is 200 times lower than the proton inelastic cross-section in air at several GeV and both cross-section have similar logarithmic increase with the interaction energy. The GeV muon content of γ-showers at 10^{15} eV is not expected to exceed 10 per cent of the proton showers, far below the Kiel measurement of 80 per cent[7].

Gamma-rays are not any more effective in producing high-energy muons. A detailed calculation by Stanev and Vankov[14], taking into account both photoproduction and creation of $\mu^+\mu^-$ pairs, shows that the total number of muons produced by the air-shower flux, if it were γ-rays, in the Soudan detector for one year is less than 0.4. Unless something goes really wrong with the electromagnetic processes at 10^{15} eV, which does not seem likely, the underground signals are not caused by primary γ-rays.

A calculation of the neutrino production at Cygnus X-3 was recently published by Gaisser and Stanev.[15] A Vestrand and Eichler type system and 10^{39} erg s^{-1} proton beam were used to calculate the neutrino production. The physical conditions at the companion star were assumed to be those of a 2.8 $M_{\odot}$, $2R_{\odot}$ main sequence star. The calculation was carried out for a monoenergetic 10^{17} eV proton beam and for a E^{-2} beam of the same luminosity. Neutrino absorbtion in the companion star was accounted for. The final product of the calculation is the neutrino induced muon flux in a deep underground detector. Calculations were also performed for stellar atmsphere of uniform density for a variety of densities and column densities.

The main conclusions from the calculation are that the ν-induced muon flux is not very sensitive to the conditions at the companion star but the flux expected from Cygnus X-3 is low. In this model neutrinos are generated during 40 per cent of the orbital period, while the pulsar is eclipsed. The upward muon flux is reduced by no more than a factor of 2 if the thickness of the stellar atmosphere is only 30 g.cm^{-2}. The expected rate for upward going muons induced by Cygnus X-3 neutrinos is 1 per 1000 m^2 per year. This number does not take into account the fraction of the time Cygnus X-3 is under the horizon and for the typical detector in the northern hemisphere it has to be divided by 5. Similar conclusions are reached in an independent estimate, reported on this conference[16], which also gives a detailed discussion of the conditions at the source.

The neutrino induced muon flux will also be independent of the zenith angle in sufficiently deep detectors, which is obviously not the case with both muon signal observations, which exibit angular distribution similar to that of atmospherically generated muons. The inconsistency of the angular dependence of the observed muon signal with that expected from neutrino induced muons, as well as the high magnitude of the signal rule out the last 'natural' candidate for signal carrier.

Propagation Scenarios

We have already established that the signal carrier has to be a neutral, long lived and have Lorentz factor of more than 10^4. It is very difficult to imagine how a neutral particle can be directly accelerated at the source. An acceleration mechanism will have most probably to include the acceleration of a charged object and stripping the charge off after the acceleration to produce the signal carrier.

Independently of the nature of the signal carrier and the acceleration mechanism one could imagine at least two scenarios to produce the air shower and underground signals. A simple, one step scenario will assume that both signals are produced by the signal carrier. A more complicated one would suggest that a long lived and very energetic signal carrier decays in the vicinity of the Earth into hadrons and γ-rays. Decay hadrons produce the underground signals and a fraction of the air showers, the other fraction being initiated by γ-rays.

Let us assume, following Ref. 17, that a signal carrier decays in the vicinity of the Earth into neutrons, which maintain its trajectory and produce the underground signal in interactions with the atmosphere. Because only a fraction of the initially acclerated signal carrier is now involved in the production of the muon flux, this will increase the power requirement at the source. The increase is proportional to the ratio of the lifetimes τ_{SC}/τ_n. The lifetime of an object with $\gamma{=}10^4$ to travel 10 kpc is $\tau_{SC} \geq 10^8$ sec and the increase of the required luminosity is

$$\frac{\tau_{SC}}{\tau_n} = 10^5,$$

which is too big a factor to be happy with on top of the 10^{39} erg s^{-1} required for the air shower signal.

The simpler 'one-step' scenario allows us to adjust the signal carrier interaction cross-section to explain the relative magnitude of the air-shower and muon signals. If the mean free path of the carrier is longer than the thickness of the atmosphere, some of those particles will interact with the rock and will still be able to generate muons without affecting the air-shower flux. If this were the case, underground muons would have to be produced promptly and not through π and K decays. As pointed out by T. K. Gaisser at this conference[8] such assumption explains better the ratio of the signals at Soudan and NUSEX.

Fig. 3 shows the angular distribution of underground muons produced by an E^{-2} primary flux in the atmosphere through π and K decays and promptly with different cross-section. The small statistics of the underground signal does not allow to distinguish the angular distribution of muons produced promptly with a 1 *mb* cross-section from this of atmospherically generated muons.

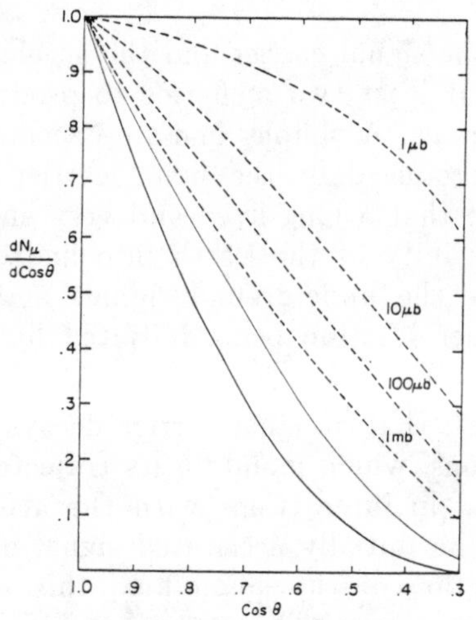

Fig. 3. Angular distribution of underground muons generated by: (a) Cosmic rays through π and K decays (solid line); (b) E^{-2} signal through π and K decays; (c) Promptly with different cross-sections by E^{-2} signal (dash lines).

The angular distributions from smaller cross-sections are, however, quite different from the observed one. Furthermore if the cross-section were much smaller than 100 μb (m.f.p. in rock 0.166 km.w.e.) the flux of signal carriers would not have been significantly attenuated and its interaction would have been seen in the large nucleon decay detectors.[18] Fig. 4 shows the expected interaction rate per kT.year in a detector at depth 2 km.w.e. as function of the cross-section.

Some Possible Signal Carriers.

One of the possible candidates for signal carrier are nuggets of quark matter[17]. There are theoretical reasons for believing that such objects might exist and be stable for certain ranges of mass.[19]. They could be abundant in the enviroment of a compact quark star and also be produced in high-energy processes in the system. Some fraction of such objects could be neutral and thus signal carrier. A high content of strange quarks, which is required for stabilization of the nugget, would lead to enhanced kaon production and correspondingly high muon yield.

It has already been suggested [20] that quark globs could penetrate deep in the atmosphere and explode to produce Centauro events. The low flux of Centauro events does not contradict this hypothesis, as Centauros might be the result of desintegration or interaction of quark globs from the tail of the mass distribution.

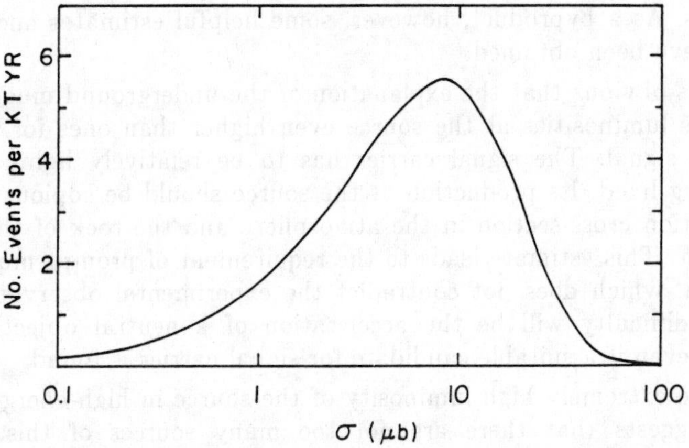

Fig. 4. Interaction rates (per kT.yr) of signal carrier with a flux of 10^{-10} $cm^{-2}s^{-1}$.

A much more specific version of a stable configuration of quark matter has been presented at this conference by L. D. McLerran[21]. This is a di-Λ, a bound dihyperon state of 2u, 2d and 2s quarks. The enhanced kaon production of such objects, however, increases the muon yield of such objects only by a factor of less than 2, which is not big enough to account for the relative magnitude of the air-shower and muon signals. The total cross-section will have to be adjusted at value significantly smaller than the nucleon cross-section in air and a prompt muon production mechanism will be quite helpful.

V. J. Stenger has studied the production of photinos in Cygnus X-3 and the flux of photino-induced muons underground.[22] Both the production and the interaction cross-sections are too low to apply to the measured muon signal even in optimistic assumptions. Photinos will also tend to produce much flatter angular distribution than the observed one.

Conclusions

A self-consistent model of the discussed phenomenon should consist of an indentification of the signal carrier, as well as of models for its production (acceleration) and interaction properties. There have been no successful attempts to construct such a model during the short time since the observation of the underground signals. Most of the work is aimed at eliminating some natural candidates for signal carriers. As a byproduct, however, some helpful estimates and guidelines have been obtained.

It is obvious that the explanation of the underground muon signal requires luminosities at the source even higher than ones for the air-shower signal. The signal carrier has to be relatively light, neutral and long lived. Its production at the source should be copious and its interaction cross-section in the atmosphere and the rock of the order of 1 *mb*. This estimate leads to the requirement of prompt muon production, which does dot contradict the experimental observations. A major difficulty will be the acceleration of a neutral object at the source even if a suitable candidate for signal carrier is found.

The extremely high luminosity of the source in high-energy particles suggests that there are not too many sources of this caliber simultaneously on duty in our Galaxy. If this is true and if at least some of these sources accelerate new particles, the flux of such particles will be highly anisotropic. Thus it is possible that not all cosmic ray detectors, with a typical angular acceptance of a steradian, will be able to see the same phenomena. A sketch of this idea[23] is shown on Fig. 5.

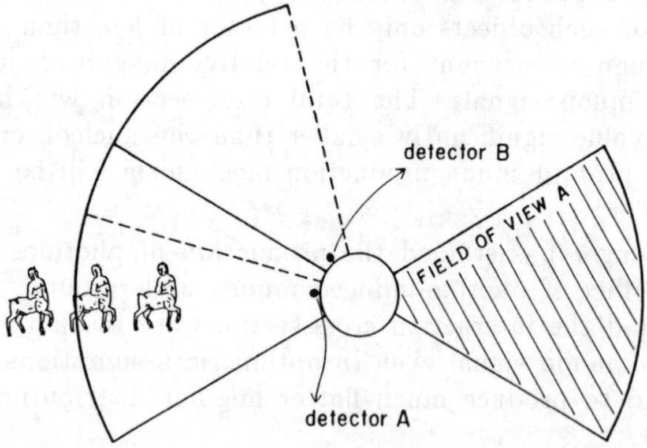

Fig. 5. Are cosmic point sources sending disturbing signals to some cosmic ray detectors?

Acknowledgments

Most of the work, upon which this talk is based, was made in close collaboration with M. V. Barnhill III, T. K. Gaisser and F. Halzen. Special thanks to T. K. Gaisser for reading the manuscript. The author has had numerous discussions and has learned a lot from J. Learned, L. McLerran, M. Marshak, K. Ruddik, E. Peterson and J. van der Velde. This work was supported in part by the National Science Foundation under Grant PHY-8410989.

References

(*) On leave of absence from the *Institute for Nuclear Research and Nuclear Energy of the Bulgarian Academy of Sciences,* Sofia 1184, Bulgaria.

(1) M. L. Marshak *et al.,* Phys. Rev. Lett. 54 , 2079 (1985).

(2) G. Battistoni *et al., Submitted to* Phys. Lett. B

(3) M. Sammorski and W. Stamm, Ap J 268 , L17 (1983).

(4) J. Lloyd-Evans *et al.,* Nature 305 , 7784 (1983).

(5) A. M. Hillas, Nature 312 , 50 (1984).

(6) Yu. I. Neshpor *et al.,* Astrophys. Space Sci. 61 , 349 (1979); S. Danaher *et al.,* Nature 289 , 568 (1981); R. C. Lamb *et al.,* Nature 296 , 543 (1981).

(7) M. Samorski and W. Stamm, in *18th International Cosmic Rays Conference, Bangalore, India, 1983, Conference papers,* edited by N. Durgaprasad *et al.,* (Tata Institute for Fundamental Research, Bombay, 1983), Vol.11, p. 244.

(8) T. K. Gaisser, These Proceedings.

(9) T. C. Weeks *et. al.,* Publ. Astron. Soc. Pac. 93 , 474 (1981).

(10) J. F. Ormes, in *Workshop on Very High Energy Cosmic Ray Interactions,* papers edited by M. L. Cherry, K. Lande and R. I. Steinberg, (University of Pennsylvania, Philadelphia, 1982), p. 466.

(11) M. V. Barnhill and T. Stanev, Bartol preprint BA-85-17.

(12) W. Thomas Vestrand and David Eichler, Ap J 261 , 251 (1982).

(13) T. Stanev, T.K. Gaisser and F. Halzen, To be published in Phys. Rev. D.

(14) T. Stanev and Ch. P. Vankov, To be published in Phys. Lett B.

(15) T. K. Gaisser and T. Stanev, Phys. Rev. Lett. 54 , 2266 (1985).

(16) E. W. Kolb, M. S. Turner and T. P. Walker, These Proceedings.

(17) M. V. Barnhill *et al.,* MAD/PH/243 (April 1985).

(18) J. van der Velde, Private Communication.

(19) For a review see e.g. F. Halzen, in *Cosmic Ray and High Energy Gamma Ray Experiments for the Space Station Era,* edited by W. V. Jones and J. P. Wefel (Louisiana State University, 1984).

(20) J. D. Bjorken and L. D. McLerran, Phys. Rev.D 20 , 2353 (1979).

(21) L. D. McLerran, R. Jaffe *et al.,* These Proceedings.

(22) V. J. Stenger, HDC -3-85 (April 1985).

(23) L. D. McLerran, Private Phone Call.

HIGH ENERGY NEUTRINOS FROM CYG X-3

T. P. Walker, E. W. Kolb, and M. S. Turner
NASA/Fermilab Astrophysics Center, Fermi National
Accelerator Laboratory, P.O. Box 500, Batavia, IL 60510 USA

ABSTRACT

Assuming that the UHE air showers from Cyg X-3 are produced by photons, we calculate the expected neutrino emission from a model which produces the γ-rays in the atmosphere of the Cyg X-3 companion. We discuss the possibility of detecting such neutrinos in underground detectors and the constraints that such a signal places on the use of this model in other particle production scenarios.

INTRODUCTION

The Cyg X-3 system has been observed at energies ranging from 10^{-4} eV to 10^{16} eV, with all but the radio emission showing a 4.8 hour period.[1] If we interpret the > TeV energy particles from Cyg X-3 to be photons,[2] then the differential flux from the system is given as[3]

$$\frac{dN_\gamma}{dE_{TeV}} = 3\times10^{-10}\ E_{TeV}^{-2.1}\ \mathrm{cm}^{-2}\ \mathrm{sec}^{-1}\ \mathrm{TeV}^{-1}\ . \tag{1}$$

The 4.8 hour periodicity appears to be associated with the orbital period of a binary system consisting of a pulsar and $\sim 4M_\odot$ companion.[4]

Vestrand and Eichler[5] have used such a system to construct a model which generates > TeV γ-ray (see fig. 1). The pulsar is considered a source of high energy protons, accelerated by large potential differences set up by the rapidly rotating magnetic field of a young neutron star.[6] These protons collide with nucleons in the atmosphere of the companion star, producing π's and K's which then decay into photons and neutrinos. If the π-production region is optically thin to TeV γ-rays, the photons escape and are observed as pulses with widths determined by the size of the optically thin region and having phase dictated by the orientation at which the observer "sees" the pulsar through the companion's atmosphere. In this way, the model gives > TeV γ-ray bursts at phase .25 and .75 with $\Delta\psi \sim 5\%$, in general agreement with the data.

THE HIGH ENERGY NEUTRINO SPECTRUM

If the > TeV events from Cyg X-3 are caused by photons, then the observed γ-ray spectrum is related to a π-spectrum which in turn can be related to a neutrino flux at the earth. Consider the observed γ-ray spectrum to originate from a source spectrum of the form

$$\frac{dS_\gamma}{dE_\gamma} \equiv AE^{-n} . \tag{2}$$

In the model of Vestrand and Eichler, each $\pi°$ decay produces 2γ's and so

$$\frac{dS_{\pi°}}{dE_\pi} = A2^{n-1}E^{-n} . \tag{3}$$

Each nucleon-proton interaction produces as many $\pi^\pm$ as $\pi°$ and so

$$\frac{ds_{\pi^\pm}}{dE} = 2\frac{ds_{\pi°}}{dE} = 2^n\frac{ds_\gamma}{dE_\gamma} . \tag{4}$$

We get a ν_μ from each charged π decay, having an energy E_ν = $1/2[1-(m_\mu/m_\pi)^2]E_\pi$. Therefore the relationship between the neutrino source spectrum and the γ-ray source spectrum is[7]

$$\frac{dS_\nu}{dE_\nu} = [1-(\frac{m_\mu}{m_\pi})^2]\frac{dS_\gamma}{dE_\gamma} . \tag{5}$$

The neutrino source spectrum is degraded by $\nu N \to \mu x$ interactions as it is propagated through the companion star. The neutrino-nucleon cross section is[8]

$$\sigma = \begin{cases} 7\times10^{-36}\, E_{TeV}\ cm^2 & (E\lesssim100\ TeV) \\ 1.2\times10^{-34}\, \ell n\, E_{TeV}\ cm^2 & (E\gtrsim100\ TeV) , \end{cases} \tag{6}$$

and the degredation at a given phase is dependent upon the amount of material a neutrino traverses at that phase (see fig. 1). We have used a ZAMS model of a 2.8 $M_\odot$ star[9] to approximate the density profile of the companion. The effect of neutrino absorption as a function of phase is shown in fig. 2 and the resulting neutrino "light-curves" are shown in fig. 3.

The derivation of the neutrino source spectrum assumed that the π's and K's decay before interacting. The condition that this assumption be valid is just that the decay lengths be shorter than a few interaction lengths:

$$\lambda_{dec} = (\gamma c\tau)_{\pi,K} \lesssim 3 \cdot \lambda_{Int} = \frac{3}{(n\sigma_{Int})_{\pi,K}} . \tag{7}$$

Taking $\sigma_{Int} \simeq 3\times10^{-26}cm^2$ at TeV energies and scaling n to stellar envelope densities (10^{-6}g cm^{-3}) we have

$$\left.\begin{array}{ll} \pi: & 5\times10^6\ E_{TeV}\ cm \\ K: & 8\times10^5\ E_{TeV}\ cm \end{array}\right\} \lesssim 6\times10^7/\rho_{-6}\ cm , \tag{8}$$

where $\rho_{-6} \equiv \rho/10^{-6}$g cm^{-3}. The inequality of equation (8) does not hold for $E > E_c \simeq 100/\rho_{-6}$ TeV. In this energy range, only neutrinos produced by charmed (or heavier flavor) meson decay will be emitted, with a flux of $\sigma_{charm}/\sigma_{\pi,K}$ relative to that produced by π and K decay. Heavy flavor production is down by a factor of $10^2 - 10^3$ relative to π-K production and thus the neutrino spectrum above 100 TeV should be down by a factor of $10^2 - 10^3$ relative to the flux expected from the decay pipe scenario. π's and K's produced with $E > E_c$ will have their energy reduced until $E \sim E_c$, at which point they produce a neutrino with $E_\nu \sim E_c$ resulting in a bump in the neutrino spectrum at E_c.

Finally, we can calculate the relationship between the observed differential energy spectra of neutrinos and γ-rays. Taking into account the fact that we only see photons from π^o decay under favorable orientations we write

$$\frac{dN_\nu}{dE} \simeq \left[1-\left(\frac{m_\mu}{m_\pi}\right)^2\right]^{2.1} \frac{\Delta\psi_\nu}{\Delta\psi_\gamma} \cdot \frac{dN_\gamma}{dE} , \tag{9}$$

where $\Delta\psi_\nu$ (~ 40%) and $\Delta\psi_\gamma$ (~ 5%) are the duty cycles of neutrino and γ-ray production. Thus

$$\frac{dN_\nu}{dE} \simeq 4\times10^{-10}\, E_{TeV}^{-2.1}\ \text{cm}^{-2}\ \text{sec}^{-1}\ \text{TeV} \quad (E<E_c) \tag{10}$$

and about $10^{-2} - 10^{-3}$ of this value for $E > E_c$.

THE NEUTRINO SIGNAL IN UNDERGROUND DETECTORS

Here we discuss the possibility of observing the nuetrino flux from Cyg X-3 in an underground water Cerenkov-type detector (e.g. the IMB proton decay detector[10] having dimensions 17m × 18m × 23m, located at a depth of 1500 m.w.e.). In figure 4, we give a generic sketch of such a detector. Muon neutrinos from Cyg X-3 can cause two types of events in these detectors. They can: (1) Interact within the detector, producing a muon which is observed by its Cerenkov radiation; or (2) Interact with the earth, producing a muon which passes through the detector. Events (1) will be called contained events and events (2) will be called external events.

The probability of observing a contained event is just the ratio of the average detector trajectory to the interaction length:

$$\ell_c(E_\nu) = \frac{\langle\ell\rangle_{det}}{\lambda_{Int}} = \langle\ell\rangle n\sigma \simeq \begin{cases} 4\times10^{-9}\ \ell_{10}\ E_{TeV} & (E\lesssim100\text{TeV}) \\ 7\times10^{-8}\ \ell_{10}\ln E_{TeV} & (E\gtrsim100\text{TeV}) \end{cases} , \tag{11}$$

where $\ell_{10} \cdot 10$ m = $\langle\ell\rangle_{det}$ (m.w.e.), $n \simeq 6\times10^{23}$ cm^{-3}, and σ is taken from equation (6).

In order to calculate the probability of observing an external event, we need to know the range of a muon in rock. Taking into account energy losses due to ionization, bremsstrallung, pair production, and inelastic collisions, we can write[12]

$$- \frac{dE}{dx} = 1.9\times10^{-6} \text{ TeV cm}^{-1} + 4\times10^{-6} E_{TeV} \text{ cm}^{-1} , \qquad (12)$$

which gives a muon range of

$$R(E_{TeV}^{\mu}) = 3\times10^{5} \ln (1 + 2E_{TeV}^{\mu}) \text{ cm}. \qquad (13)$$

The fact that muons made within R(E) of the detector extends the size of the detector towards Cyg X-3 and results in a probability for an external event of

$$P_e(E_\nu) \simeq \frac{R[(1-y)E_\nu]}{\lambda_{Int}} = 1\times10^{-6} E_{TeV}^{\nu} \ln[1+2(1-y)E_{TeV}^{\nu}] \quad (E_{TeV}<100) \qquad (14)$$

where (1-y) is the fraction of energy carried off by the muon in a $\nu N \to \mu x$ interaction. For simplicity, we take $y \simeq 1/2$.[12] Note that P_e/P_c increases with energy, making contained events rare in a signal dominated by > TeV neutrinos.

The rate of external events due to a neutrino spectrum of the form $dN_\nu/dE = aE^{-n}$ incident upon a detector of cross sectional area A is

$$\Gamma_e = A \int P_e(E) \frac{dN_\nu}{dE} dE$$

$$= Aa\ 1\times10^{-6} \int E_{TeV}^{-n+1} \ln(1+E_{TeV}) dE \qquad (E_{TeV}<100) . \qquad (15)$$

For a spectral index n < 3, the integral of equation (15) is dominated by 10 - 100 TeV events, cut off by the logarithmic range dependence. In this energy regime we expect external events to be 1000 times more likely than contained events.

Using the neutrino spectrum derived from the observed photon spectrum in the previous section ($a = 4\times10^{-10} \text{cm}^{-2}\text{sec}^{-1}$ and $n = 2.1$) and scaling A to the IMB detector ($A_{IMB} \simeq 400 \text{ m}^2$), the predicted event rate for Cyg X-3 is

$$\Gamma^{Cyg\ X-3} \simeq 2\times10^{-8} \text{ sec}^{-1}/400 \text{ m}^2 , \qquad (16)$$

or slightly less (.6) than 1 event/year.

This signal must be picked out from the muon background due to cosmic ray interactions in the atmosphere. Since muons have a finite range in earth (see equation (13)), there is a zenith angle dependent energy threshold for a background muon to reach the detector,

$$E_{TeV}^{TH}(\gamma) = \frac{1}{2} \left[\exp \frac{x(\gamma)}{3\times10^5} - 1 \right] , \qquad (17)$$

where γ is the zenith angle and from figure 4,

$$x(\gamma) = R\left\{(\frac{d}{R} - 1)\cos\gamma + [(\frac{d}{R} - 1)^2 \cos^2\gamma - \frac{d}{R}(\frac{d}{R} - 2)]^{1/2}\right\} . \quad (18)$$

Using the fact that the integrated muon background goes as E^{-2},[13] the background rate in IMB is

$$\frac{d\Gamma^{BCKGRND}}{d\Omega} \simeq 5\times10^{-4}[\exp(\frac{.5}{\cos\gamma}) - 1]^{-2} \sec^{-1} \deg^{-2} , \quad (19)$$

which must be multiplied by the solid angle corresponding to the detector's acceptance cone.[14] The muon background as a function of zenith angle is shown in figure 5.

In the case of Cyg X-3, one cannot use the zenith angle dependence of the muon background to full advantage because of Cyg X-3's location for northern hemisphere observers. With a declination of 40.8°, the zenith angle of Cyg X-3 is restricted to the range $\theta - 40.8° \lesssim \gamma \lesssim 139.2° - \theta$ for an observer at latitude θ. This confinement reduces northern latitude exposure times to a few hours/day (e.g. IMB has $\gamma > 86°$ for ~ 5 hours/day).

NEUTRINO PRODUCED MUONS AND CONSTRAINTS ON CYG X-3

Observations of the muon-content and zenith angle dependence of the air showers from the direction of Cyg X-3 appear to indicate that the showers cannot be caused by photons or neutrinos.[15] The phase correlation and directionality of the signal and the distance to Cyg X-3 constrain the air shower particle to be neutral, less than a few GeV in mass, and metastable ($\tau \gtrsim$ months).[15] Several authors[16] have tried to construct scenarios in which these "cygnets" are produced. We would like to point out that even if the air showers are not due to photons, the neutrino flux accompanying the creation of cygnets and the fact that cygnets have not been seen in accelerator searches places severe constraints on using a Verstrand and Eichler-type mechanism for cygent production.

This model would use pN → Cygnet, (π,K) interactions in the companion's atmosphere to produce the cygnet flux. The ratio of cygnets to muon neutrinos is just

$$\frac{N_{\nu\mu}}{N_c} \simeq \frac{N_\pi}{N_c} = \frac{\sigma_\pi}{\sigma_c} . \quad (20)$$

If we want cygnets to have the same flux as indicated by air showers, then the rate of external events in IMB scales as

$$\Gamma^{Cyg\ X\text{-}3} \simeq \frac{\sigma_\pi}{\sigma_c} \cdot yr^{-1} . \quad (21)$$

The fact that we don't see neutrinos from Cyg X-3 implies that $\sigma_c \gtrsim 10^{-2}\ \sigma_\pi$ (i.e. $\Gamma^{Cyg\ X\text{-}3} \lesssim 10^2\ yr^{-1}$). A GeV particle with this large a production cross section would be difficult to hide in accelerator searches. Therefore the model for producing anything but neutrinos and γ-rays with a proton beam hitting the companion star seems to be inconsistent with our calculation of the neutrino signal and we conclude that such new particles must come directly from the compact object.

SUMMARY

Using a model which produces > TeV γ-rays, we have calculated the expected neutrino flux from Cyg X-3, normalized to the photon flux (see equation (10)). The source of spectrum of neutrinos is modified by passage through the companion star, resulting in the neutrino "light curves" of figure 3. The event rate in an underground detector is about 1/yr/400 m^2. Although the rate is not measurable with current detectors, this calculation can be used to put limits on the production of other types of particles in the Cyg X-3 system.

ACKNOWLEDGEMENTS

We wish to thank Chris Hill, J. D. Bjorken, David Seckel, David Schramm, John Learned, and Jim Stone for useful conversations. This work was supported by NASA (at Fermilab), the DOE (at Fermilab, Chicago, and Indiana University), and by MST's A. P. Sloan Fellowship.

Note added: After our work of ref. 9 was completed we learned of similar work by: T. Gaisser and T. Stanev, Phys. Rev. Lett., in press (1985); V. S. Berezinsky, C. Castagnoli, and P. Galeotti, preprint (1985); G. Cocconi, CERN preprint (1985). These groups reached similar conclusions to ours.

REFERENCES

1. For a review of the e.m. observations of Cyg X-3, see N. Porter, Nature 305, 179 (1983) and references therein.
2. See T. Stanev's contribution to these proceedings, as well as T. K. Gaisser, T. Stanev, and F. Halzen, University of Wisconsin Preprint, MAD/PH/243 (1985).
3. M. Samorski and W. Stamm, Ap. J. 268, L17 (1983); J. Lloyd-Evans, etal., Nature 305, 784 (1983).
4. D. R. Parsignault, J. Grindlay, H. Gursky, and W. Turker, Ap. J. 218, 232 (1977); W. T. Vestrand and D. Eichler, Ap. J. 261, 251 (1982).
5. Vestrand and Eichler, ref. 4.; see also A. M. Hillas, Nature 312, 50 (1984).
6. M. A. Ruderman and P. G. Sutherland, Ap. J. 196, 51 (1975).
7. See also F. W. Stecker, Ap. J. 228, 919 (1979); V. J. Stenger, Ap. J. 284 (1984).
8. Yu. M. Andreev, V. S. Berezinsky, and A. Yu. Smirnov, Phys. Lett. 84B, 247 (1979).
9. See, for example, D. Clayton, Principles of Stellar Evolution and Nucleosynthesis (McGraw-Hill: New York, 1968); for a discussion on the effect of varying ρ(r) see E. Kolb, M. Turner, and T. Walker, to appear in Phys. Rev. D (1985).

10. R. M. Bionta, etal., AIP Conference Proceedings 96, 138 (1983).
11. L. B. Bezrukov and E. V. Bugaev, ICRC 17, MN 102.
12. See E. W. Kolb, etal., ref. 9 for further discussion.
13. T. K. Gaisser, T. Stanev, S. A. Bludman, and H. Lee, Phys. Rev. Lett. 51, 223 (1983).
14. For a cone of angle a, $\Delta\Omega \approx 2\times10^4$ [1-cos(a/2)]deg^2.
15. See the discussions of ref. 2; also M. L. Marshak, etal., Phys. Rev. Lett. 54, 2079 (1985).
16. T. K. Gaisser, T. Stanev, and F. Halzen, ref. 2; G. Baym, E. W. Kolb, L. McLerran, and T. P. Walker, Fermilab Preprint (1985); V. J. Stenger, University of Hawaii Preprint, HDC-3-85 (1985); G. L. Shaw, G. Benford, and D. J. Silverman, U.C. Irvine Preprint 85-14 (1985).

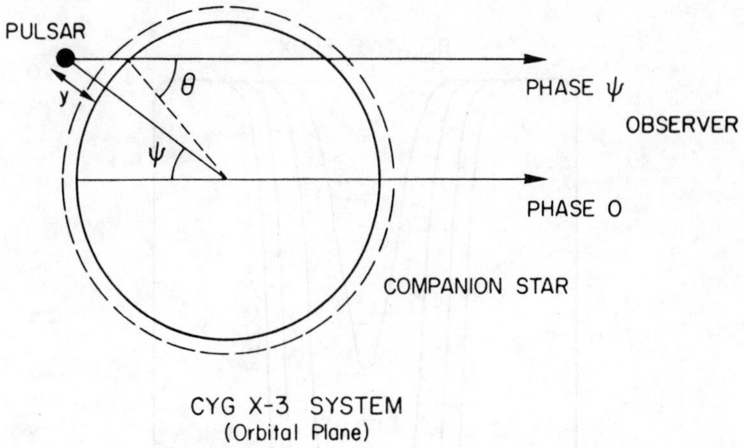
PULSAR
y
θ
ψ
PHASE ψ
OBSERVER
PHASE 0
COMPANION STAR
CYG X-3 SYSTEM
(Orbital Plane)

Fig. 1

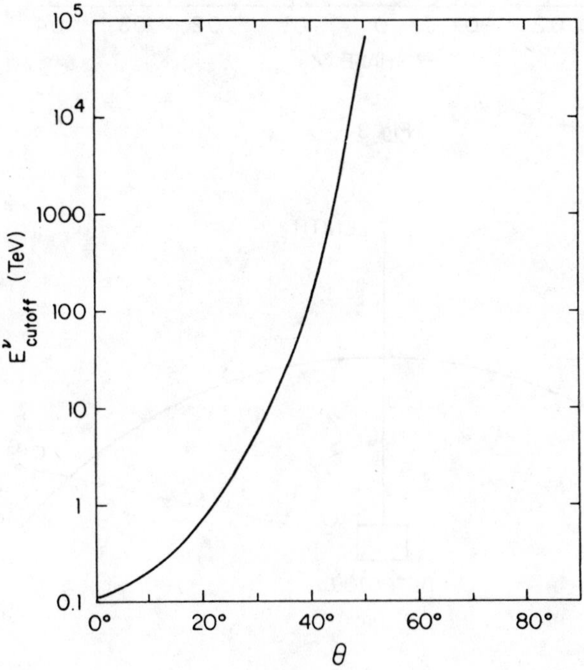
10^5
10^4
1000
100
10
1
0.1
E^ν_cutoff (TeV)
0°
20°
40°
60°
80°
θ

Fig. 2

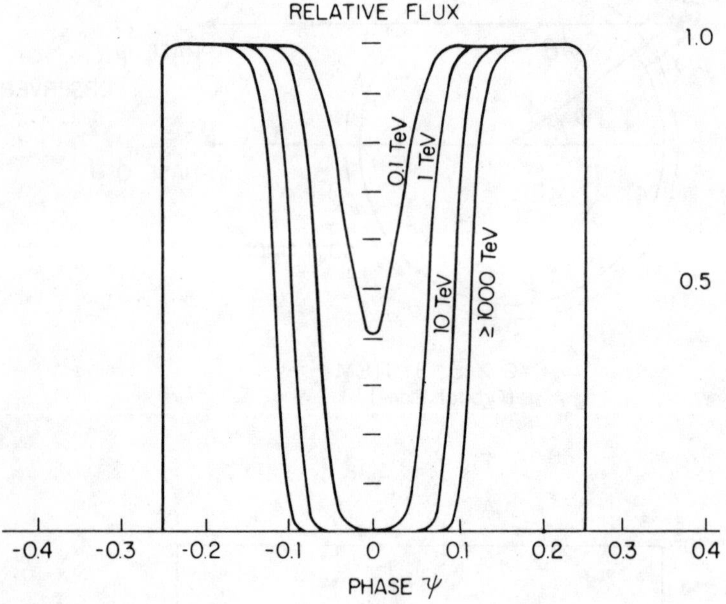
RELATIVE FLUX
1.0
0.5
0.1 TeV
1 TeV
10 TeV
≥ 1000 TeV
-0.4
-0.3
-0.2
-0.1
0
0.1
0.2
0.3
0.4
PHASE ψ

Fig. 3

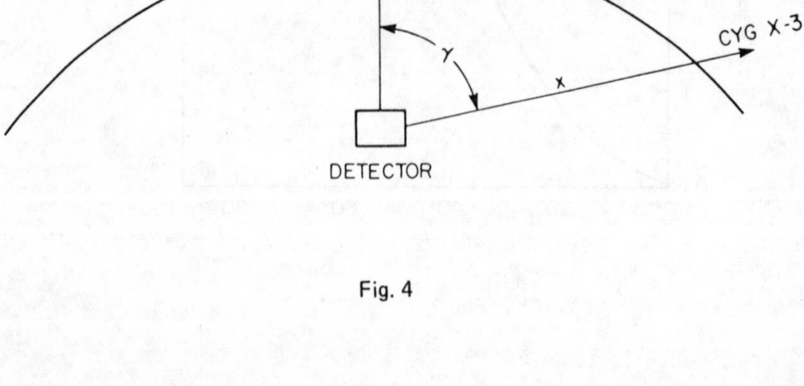
ZENITH
d
γ
x
CYG X-3
DETECTOR

Fig. 4

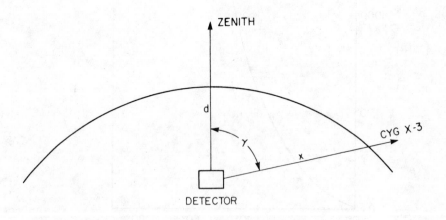

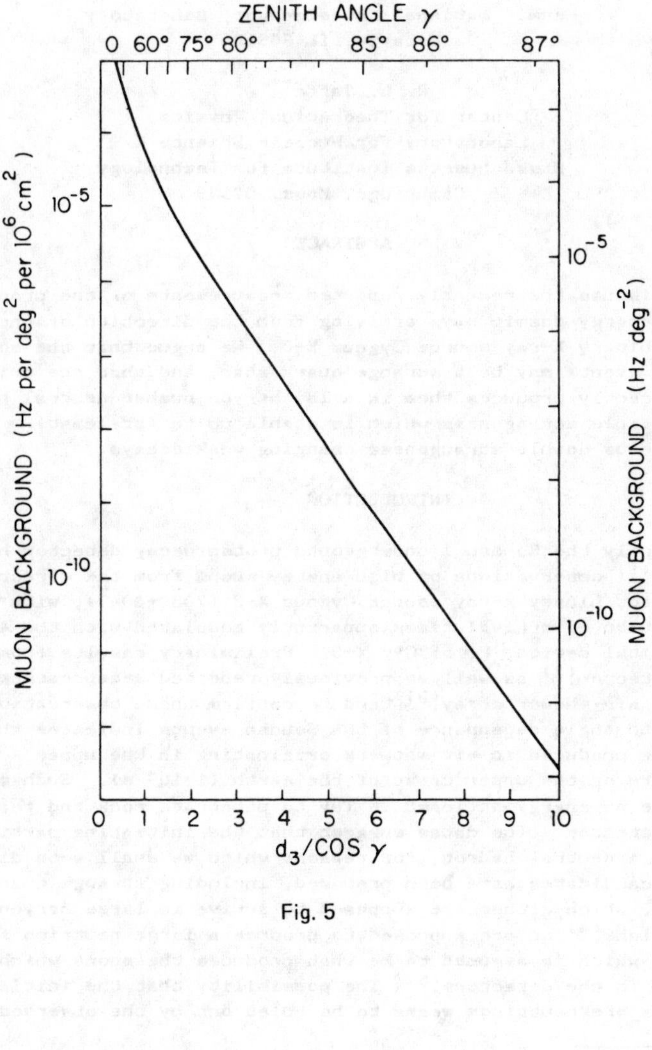
ZENITH ANGLE γ
0
60°
75°
80°
85°
86°
87°
MUON BACKGROUND (Hz per deg² per 10⁶ cm²)
10⁻⁵
10⁻¹⁰
MUON BACKGROUND (Hz deg⁻²)
10⁻⁵
10⁻¹⁰
0
1
2
3
4
5
6
7
8
9
10
d₃/COS γ

Fig. 5

IS CYGNUS X-3 STRANGE?

Gordon Baym,*‡ Edward W. Kolb, Larry McLerran, T. P. Walker
Theory Department
Fermi National Accelerator Laboratory
Batavia, IL 60510

R. L. Jaffe**
Center for Theoretical Physics
Laboratory for Nuclear Science
Massachusetts Institute for Technology
Cambridge, Mass. 02139

ABSTRACT

We discuss the recently reported measurements of the properties of high energy cosmic rays arriving from the direction of the compact binary X-ray source Cygnus X-3. We argue that the source of these events may be a strange quark star, and that the primary which directly produces them is a low baryon number neutral hadron with multiple strangeness which is stable up to (at least) simultaneous double strangeness changing weak decays.

INTRODUCTION

Recently the Soudan 1 underground proton decay detector has reported[1] observations of high energy muons from the direction of the compact binary x-ray source Cygnus X-3 (2030+4047), with a distribution of arrival times apparently modulated with the 4.8 hour orbital period, P, of Cyg X-3. Preliminary results from the Nusex detector,[2] as well as previously reported measurements from the Kiel air shower array,[3] tend to confirm these observations. The zenith angle dependence of the Soudan events indicates that the muons are produced in air showers originating in the upper atmosphere or the upper crust of the earth (1-10^3 m). Such muons must have an energy at least .6 TeV to penetrate rock and reach the Soudan detector. The datas suggest that the initiating particle might be a neutral hadron, for reasons which we shall soon discuss. Several candidates have been proposed, including strange quark droplets, which either are supposed to arrive in large baryon number globs,[4] or are supposed to produce a large neutrino flux at the star which is assumed to be what produces the muons which are observed in the detectors.[5] The possibility that the initiating particles are neutrinos seems to be ruled out by the observed

---

* Research supported through NSF grant PHY 84-15064

‡ Permanent address: Physics Dept. University of Illinois at Urbana-Champaign.

**This work is supported in part through funds provided by the U. S. Department of Energy (D. O. E.) under contract #DE-AC02-76EER03069.

of muons arriving in detectors. In this paper, we shall assume that the experimental observations are correct and explore unconventional sources for the muons. We suggest that the muons are produced by metastable neutral low-baryon number strange hadrons originating from the condensed star in Cyg X-3, argue that for this to be the case most probably the condensed star must be entirely composed of matter with a large strangeness fraction, and describe how such a strange star might be produced in a supernova explosion.

As several authors have already noted,[4] the observation that the absolute flux of muons at the Soudan detector, 7×10^{-11} $cm^{-2}sec^{-1}$ is comparable to that of air showers from Cyg X-3 themselves extrapolated into this energy range indicates that the initiating particle is unlikely to be either a photon or neutrino, if the photon or neutrino cascade by conventional processes. The flux of muons is two to three orders of magnitude too large to be produced by such air showers if the showers are initiated by high energy gamma rays. On the other hand, since the overall neutrino and gamma ray fluxes from Cyg X-3 should be comparable at high energies, if they are both produced by high energy hadronic interactions, the absolute neutrino flux at the Soudan detector should be below the limit of detectability.[6] If the muons are produced by neutrino interactions, which do not generally produce air showers, the zenith angle distribution of such produced muons would be different from that observed. The inference therefore is that the most likely candidate for the initiating particle is a hadron, and in order for interstellar magnetic fields not to alter its direction relative to that of the source, the hadron must be neutral. We shall refer to this particle here as the "cygnet."

In order that the cygnet may arrive from Cygnus X-3 without a tremendous reduction in flux due to decays, the cygnet must have a minimum lifetime $> d/\gamma c$, where $d \sim 12.5$ kpc. (This lower limit might be violated by perhaps as much as a factor of two and still allow for a reasonable flux of cygnets, but much more reduction forces the production rate of cygnets at the source to be unrealistically large.) The energy threshold for muons in the Soudan detector is about .5 Tev, corresponding to hadron primary energy of 5-10 Tev, or to $\gamma \sim$ 5-10 Tev/m, with m the mass of the hadron. This value of γ is about a factor of 5-10 higher for the NUSEX data. In the Soudan detector, the pulses appear in an interval of about .3 units of phase, and in order that the dispersion in travel times not smear out the observed orbital modulation of the signal, γ must be $> (d/.6Pc)^{1/2} \sim 10^4$. For NUSEX, the signal is in .1 units of phase, and the corresponding limit on γ is 10^3. The minimum mass for these energies is therefore $m \lesssim 1\text{-}2$ Gev. The minimum lifetime is ~ 10 yr. These considerations rule out the possibility that cygnets might be neutrons. A possible cygnet candidate is the doubly strange H dibaryon first proposed in Ref.(7). This hadron has the quantum numbers of two lambdas, and is a tightly bound 6-quark bag. If $m_H < m_p + m_\Lambda$ then the particle can only decay by doubly weak processes, and can have a lifetime in the requisite range.

The observed 4.8 hr period of Cygnus X-3 has been associated with the orbital period of a binary system, thought to consist of a young pulsar and a 4 M_{sun} companion.[8] Vestrand and Eichler[8] have used these components to generate short pulses of E > 1 Tev radiation assuming the magnetosphere of the pulsar to be an efficient accelerator of charged particles[9] which then collide with the atmosphere of the companion star.

The cygnets may either be present in some form in the beam which originates from the neutron star, or arise from conventional nucleon interactions of nucleons from the neutron star with the companion star. In the latter case, the production cross section for cygnets is large, and it is difficult to imagine that such a particle would have escaped detection in accelerator experiments. We shall therefore assume that the cygnets are components of the beam of particles originating from the neutron star. We picture then the cygnets originating tightly bound to charged hadrons and being stripped by the interactions of such complexes with matter in the companion star.

If cygnets or their primaries are already present in the flux leaving the neutron star then they must be present in some form in the neutron star itself. The alternative, that they are created by surface bombardment by counter-accelerated particles in the high energy beam, probably requires too large a production cross section to be consistent with laboratory data. It has long been suggested that neutron stars could be largely composed of strange quark matter.[10] If this is true they provide a copious source of H's. A possibility is that strange quark matter exists in the core of the star and is somehow brought up to the surface.[5] We are unable to find any mechanism which efficiently transports matter from the core to the surface by diffusion, convection, or excavation. The other possibility, which we now pursue, is that exotic matter exists stably up to the stellar surface in sufficient quantity to form a rich component of the emitted cosmic rays.[11]

The scenario we describe is based on the possibility, suggested by Witten,[11] that high density strange quark matter is absolutely stable with respect to ordinary nuclear matter, i.e., the energy per baryon of strange quark matter, with strangeness per baryon f_s, around a density ρ_s is less than that of ordinary nuclear matter at $\rho_{nm} \sim 0.16\ fm^{-3}$. In this scenario, ordinary nuclei cannot decay into strange matter because of the presence of a barrier as a function of strangeness fraction. Conversion to strange quark matter would require a very high order weak interaction,[11-12] with a lifetime far in excess of the age of the universe. If strange matter is stable at all it is stable in bulk and for all baryon numbers above some A_{min}.[12] A_{min} cannot be too small lest light nuclei decay by first order or second order weak processes into strange matter. A strangelet (a droplet of strange matter of nuclear dimensions) with $A < A_{min}$ would decay by sequential first order weak processes and by α particle and nucleon emission. Strange matter must have positive electric charge on the quarks. Otherwise, without a Coulomb barrier, a single strangelet would rapidly gobble up all ordinary matter with which it came in contact.[12] For generic values of model parameters the Coulomb barrier of strange matter, while lower than that of ordinary matter, suffices to prevent strangelets from absorbing ordinary nuclei at ordinary stellar temperatures.

In this scenario the lowest energy configuration of the star would have to contain strange matter out to its edge; the problem is to understand how it forms in the star. Let us first look at the high density core. In the cases discussed,[11-12] non-strange quark matter becomes stable relative to nuclear matter at some sufficiently high density, ρ_{crit}. If, in the formation of the neutron star during a supernova, the central density exceeded ρ_{crit}, then the core would form quark matter over strong interaction time scales and later relax to finite strangeness fraction via single weak processes. These processes should be rapid [~ 10^{-10}sec, or longer, for strangness-changing hadronic weak decays in dense matter] compared with deleptonization times for the supernova core, and so the rate at which the initially degenerate neutrinos leave the core determines the timescale for formation of a strange core. (Until the neutrinos have left, the matter is diffuse and not at sufficiently high density to rapidly form quark matter) On the other hand, if the mean baryon densities reached are not high enough to overcome the barrier, a strange quark region can still form through fluctuations in the local density. Once formed they would not only be absolutely stable, they would proceed to expand by converting the neighboring normal matter -- either quark matter or neutrons -- to strange quark matter. For example, if a neighboring neutron crosses the surface of the strange region, it would disassemble into its component quarks; the up quark would be converted to strange either by a direct semi-leptonic process or by purely hadronic weak interactions.

Eventually the entire core would be turned into a strange quark core. The conversion process releases an energy of order tens of MeV per baryon; however, because the core after deleptonization is bound with ~ 100 MeV per baryon, the burning should not lead to explosive disassembly of the core.

After the core has been converted to strange quark matter, the strange matter begins to eat its way out from the core. At the interface between the strange matter and ordinary nuclear matter, ordinary matter is absorbed through the interface and is converted to quark matter, if the matter being eaten through is sufficiently neutron rich, or if the temperature of the star remains sufficiently high that there are some particles in the nuclear matter with kinetic energies sufficient to overcome the Coulomb barrier of strange matter. As the strange quark matter burns its way to the surface, it may also preheat the matter in front of the conversion region and generate enough particles with kinetic energies above the Coulomb barrier to maintain burning. Our estimates indicate that this is indeed possible, and that a time of about a year is necessary to burn a neutron star into a quark star.

The matter emitted from the surface of the strange quark star would consist of strangelets of relatively low baryon number (including H's) in addition to the expected mix of nucleons, hyperons, and mesons. Strangelets that are so light as to be unstable via strong interaction processes decay rapidly away. Those that are stable and many which decay weakly survive long enough to be stripped in the atmosphere of the companion yielding nucleons, hyperons, H's in addition to high energy photons and neutrinos (from meson decay). The strangelets which decay in flight from the surface of the strange quark star might provide a source of high energy neutrinos with a flux not constrained by the photon flux, and might be measurable.[5] Stable strangelets which miss the companion seed the galaxy with stable quark matter. Such matter, in the form of large baryon number globs may be the source of Centauros[13-14] and explain a number of cosmic ray anomalies.[15] In addition the stable quark matter produced in this manner by Cyg X-3 like objects in the past would have seeded the solar nebula and led to a substantial abundance of strange matter, which might be detectable in terrestial searches.

In order to relate the flux of cygnets to the flux of muons, the detailed properties of cygnet-hadron interactions must be understood. To a first approximation, if the cygnet is an H particle, its interactions are similar to those of a proton with an energy equal to that of an H. The properties which might allow for a more sophisticated computation of the showering of the H particle are the following. The cross section for H-p interactions is somewhere midway between that of a proton and that of a deuteron. The fractional energy loss for the H, so long as the H holds together should be roughly half that of a proton, since the mass is twice as large and the energy loss per collision should be about the same. After several interactions, the H particle might fall apart into two Λ's or protons and kaons. The Λ's and kaons are in the projectile fragmentation region, and should quickly decay, generating fast muons. When the Λ decays into πp, the pion is faster than is typical for centrally produced π's. and should increase the fast muon flux. Kaons have smaller cross sections and mean free paths than pion and therefore decay before interacting as much as pions, and therefore produce fast muons.

Estimates of the total flux of hadronic primaries which would produce a flux of muons corresponding to the rate seen at Soudan give a flux which is about a factor of twenty times larger than the rate observed in air showers extrapolated into this energy range. Because the H particle may shower somewhat differently than a proton primary, this discrepancy may not be so severe. Also, since the flux from Cyg X-3 may be variable, and the measurements were not made at the same time, this discrepancy, which deserves more study, does not yet appear to rule out the H particle hypothesis.

There is a puzzling feature of the NUSEX data which is not explained by our proposal. The NUSEX detector sees the muon signal from a region many degrees on a side around Cygnus X-3. We have no mechanism for dispersing H particles over such a large angular range. This result seems difficult to explain by any mechanism since the NUSEX experiment sees a wider angular dispersion than

Soudan, and since NUSEX measures higher energy particles than Soudan, one would expect that the dispersion at Soudan would be larger. For example, if there was production of a new particle high in the atmosphere with high transverse momentum, the spread at Soudan should be larger by a factor of about 5-10 compared to NUSEX since the energy of particles detected is lower by about this amount. Two plausible, although unattractive explanations for the increased angular broadening would be either multiple muon scattering in the rock beyond that computed using multiple Coulomb scattering Monte-Carlo computations, or an improper determination of the detector orientation.

ACKNOWLEDGEMENTS

We all thank C. Alcock, J. D. Bjorken, S. Errede, T. Gaisser, M. Gell- Mann, F. Halzen, S. Kahana, K. Ruddick, T. Stanev, M. Turner and G. Yodh.

REFERENCES

(1) M. L. Marshak, J. Bartlet, H. Courant, K. Heller, T. Joyce, E. A. Petersson, K. Ruddick, M. Shupe, D. S. Ayres, J. Dawson, T. Fields, E. N. May, L. E. Price and K. Sivaprasad, U. Phys. Rev. Lett. 54, 2079 (1985).

(2) NUSEX Collaboration, Talk presented at the 1'st Symposium on Underground Physics, Saint Vincent, Italy, April (1985).

(3) M. Samorski and W. Stamm, Astrophys. J. 268, L17, (1983).

(4) M. V. Barnhill, T. K. Gaisser, T. Stanev and F. Halzen, U. Wisc. preprint MAD/PH/243 (1985).

(5) G. L. Shaw, G. Benford, and D. J. Silverman, U. C. Irvine preprint 85-14, (1985).

(6) E. W. Kolb, M. S. Turner, and T. P. Walker, Fermilab preprint, 1985. (1985); T. Gaisser and T. Stanev, Bartol Preprint BA-85-12 (1985),

(7) R. L. Jaffe, Phys. Rev. Lett. 38, 195, (1977).

(8) W. T. Vestrand and D. Eichler, Ap. J. 261, 251 (1982).

(9) M. A. Ruderman and P. G. Sutherland, Ap. J. 196, 51 (1975).

(10) G. Baym and S. A. Chin, Nuc. Phys. A262, 527, (1976); G. Chapline and M. Nauenberg, Nature, 264, 23, (1976); B. A. Freedman and L. D. McLerran, Phys. Rev. D16, 1169, 1976 and Phys. Rev. D17, 1109, (1978); K. Brecher Astrophys J. 215, 117 (1977); W. Fechner and P. Joss, Nature, 274, 347, (1978).

(11) E. Witten, Phys. Rev. D30, 272 (1984).

(12) R. L. Jaffe and E. Farhi, Phys. Rev. D30, 2379, (1984).

(13) J. D. Bjorken and L. D. McLerran, Phys. Rev. D20, 2353, (1979); S. A. Chin and A. Kerman, Phys. Rev. Lett. 43,1292 (1979); A. K. Mann and H. Primakoff, Phys. Rev. D22, 1115, (1980); F. Halzen and H. C. Liu, Phys. Rev. Lett. 48, 771, (1982); F. Halzen and H. C. Liu, Phys. Rev. D25, 1842, (1982); J. Elbert and T. Stanev University of Utah Preprint UU/HEP82/4 (1982).

(14) C. M. G. Lattes, Y. Fujimoto, and S. Hasegawa, Phys. Rept. 65, 151, (1980).

(15) H. C. Liu and G. Shaw, Phys. Rev. D30, 1137, (1985).

COMMENTS ON COSMIC-RAY SIGHTINGS OF CYGNUS X-3*

T.K. Gaisser
Bartol Research Foundation of the Franklin Institute
University of Delaware
Newark, DE 19716 USA

Two experimental groups[1,2] working at different minimum energies have reported underground muons coming from the direction of Cygnus X-3 with rates that vary with its binary period. At the Mont Blanc detector[2] the events are, within statistics, uniformly spread over a 5-degree circle around the position of Cygnus X-3, even though the angular resolution is significantly better than this. Previous speakers[1,3] have given the reasons that the underground muon signals cannot be induced by charged particles or by any known neutral particles that have travelled all the way from Cygnus X-3, a distance greater than 10 kpc. In addition the requirement that the phase information not be lost imposes a constraint on the Lorentz factor of the signal carrier.

Clearly, it is difficult to account for these data. One possibility that has been suggested[4] is that the parent primaries might be nuggets of quark matter. There are theoretical reasons for believing that such objects might exist and be stable for certain ranges of mass.[5] Furthermore, they might be produced in high energy processes around a compact quark star. One would then expect comparable numbers of up, down and strange quarks. Some fraction of such nuggets would be neutral and thus a possible signal-carrier. A high content of strange quarks would lead to enhanced kaon production in the atmosphere and thus to a relatively high yield of muons. Quark globs of the right mass could penetrate deep in the atmosphere and explode to give rise to Centauro events.[6] At this conference McLerran dscussed a specific version of a stable ensemble of quarks that could be relevant in the context of underground signals from Cygnus X-3.[7] This is the di-lambda, a bound dihyperon state of 2u, 2d, and 2s quarks.

Hillas[8] has pointed out, however, that the surface air shower signal from Cygnus X-3 puts a significant constraint on models which would produce the muons by interactions of nucleon-like objects in the atmosphere: Assume such parent "nucleons" are bound in aggregates of mass number A. These particles will also produce air showers.

---

*Invited talk given at the Madison Workshop on New Particles, May 11, 1985.

To be consistent with the observed air shower signal, $dF_{surface}/dE$, one then requires

$$\text{Cyg X-3 underground signal} < \int_{AE_\mu}^{\infty} N_\mu(>E_\mu)[dF_{surface}/dE]dE,$$

where N_μ is the number of muons per primary that have sufficient energy to penetrate to the detector. The differential surface flux is roughly $4\times10^{-8}/E^2\ cm^{-2}s^{-1}GeV^{-1}$. Using an Elbert formula[9] for underground muon yield from incident nuclei[10], we find a bound on the underground signal from Cyg X-3 of $1.3\times10^{-6}(E_{GeV})^{-2}cm^{-2}s^{-1}$, where E_{GeV} is the minimum muon energy for the underground detector. For Soudan (E_{GeV}=650) this bound is $3\times10^{-12}\ cm^{-2}s^{-1}$ and for Mont Blanc (E_{GeV}=3400) $10^{-13}\ cm^{-1}s^{-1}$. In contrast, the reported signal at Soudan is about 7×10^{-11}. A flux is not stated for the signal at Mt. Blanc, but an estimate can be obtained from a comparison of signal/background ratio with the background flux of single atmospheric muons in the angular region around Cygnus X-3. Such an estimate gives of order 10^{-11} $cm^{-2}s^{-1}$. Thus the underground signal appears to be at least a factor 20 too high to be induced by nucleons. Conversely, the parent hadrons must be at least 10-20 times more prolific at producing muons relative to air showers than nucleons are. In view of the quark matter suggestion (for which kaon and hence atmospheric muon production should be enhanced), we ran the cascade simulation of Ref. 10 for incident lambda hyperons, forcing production of a leading kaon at each lambda interaction. The muon production was enhanced by a factor less than two relative to nucleons, so even in this case there is a problem of consistency with the surface air shower fluxes.

A conceivable way out is to arrange the interaction length of the parent to be comparable to or greater than the thickness of the atmoshpere so that production of the signal occurs too low for air shower production (i.e. mostly in the Earth). In this case, however, muon production must be prompt. In addition, as pointed out by Vander Velde, the cross section must be tuned to avoid producing too many contained events in large underground detectors. See Ref. 3 for a further comment on this point.

One can in principle use the energy-dependence of the signal implied by the different depths of the experiments to determine whether the muon production is prompt or atmospheric via pion and kaon decay. In the latter case the signal should be suppressed by an extra power of E_{GeV} as the depth increases due to time dilation of the parent pions and kaons. If the spectrum of the carrier from the source is E^{-2} (differential) one would expect the ratio Soudan/NUSEX underground signal = 5 for prompt and = $(5)^2$ for atmospheric pion and kaon decay. The ratio of the observed fluxes quoted above is closer to 5, but the analysis is not conclusive because we have not taken account of the complex variation of the overburden in the line of sight to Cygnus X-3 as it passes across the sky at Mont Blanc.

Both the observed angular dependence of the underground signal and calculations of the expected neutrino flux [11,12] are inconsistent with the possibility that the apparent underground signal from Cygnus X-3 is neutrino induced. It is well known that Southern hemisphere sources are more promising for doing neutrino astronomy with underground detectors in the Northern hemisphere. One such source is Vela X-1, which has been observed by the Adelaide group[13] in air showers above 10^{15} eV. (See Elbert's talk at this conference[14] for a review of air shower observations of point sources.) Another[15] is LMC X-4 for which a flux of 4.6×10^{-15} air showers $cm^{-2}s^{-1}$ with energies above 10^{16} eV is reported. This is about equal to the flux of air showers from Cygnus X-3.[14]

Cocconi recently pointed out[16] in addition that LMC X-4 might give a neutrino flux at Earth significantly greater than that expected from Cygnus X-3 because it appears to give the same photon flux despite significantly greater attenuation due to electron-positron production on microwave background photons. The estimated distance to this source is 50 kpc. In Fig. 1 I show the transmission coefficient as a function of energy for photons that have traversed 10 and 50 kpc through the microwave background.[17] The energy range of the air shower observations of Cyg X-3[18,19] and of LMC X-4[20] are marked on the respective curves. To check the estimate of the ratio of neutrino fluxes from the two sources more precisely, I assumed an E^{-2} differential photon spectrum at the source and folded in the energy-dependent attenuation factor to get

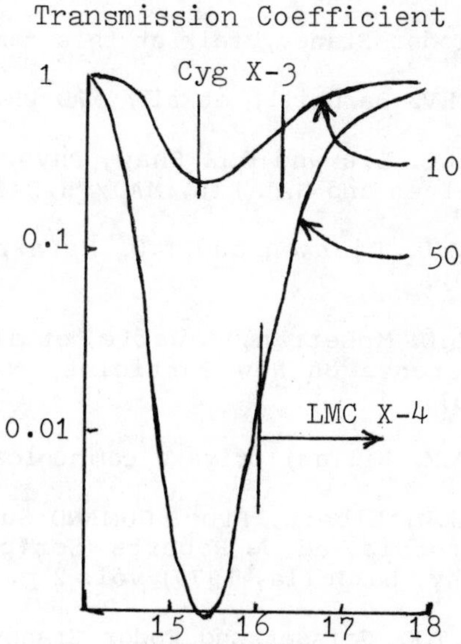

Fig. 1

the observed photon fluxes at Earth. The integrated signal from LMC X-4 is attenuated by a factor three more than that from Cygnus X-3[21]. (Since attenuation is exponential, this statement obviously depends crucially on the assumed distances.) Thus <u>if the air showers are photon-induced and if the relation between neutrino production and photon production at the two sources is the same</u> one would expect a factor three[20] higher neutrino flux from LMC X-4. Moreover, its duty factor for a Northern hemisphere detector would be 100% rather than about 25% as for Cyg X-3.

ACKNOWLEDGMENTS

Much of this work was done in collaboration with M.V. Barnhill III, Francis Halzen and Todor Stanev. I am grateful to R.J. Protheroe, A.M. Hillas, G. Cocconi and J. Vander Velde for useful discussions and correspondence. Work supported in part by the U.S. Department of Energy.

REFERENCES

[1] M.L. Marshak, et al., Phys. Rev. Letters 54, 2079 (1985) and talk at this conference.

[2] G. Battistoni et al., submitted to Phys. Letters B (May 1985).

[3] Todor Stanev, talk at this conference.

[4] M.V. Barnhill, et al., MAD/PH/243 (April 1985).

[5] H.C. Liu and G.L. Shaw, Phys. Rev. D30, 1137 (1984) and F. Halzen and H.C. Liu, MAD/PH/240 (1985).

[6] J.D. Bjorken and L.D. McLerran, Phys. Rev. D20, 2353 (1979).

[7] L.D. McLerran, R. Jaffe, et al., paper presented at the Conference on New Particles, Madison, Wisconsin, April (1985).

[8] A.M. Hillas, private communication.

[9] J.W. Elbert, Proc. DUMAND Summer Workshop, La Jolla, California, ed. A. Roberts (Scripps Institution of Oceanography, La Jolla, 1979) vol. 2 p. 101.

[10] T.K. Gaisser and Todor Stanev, Nuclear Instruments and Methods in Physics Research, A235, 183 (1985).

[11] T.K. Gaisser and Todor Stanev, Phys. Rev. Letters 54 2265 (1985).

[12] E.W. Kolb, M. Turner and T. Walker, FNAL preprint and T. Walker talk at this conference.

[13] R.J. Protheroe, R.W. Clay and P.R. Gerhardy, Ap. J. 280, L47 (1984).

[14] J.W. Elbert, this conference.

[15] R.J. Protheroe and R.W. Clay, Nature 315, 205 (1985).

[16] G. Cocconi, preprint, April 1985 and private communication.

[17] R.J. Gould and G.P. Schreder, Phys. Rev. 155, 1404 (1967) and R.J. Gould, Ap. J. 274, L23 (1983).

[18] M. Samorski and W. Stamm, Ap. J. 268, L17 (1983).

[19] J. Lloyd-Evans et al., Nature 305 784 (1983).

[20] In evaluating the attenuation from LMC X-4 it was assumed that showers up to at least 10^{18} eV contribute to the observed signal. If the upper cutoff is significantly lower than this, then the estimate of source power per decade of energy would increase and so would the estimated neutrino flux.

[21] I have assumed that photons that interact with the background radiation are lost from the beam. Protheroe (paper OG2.7-13, submitted to the XIX Int. Cosmic Ray Conf. at La Jolla, August, 1985 and in Ref. 15) has pointed out that the attenuation is reduced when regeneration of photon flux by inverse Compton scattering of produced electrons is considered. He in fact concludes that the luminosities of LMC X-4 and Cyg X-3 are approximately proportional to the square of their respective distances, which would suggest equal neutrino fluxes at Earth. However, for a mean magnetic field less than about 10^{-9} Gauss, the electron gyroradius is less that the interaction length of electrons in the microwave background. In this case regeneration should not be important.

List of Participants

Albrow, Mike	Rutherford Appleton Lab, England
Arnowitt, Richard	Northeastern University
Baer, Howard	Argonne National Lab
Band, Henry	Stanford Linear Accelerator Center
Banner, Marcel	CEN/Saclay, France
Barger, Vernon	University of Wisconsin
Bellettini, Giorgio	INFN–Firenze, Italy
Bellinger, James	University of Wisconsin
Beretvas, Andrew F.	Fermilab
Berger, Edmond L.	Argonne National Lab
Boudreau, Joe	University of Wisconsin
Branson, James G.	Massachusetts Institute of Technology
Ceradini, Filipo	University of Rome, Italy
Chiu, Charles	University of Texas–Austin
Cline, David	University of Wisconsin
Cochet, Christian	CEN/Saclay, France
Colas, Paul	CEN/Saclay, France
Conti, Antonio	University of Florence, Italy
Cudell, Jean-René	University of Wisconsin
Dau, Deiter	University of Kiel, Germany
de Brion, J. P.	CEN/Saclay, France
Delfino, Manuel	University of Wisconsin
Denegri, Daniel	CEN/Saclay, France
Deshpande, N. G.	University of Oregon
Di Caporiacco, Giuliano	INFN–Firenze, Italy
Dowell, John	University of Alabama–Birmingham
Duck, Ian	Rice University
Durand, B.	University of Wisconsin
Eggert, Karsten	University of Aachen, Germany
Elbert, Jerome	University of Utah–Salt Lake City
Ellis, Nick	University of Birmingham, England
Ellis, R. K.	Fermilab
Erhard, Peter	University of Aachen, Germany
Fontaine, Gerard	College de France, France
Gaisser, Tom	Bartol Research Foundation
Goldberg, Haim	Northeastern University
Haber, Howard	University of California–Santa Cruz
Hagiwara, Kaoru	DESY, Germany
Hahn, Stephen	Fermilab
Halzen, Francis	University of Wisconsin
Hasan, Abul	University of Wisconsin

Hedin, David	SUNY–Stony Brook
Hikasa, Ken-Ichi	University of Wisconsin
Hollebeek, Robert	Stanford Linear Accelerator Center
Honma, Alan	Queen Mary College, England
Introzzi, Gianluca	University of Colorado
Izen, Joseph	University of Wisconsin
Jaffe, Robert	Massachusetts Institute of Technology
Kalmus, Peter	Queen Mary College, England
Kane, Gordon	University of Michigan
Keung, Wai-Yee	University of Illinois–Chicago
Kienzle, Werner	CERN, Switzerland
Kitazawa, Yoshihisa	Enrico Fermi Institute
Kondo, Kunitaka	University of Tsukaba, Japan
Kozaneki, Wijold	Stanford Linear Accelerator Center
Kunszt, Z.	CERN, Switzerland
Langacker, Paul	University of Pennsylvania
Levi, Mike	CERN, Switzerland
Lichtenberg, Don	Indiana University
Markeloff, Richard	University of Wisconsin
Markiewicz, Thomas	University of Wisconsin
Marshak, Marvin	University of Minnesota
Martin, A. D.	University of Durham, England
Maruyama, Takashi	Stanford Linear Accelerator Center
Maurin, Guy	CERN, Switzerland
McLerran, Larry	University of Washington
Menzione, Aldo	INFN–Firenze, Italy
Meshkov, Sydney	National Bureau of Standards
Messner, Robert L.	Stanford Linear Accelerator Center
Mestayer, Mac	Vanderbilt University
Meyer, Olivia	CERN, Switzerland
Meyer, Thomas C.	CERN, Switzerland
Miller, Marshall	University of Pennsylvania
Muller, Thomas	CERN, Switzerland
Nandi, Satyanarayan	University of Texas–Austin
Nanopoulos, D.	CERN, Switzerland
Naumann, Lutz	CERN, Switzerland
Norton, Alan	CERN, Switzerland
Olsson, Martin G.	University of Wisconsin
Pakvasa, Sandip	University of Hawaii
Parrini, Giuliana	INFN, Italy
Pauss, Filicitas	CERN, Switzerland
Phillips, R. J. N.	Rutherford Appleton Lab, England
Piano-Mortari, G.	University of Rome, Italy

Placci, Alfredo	CERN, Switzerland
Pondrom, Lee	University of Wisconsin
Prepost, Richard	University of Wisconsin
Proudfoot, James	Argonne National Laboratory
Radermacher, Ernst	University of Aachen, Germany
Raja, R.	CERN, Switzerland
Reeder, D.	University of Wisconsin
Repko, Wayne W.	Michigan State University
Revol, Jean-Pierre	M. I. T.
Ryssenbeck, Michael	CERN, Switzerland
Sajot, Gerard	CERN, Switzerland
Salvini, Giorgi	University of Rome, Italy
Savoy-Navarro, A.	Saclay, France
Schoessow, Paul	Argonne National Laboratory
Shapiro, Anatole	Department of Energy
Stanev, Todor	Bartol Research Foundation
Stenzler, Mark	University of Wisconsin
Stevenson, Paul	Rice University
Stump, Daniel	Michigan State University
Summers, Don	University of Wisconsin
Sumurok, Sham	CERN, Switzerland
Suzuki, Mahiko	University of California–Berkeley
Swartz, Morris	CERN, Switzerland
Touminiemi, Jorma	University of Helsinki, Finland
Tscheslog, Evelin	CERN, Switzerland
Tung, Wu-Ki	Illinois Institute of Tehcnology
Turkot, Frank	Fermilab
Underwood, David	Argonne National Lab
Van Eijk, Robert	CERN, Switzerland
Vankov, Christofor	INRNE, Bulgaria
Vlassopulos, S. D. P.	McGill University, Canada
Weekes, Trevor	Smithsonian Institute
Weiler, Thomas	Vanderbilt University
Whisnant, Kerry	Florida State University
Wicklund, A. B.	Argonne National Laboratory
Wilcke, Rainer	CERN, Switzerland
Wu, Sau Lan	University of Wisconsin
Wulz, Claudia-Elizabeth	CERN, Switzerland
Yasuoka, Kiyoshi	Fermilab